FUNCTION AND EVOLUTION OF BEHAVIOR

An Historical Sample from the Pens of Ethologists

Edited by
PETER H. KLOPFER
Duke University
and
JACK P. HAILMAN
University of Wisconsin

ADDISON-WESLEY PUBLISHING COMPANY
Reading, Massachusetts · Menlo Park, California · London · Don Mills, Ontario

 Printed in the United States of America. Published simultaneously in Canada. Library of Congress Catalog Card No. 74-162466.

PREFACE

This is a collection of especially chosen research papers about animal behavior. We assembled it because many communities and colleges can provide only a minimum of library resources, and all the desired journals may not be available. While one can learn about the concepts of ethology—the study of animal behavior—through any one of many recent textbooks (e.g., Hinde, 1966; Marler and Hamilton, 1966; Klopfer and Hailman, 1967), one cannot get the flavor of the original studies from texts: For this the journal literature is essential.

No random sample is this. The papers or excerpted papers included here either represent important milestones in the development of ethology, or they are representative papers about important issues. There are certain biases in the selection. First, we have selected papers that interest us personally, as is generally the case with personal selections. Second, shorter papers were taken before longer papers of similar content. Third, English language papers were selected over those in other languages. Fourth, more papers are about birds than other animals, because, simply, there are far more papers about birds from which to choose. Finally, we intended that this selection complement our own book (*An Introduction to Animal Behavior: Ethology's First Century*, Klopfer and Hailman, Prentice-Hall, 1967), though we expect it will prove valuable to users of other texts as well.

The selections are presented with a minimum of comment, although they are grouped and subgrouped. The four major groups correspond to the four major causes and origins of behavior (Tinbergen, 1963; Hailman, 1967; Klopfer and Hailman, 1967): function, evolution, control, and ontogeny. Each group is preceded by a short introduction to the concept that the papers are intended to illustrate. The collection begins with some early theoretical papers, followed by empirical studies comprising the bulk of the volumes, and ending with some theoretical overviews. For empirical papers of substantial length, we have moved the author's summary, where one exists, to the beginning of the article to serve as an abstract for the reader; where a summary did not exist, we have written one and so indicated so that any errors therein will be charged to us and not the original authors. Some long papers in another language have been presented only as lengthy English summaries, and many papers have been edited to eliminate tangential material (omissions indicated by ***). Some contributions are merely excerpts from much longer works, as indicated.

Within each of the four major groups of empirical papers there are subgroups of papers on the same problem or issue. In many such subgroups the reader can see the thesis, antithesis, and synthesis of a topic unfold; in other cases there is no convenient resolution of an issue raised, even today. It is the grouping, as well as the selection, that gives distinction—and, we hope, value—to this collection of readings for students.

Durham, North Carolina P. H. K.
Madison, Wisconsin J. P. H.
January 1972

CONTENTS

ABRIDGED CONTENTS OF CONTROL AND DEVELOPMENT OF BEHAVIOR

PROLOGUE

In this first volume, Konrad Lorenz presents the initial thesis of ethological theory in his paper on the comparative method in studying innate behavior patterns. Then, J. S. Kennedy and T. S. Schneirla present general objections to the viewpoint: an antithesis. These fascinating papers serve to set the stage for the four groups of empirical papers to come, the first two of which appear in this volume. The selective function and historical origins of behavior are discussed, before the close of the volume, providing material from which an ultimate synthesis of ethological theory is formulated in the second volume.

INTRODUCTION TO ETHOLOGY

The science known as "ethology" owes much of its immediate impetus to the writings and discoveries of a small group of European zoologists, of which Konrad Lorenz and Niko Tinbergen are foremost. But the theories that had been developed by the end of ethology's first century—dating from when Charles Darwin began to publish his many works touching on animal behavior—were not to go unchallenged. To be useful, any collection of readings must begin by setting the scene, and that is precisely what this chapter intends.

A. *THE FORMULATIONS OF LORENZ AND TINBERGEN*

In his own words Lorenz gives his views of animal behavior, as they were about 1950, in a sweeping epistemological treatise. The conclusions of Tinbergen are similar: Most of animal behavior consists of highly organized, unlearned, stereotyped behavior patterns that may be studied as objectively by the ethologist as a bone is by the anatomist.

The Comparative Method in Studying Innate Behaviour Patterns

KONRAD Z. LORENZ

INTRODUCTION

Like many other branches of biological research, comparative ethology owes its existence to a discovery. The discovery of a new *particulate* process, the function of which is comparatively independent from the function of the organism as a "whole," always opens a new line of analytical approach. The nature of the process then determines the method most favourable for its experimental investigation, and this method, in its turn, determines the direction in which analytical research continues to proceed. There are many cases in which the development of a new and independent branch of biological science has thus been determined by the character of one particularly favourable object, modern genetics being the most obvious example.

The distinct, particulate physiological process whose discovery may be identified with the origin of comparative ethology as an independent branch of science is represented by a certain type of innate, genetically determined behaviour patterns. Charles Otis Whitman, who was the first to discover them, called them simply "Instincts"; Oskar Heinroth, who, ten years later, independently rediscovered them, spoke of "*arteigene Triebhandlungen*"; I myself have called them "*Instinkthandlungen,*" "*Erbkoordinationen*" and "*Instinktbewegungen*" successively, thus further mixing up

Reprinted in part from the *Symposia of the Society for Experimental Biology*, volume IV (Physiological mechanisms in animal behaviour), pages 221–268 (1950).

terminology. I shall, in the conference on terminology, included in the programme of this Symposium, propose the term *endogenous movements* for this type of innate behaviour pattern. Endogenous movements were not only discovered and recognized as a very distinct phenomenon by C. O. Whitman as early as 1898, but also systematically studied and evaluated as taxonomic characters. Nevertheless, their extreme importance as independent elements of behaviour was not generally realized until much later. It was only about 1930 that the discovery of endogenous activities began to act as a centre of crystallization, about which concerted efforts to analyse innate behaviour patterns began to become organized. From then on, however, a quickly growing number of investigators began to apply the methods dictated by the nature of the favourable object which had been discovered so much earlier by Whitman and Heinroth. Thus comparative ethology developed, and therefore it is distinguished by a particular set of methods from all other schools also occupying themselves with the innate behaviour of animals and men. To convey some idea of this particular set of methods is the task I have set myself in this paper.

I shall set about this task from two sides. Before trying to explain what comparative ethology is trying to do, I propose to state, at least as clearly, what it is trying *not* to do. I deem it most necessary to show, what particular errors of method, committed by the vitalistic and by the mechanistic schools of behaviour study, we are meticulously trying to avoid, also to make it quite clear why these errors have prevented some fundamental facts about innate behaviour from being discovered half a century sooner than they actually were. After thus defining our own methods *per exclusionem*, I shall proceed to describe them in some detail.

This plan involves some serious criticism of both vitalistic and mechanistic schools, on "purposive psychology" and some branches of "Gestalt" psychology, as well as on behaviourism and Pavlov's school of reflexology. I confidently assert that this criticism is irrefutable from the methodological point of view. But I want to make it absolutely clear at once that comparative ethology has no right whatsoever to claim credit for not having repeated the methodical errors of which we accuse vitalists and mechanists. A preconceived idea is a very terrible danger indeed to all inductive research, the very essence of which is unprejudiced observation. The narrow path of inductive natural science is hemmed in on both sides by methodological pitfalls, and is, therefore, by far easiest to walk along blindfolded. And this is exactly what Whitman and Heinroth did. Happily ignorant of the great battle waged by vitalists and mechanists on the field of animal behaviour, happily free from even a working hypothesis, two "simple zoologists" were just observing the pigeons and ducks they loved, and thus kept to the only way which leads to the accumulation of a sound, unbiased basis of induction, without which no natural science can arise. Small tribute is due to them for not tumbling into the methodological traps, into which vitalists and mechanists succeeded in shouldering each other, as I shall at once proceed to explain. No very great tribute either is due to them for applying the phyletic method to the study of innate behaviour. Being comparative morphologists, it was only natural for them to do so. But by doing it, they discovered a fact which could not be discovered from any but the phyletic point of view; they discovered an independent, particulate function of the central nervous system which, as an element of animal and human behaviour, is, at the very least, equally important as the reflex or the conditioned reaction.

CONSEQUENCES OF THE DISPUTE BETWEEN VITALISTIC AND MECHANISTIC SCHOOLS OF BEHAVIOUR STUDY

Antagonistic exaggerations

Without any doubt the dispute between vitalism and mechanism has created and exaggerated a number of erroneous preconceptions which subsequently developed into a serious obstacle to analytical research. It would be a difficult task for the historian to decide which side originally sent the pendulum swinging and was responsible for the whole dispute. One must not forget, however, that vitalism is much older than mechanism, that the concept of entelechy comes from Aristotle, and that Johannes Müller and Claude Bernard were quite decidedly vitalists, though they also undoubtedly were the pioneers of analytical physiology. Furthermore, the quarrel has had particularly disastrous results for the analytical investigation of such life processes as embryonic development, regeneration and instinctive behaviour, which, because of their "wholeness," their "directedness" and their apparent spontaneity, have, since the very beginning of the dispute, constituted the stronghold of vitalism. Last, but not least, the theoretical and philosophical position of vitalism is much more fundamentally wrong than the few much less portentous methodical errors, of which we accuse the mechanistic schools. For all these reasons I think that we are justified in regarding the vitalistic errors as primary and the reciprocal errors of the mechanists' reactions justified in themselves and only erring through exaggeration.

Vitalists base all their arguments on the purely dogmatic assertion that all life processes are governed by an essentially non-explainable, preternatural "factor," termed vital force, entelechy, whole-producing factor, and so on. The legitimate answer to this is obviously that no man can know to what extent causal analysis is going to succeed in explaining the processes of life, and that, if a man prefers to meditate upon the supernatural instead of trying to explain things in a natural way, a lamasery is a much more suitable place to do so than a research institute. Instead of which, mechanists replied that all life processes could be explained on the foundation of laws of classical mechanics. If they relinquished this position later, they only did so because modern physics did.

If vitalists asserted that the existence of organic systems was due to a whole-producing factor (Müller's *ganzmachende Kraft*), and therefore not explainable by way of synthesis, mechanists answered that organic systems could be explained very simply as a sum of very few sorts of elements. They even grew quite dogmatic, in some cases, as to what these elements were. If vitalists asserted that life was the result of a directing force, mechanists altogether neglected the fact that life processes do, in some cases, continue to run in a certain direction. If vitalists regarded the spontaneity of some life processes as a wonder and as the direct effect of their preternatural factor, mechanists replied by denying the very existence of spontaneous processes. If all the vitalistic factors, vital forces, directing forces, entelechy, instinct and however they were called, were, if we come down to fundamentals, nothing else than the old, preternatural, immortal *soul* in a new guise, mechanists tried to build up a psychology without a soul, and went so far as to neglect the existence of subjective phenomena altogether, even where introspection can tell us some most important facts about our own ways of acting and reacting.

The very nature of these antagonistic exaggerations, into which vitalism and mechanism were forcing each other, makes it intelligible that of all fields of analytical research the investigation of *behaviour* suffered most. Its well-defined character as a "whole" or system, its directedness and purpose and, in some cases, its evident spontaneity, brought animal behaviour into the very focus of the great dispute. Especially innate behaviour suffered from this, mainly because "instinct" was one of the cardinal points of dissension. To make clear the methodical point of view held by comparative ethology, I think it necessary to explain in some detail the reciprocal errors which both the vitalistic and the mechanistic schools of behaviour study committed regarding the items mentioned above.

Reciprocal errors in the treatment of systems

The conception of the "whole" was to vitalists the very essence of what never could be explained, being the direct consequence of a preternatural factor. Particular stress was laid on the thesis that a whole could never be explained by synthesis of particulate elements. To this fatal resignation of analysis on the vitalistic side mechanists replied by trying to synthetize all the systems of behaviour on the basis of too few elements. This "atomism" is an error very characteristic of mechanistic schools. This optimistic attempt was, of course, doomed to failure, but not before it had brought valuable results. Even very crude simplisms have done less damage to the progress of science than resignation.

One of the most deplorable consequences which resulted from the "atomistic" attempts at explaining behaviour was the counter-recoil of "Gestalt" psychology. It is an everlasting merit of Gestalt psychologists to have been the first consciously to introduce the method of a truly correlative analysis of organic systems. They were the first to prove the inherent fallacy of the atomistic method of artificially isolating a single chain of causation where in reality there was an immeasurably complex system of interaction. Gestalt psychology should have undertaken to correct the atomistic errors of behaviourism and of reflexology and to introduce the method of a correlative analysis into the study of behaviour. But it did not. Instead of correcting the error mentioned, it merely committed the reciprocal errors. That is to say, the conception of "Gestalt" has assumed, with a good number of Gestalt psychologists, a character dangerously akin to that of a vitalistic factor. To very many authors "Gestalt" is something that neither stands in need of, nor is susceptible to, a natural, physiological explanation. Also, "Gestalt" has been very badly overrated as an explanatory principle. The characters of "Gestalt" were uncritically attributed to all "wholes" and therewith to all organic systems in general. I want to show that this is just as much of an error as it is to regard organism as "sums" or "mosaics" consisting of only a few elements.

There is not a single organic system which is really a "Gestalt," nor one that is a mosaic. Every real organic system is built up on *both* principles, mutual interaction of correlated parts playing a role as well as particulate elements influencing the whole in one-way causation. It is extremely important to know, in every single instance, with what kind of relation analysis has to deal as its method has to conform to the actuality of its object. It is as much a methodological error to treat a one-way causality as if it were an "amboceptoric" interaction, as to mistake mutual interaction for linear one-way causation.

* * *

The behaviour of the sea-urchin, for instance, is quite literally a mosaic of the independent functions of its organs and can be as literally "synthesised" by tying a broken-up sea-urchin shell together again with a piece of string. Uexküll, in his drastic manner, has said: "When a dog runs, the dog is moving his legs, when a sea-urchin runs, the legs are moving the sea-urchin"—which beautifully expresses the difference between the mosaic system and the system of universal interaction.

I do not think that there are any other organic systems, in which independent, mosaic-like constituents and regulative systems of mutual interaction alternate and interlock in so utterly incalculable a manner as they do in the behaviour systems of higher animals. It is, therefore, of superlative importance to ascertain, as early as possible, to which type of system a given object of investigation belongs. It is an extremely pertinent question, whether the dog moves his legs, or the legs the dog, quite especially in group and social psychology. We may summarize the reciprocal methodological errors of vitalists and mechanists by saying that the former dogmatically assume that it is exclusively the whole which is influencing its parts, while mechanists as dogmatically confine their attention to one-way causalities, leading from the part to the whole. Vitalists overrate the "whole" and totally neglect the particulate element, mechanists neglect regulative interaction and overrate those particulate independent elements which they happened to succeed in isolating.

* * *

If vitalists persistently ignored the question *why* behaviour develops survival value, or why it is purposive, one should expect mechanists to stress the urgent necessity of putting just these questions. Instead of doing this, the mechanistic schools treated the very fact of survival value and purpose exactly as if it were just another vitalistic superstition like entelechy, instinct and the like, that is to say, they flatly ignored it.

* * *

In saying above that no mechanist has ever raised the question after a causal explanation of directedness and purposivity of behaviour, I have neglected one memorable instance. No less a man than I. P. Pavlov has not only raised this question, but actually tried to give an answer to it. In a paper published in 1916 he writes: "Analysis of the activities of animals and men has led me to the conclusion that, amongst other reflexes, one particular reflex, the reflex of the goal (рефлекс целм) must be conceived, a striving for the attainment of one definite object sending out stimuli" (translated from the Russian). Later on, he adds that this reflex is standing in a close relation to the unconditioned reflexes of taking food and of copulation. It is quite obvious that here purposive behaviour, in the special form of appetitive behaviour, has obtruded itself to Pavlov's notice, also that, observing it chiefly in dogs and humans, he believed that food appetite and sexual appetite were the only ones in existence. It is a subject for deep meditation that the great mechanist did not see how closely akin this new "reflex," created with the sole intent of explaining something hitherto unexplainable, is to any vitalistic factor introduced for the very same purpose.

Reciprocal errors in the treatment of spontaneity

In a strictly analogous way, as the vitalistic-mechanistic dispute prevented a sensible analytical treatment of the whole, and of survival value and purpose of behaviour, it also obstructed the way to a causal explanation of spontaneity. Exactly as in the cases already discussed, the vitalists saw the phenomenon in itself, but regarded it as a "wonder" and refused to ask for its natural explanation. William McDougall has certainly gained very great merit by stressing the importance of spontaneity of behaviour as an argument against the chain-reflex theory of the mechanistic schools. His sentence that "it is evidently inadmissible to speak of a *re*-action to a stimulus not yet received" is something we ought to keep well in mind when formulating our modern terminology of behaviour. Also, his slogan, "The healthy animal is up and doing," which he threw into the teeth of the chain-reflex theory, is worth remembering. But he and other purposive psychologists did not confine themselves to using the fact of spontaneity as an argument against the chain-reflex theory, but regarded it also as an argument against the general assumption that behaviour is causally explicable. At least none of them ever made the least attempt to offer a physiological explanation of spontaneity.

Instead of unrelentingly demanding this explanation, behaviourists as well as reflexologists evaded the problem by the simple means of ignoring the fact. In my opinion it is the most serious accusation against the working hypotheses of mechanists that they conducted only such experiments as were beforehand destined to confirm the theory. This is about the worst fault a working hypothesis can have. With exceedingly few exceptions, the experiments of mechanists confined themselves to letting some sort of stimulation impinge upon the organism and then to record its answering reaction to this stimulation. This kind of experiment could not but create and confirm the opinion that the function of the central nervous system was restricted to receiving and reflecting external stimuli. No mechanist ever thought it worth while to observe what healthy animals do when left to themselves. So the central nervous system, poor thing, never got the opportunity to show that it could do more than answer to stimulation.

Again there is one exceptional case, in which a reflexologist did take notice of spontaneity, and again it is I. P. Pavlov himself who did so. In describing his experiences with a certain small and very lively dog who would on no account keep still when tied to the framework, to which reflexologists attach their dogs, Pavlov actually has introduced a new reflex to meet the case, the "reflex of freedom." The most confirmed vitalist could not have done better by introducing an instinct of freedom, and, curiously enough, he would have been far nearer the truth than the great mechanist. It is very far from my mind to make fun of Pavlov. On the contrary, I regard this jarring *contradictio in adjecto*, the "reflex of freedom," as a reminder of the limitations of the human mind!

Reciprocal errors in the treatment of innate behaviour

The conception of "instinct" is one of the doubtful legacies left to us by the scholastic philosophy of the middle ages. From the first, this conception was introduced to offer what appeared to be an explanation for such types of animal behaviour as were not explicable on the familiar terms of insight and intellect and yet had a definite survival value to the animal. Medieval science was very prodigal with just this kind of "explanation." The *horror vacui*, the "phlogiston" and many other conceptions are further examples. John Dewey, in his book *Human Nature and Conduct*, has given so irrefutable a critique of this type of explanation that it is really superfluous to add even one word. If we accept a "homing instinct" as an explanation for the fact that a pigeon flies back to its loft, then, next time a child asks me what makes the train go, I shall answer that this is caused by a special factor, called the locomotive force. The justification in both cases is absolutely the same. If we concede it as a scientifically legitimate sort of proceeding, to bridge any arbitrarily chosen gap in our present understanding of nature by assuming an obliging little wonder, then all scientific research becomes a pure farce.

* * *

The few reflexologists who deign to mention innate behaviour patterns at all, confine themselves to the rather obvious statement that they consist of chains of unconditioned reflexes, while some behaviourists, chief among them J. B. Watson, go so far as to deny the existence of any more highly differentiated forms of innate behaviour, conceding only that of simple "motor mechanisms."

It is hardly an exaggeration to say that the large and immeasurably fertile field which innate behaviour offers to analytical research was left unploughed because it lay, as no man's land, between the two fronts of the antagonistic opinions of vitalists and mechanists. Small wonder, indeed, if it became "the sporting ground of fruitless philosophical speculation," as Max Hartmann said on a memorable occasion.

Lack of observational basis

* * *

It is an inviolable law of inductive natural science that it has to *begin* with pure observation, totally devoid of any preconceived theory and even working hypothesis. This law has been broken by one and all of the great schools of behaviour study, and this one fundamental, methodical fault, is at the bottom of all the errors of which we accuse vitalists and mechanists. To put it crudely in two examples. If William McDougall had known all H. Elliot Howard knew about "reactions incomplete through lack of intensity," he would never have confounded survival value and purpose. If J. B. Watson had only once reared a young bird in isolation, he would never have asserted that all complicated behaviour patterns were conditioned. It was a really crushing blow to cherished ideals when, as a young student, I first realized that the great authorities on "instinct," such as Lloyd Morgan and W. McDougall, *did not know* the relevant facts about innate behaviour with which I, ignorant boy though I was, was mentally struggling even then; reactions not attaining their goal because of lack of intensity, vacuum activities and the innumberable ways in which innate behaviour patterns were miscarrying, were evidently unknown to the great theorists.

I know of only two students of animal behaviour who really *knew* the animals about which they wrote and who were, at the time of their writings, considered "scientists" and not just "bird lovers." One of them is H. S. Jennings, who may, in a loose sense, be considered as a behaviourist; the other is Jakob von Uexküll, who certainly was a vitalist. In spite of the world-wide discrepancy of their philosophical positions they have one fundamental point of departure in common: both of them hold the observation *of all there is to be observed* in the behaviour of a species must go *before* the quest for explanation of the single items of behaviour. This means that both of them treat the behaviour of a species as it must unconditionally be treated: as an organic system. To Jennings we owe the conception of the *system of actions* of a species, and though J. v. Uexküll gave a vitalistic name (*Bauplan*) to the same reality, he also approached it with the same method as Jennings, with the method of a truly correlative analysis, proceeding on a broad observational front.

* * *

But all other professional students of animal behaviour have been guilty of the one unpardonable offence against the most fundamental law of inductive natural science: they have one and all formed a hypothesis *first* and proceeded to look for examples to confirm it *afterwards*. The protean multiformity of organic nature and quite particularly of the behaviour of higher animals is such that a circumspect search for examples can never fail to detect a wealth of evidence for literally *any* theory, however arbitrarily you chose to invent one. The *facts* in themselves may be quite correct, but *choice* of facts in itself is ever a falsification of the inductional basis. And this statistical falsification of the very foundation of natural science can only be prevented in one way: unprejudiced observation of all there is to be observed must go temporarily *before* the building up of hypotheses.

METHODS PECULIAR TO COMPARATIVE ETHOLOGY

Observational basis of induction

It is a fact worthy of very serious contemplation that very nearly all the relevant details of innate behaviour, the ignorance of which so seriously impaired the theories of the two great schools of behaviour study, mechanism and vitalism, have long been known to and fully appreciated by *bird lovers*. H. Eliot Howard, Edmund Selous and many other men who would have been considered mere "amateurs" by the "scientific" schools, have been conversant with these facts, without, however, knowing how badly these facts were needed by psychology. Also the professional zoologists whom we consider as the pioneers of comparative ethology, such as C. O. Whitman, O. Heinroth, Julian Huxley and Jan Verwey, at first regarded their observational work more as a hobby or at best as a very secondary occupation. Indeed, sitting in a blind observing birds, or just staring into an aviary with pigeons, or on a pond with collection of ducks, or into an aquarium, does not, at first sight, appear to be as "scientific" as the tremendous experiments of mechanists or the deep philosophical speculations of vitalists. Yet it is very far from being accidental that just this kind of observation yielded facts which remained unknown to the great schools.

The development of any inductive natural science proceeds through three indispensable stages: the purely observational recording and describing of fact, the orderly arrangement of these facts in a system, and finally the quest for the natural laws prevailing in the system. These three steps have been termed the idiographic, the systematic and the nomothetic stage by Windelband. For rather obvious reasons, which I need not expound here, the relative importance of the strictly observational basis of any natural science is in direct proportion to the degree in which its object bears the character of a whole or system: the greater the number of constituent parts of a system, the more complicated their interaction, the more necessary it evidently becomes to get an *inventory* of these parts, before beginning the analysis of their correlations. Now the system of actions of any animal is at least as much of a "whole" and possesses at least as many particulate constituent parts as its body. Therefore the only legitimate manner of proceeding in the study of the behaviour of any species of living organisms is to begin, quite exactly as H. S. Jennings has done, with what I would call a morphology of behaviour, by a thorough observation and description of *all* the behaviour patterns at the disposal of the species.

* * *

The keeping of animals as a scientific method

In many natural sciences, such as botany, zoology, mineralogy and others, scientific activity has begun with collecting, and I do not doubt that this collecting was, in most cases, at first motivated much more by "fancy" for attractive objects than by cognitive purpose. No comparative ethologist of my acquaintance has ever denied that his scientific career began in the same manner. Heinroth was ten when he started collecting and keeping ducks and geese; I myself was still younger when I did the same. Whitman is said to have been addicted to pigeon keeping at about the same age. Though having done this undoubtedly is not a merit, it nevertheless was the methodologically correct thing to do, and I think I am justified in ranging the keeping of animals in captivity first among the methods peculiar to comparative ethology.

By "keeping" an animal I do not mean the mere attempt to keep it alive in captivity for as long as possible, but the endeavour to make a given species unfold its whole system of actions under constant, controllable conditions. Keeping, therefore, always implies the attempt to *breed* the species in question.

It is far from me to disparage the advantages of field observation and the merits of field observers. But for the purposes of correlative analysis it is, as I have already stressed, of the first importance to get an inventory of the particulate behaviour patterns of species and to get some notion to what extent these constituents of behaviour patterns are woven into a regulative "whole." I have already emphasized the necessity of ascertaining, at an early stage of analysis, how far a system of actions is a regulative whole and how far it is a mosaic of relatively independent elements. This question is extremely difficult to decide in field work, because in the natural surroundings of a species its system of action is apt to *seem* much more adaptive and its several activities much more purposive than they really are. Thus adaptedness is easily confounded with adaptability. *It is only by miscarrying that any behaviour pattern can show its nature of a particulate element.* Now, of course, it is possible to notice miscarrying behaviour patterns in field observation, as H. Eliot Howard has sufficiently proved. But it is quite impossible *not* to see it in observing animals kept in captivity.

It is hardly possible to exaggerate the importance of miscarrying behaviour patterns as a source of our knowledge. W. Craig's great discovery that it is the discharge of consummatory actions and not the survival value which is the goal of appetitive behaviour was exclusively due to the observation of behaviour patterns being executed without attaining their "normal" survival value. The existence of innate releasing mechanisms would never have been discovered and their function would not have been possible to analyse if it had not been for the observation of cases, in which they were activated "erroneously" by stimulus situations, in which the released activity was thoroughly miscarrying. The whole set of phenomena connected with the process of accumulation of action-specific energy (such as lowering of threshold, explosion activity, etc.) also could only be discovered by observing innate behaviour patterns miscarrying when discharged at an inadequate object.

It is not only the miscarrying of reactions caused by the absence of normal environmental stimulation which makes the observation of animals kept in captivity so valuable. The physiology of the central nervous system has gained much of its knowledge about particulate functions by studying cases in which some functions had dropped out, because of pathological processes, or had been experimentally destroyed. Very much the same manner of proceeding has led comparative ethology to analogous results. Very slight disturbances of the general "condition" of captive animals are apt to cause surprisingly far-reaching disintegrations of their systems of actions. Endogenous activities lose much of their normal intensity or drop out altogether; innate releasing mechanisms lose much of their selectiveness. The general disintegration of the species' system of actions caused by these two main processes is essentially *reversible*. If one starts keeping a *new* species, with whose needs one is not yet familiar, one automatically gets an opportunity to observe this particular form of disintegrated behaviour, very often a jumble of unconnected behaviour patterns whose normal survival value is often very difficult even to guess. Yet this jumble of disconnected behaviour patterns is extremely valuable for the purpose of taking stock of particulate, genetically fixed "mosaic stones" of behaviour. When, later on, one learns how to keep the species in question and when, with the improvement of technique, the animals are got back into the pink of condition, then the jumble of disintegrated parts is slowly reassembled into a functioning system which fully delivers its normal survival value. I hardly know a more *instructive* object of observation than just this type of disintegration and reassembling of the system of actions in animals kept in captivity. It is, in fact, an actualized example of analysis and resynthesis of behaviour!

If disintegration is a valuable source of our knowledge about behaviour and, indeed, the most valuable fulcrum for its analysis, it must also forever be remembered as a possible source of error. I want to give one warning example of this. The red-backed shrike, *Lanius collurio* L., has got a very beautifully differentiated endogenous movement for impaling insects on thorns. This activity has a taxis component generally directing it to branches or twigs, but apparently none that directs it to a thorn. The reaction to the thorn was indubitably acquired by the trial-and-error method by the shrikes I observed as early as 1932, and in a paper then published I enlarged upon this particular case of innate behaviour patterns interlocking with trial and error and conditioning. Now G. Kramer has recently reared red-backed shrikes with a view to studying their migration activities and incidentally observed the development of their reaction to the thorn. What he saw at first was in perfect accord with what I had written in 1932; the shrikes showed the movement first and got conditioned to the thorn by trial and error. Kramer is a past master in rearing birds, and you may rest assured that these shrikes were in the best of condition. And yet, when he tried still further to improve his feeding technique and started breeding silkworms for the young shrikes, the birds thus reared needed no conditioning to the thorn, but most amazingly proved to have an innate directive mechanism reacting to the specific optical sign stimuli emanating from a sharp end or thorn. This particular source of error must be kept in mind.

As a method of analysis of animal behaviour, *keeping* truly deserves the adjective "correlative," because it very strictly compels the investigator to consider the behaviour patterns of a species in correlation to the environmental conditions to which they are adapted. Again, it is the observation of miscarrying behaviour patterns, from which the keeper must be able to deduce what particular conditions of the artificial environment of the animal he must *change*, to make those behaviour patterns attain their normal survival value. Just as he would deduce from the morphological characters of a mole's forepaws that this species needs earth to dig in, so he must, from slight "hints" of miscarrying behaviour patterns, be able to deduce the corresponding environmental exigencies of the species. Perhaps the most important cognitive value of this proceeding lies therein, that it forcefully compels the observer to treat the several behaviour patterns of a species exactly as he would treat organs. The observer is very convincingly taught that behaviour patterns are not something which animals may do or not do, or do in different ways, according to the requirements of the occasion, but something which animals of a given species "have got," exactly in the same manner as they "have got" claws or teeth of a definite morphological structure.

From the recognition of this fact it is only a very short step to the systematic *comparison* of the innate behaviour patterns characteristic of allied species. It is, indeed, only in keeping animals that this comparison obtrudes itself to the observer's notice. The discovery of *phyletic homology* between innate behaviour patterns was unavoidable, once a scientist conversant with the methods of comparative morphology got the opportunity to observe closely a number of species belonging to the same systematic group. It is there that C. O. Whitman's "fancy" for pigeons and O. Heinroth's partiality to Anatides have played a very decisive part. A man knowing the single system of actions of every species as they knew it could not fail to notice that certain innate behaviour patterns were just as characteristic of a species, a genus or a family, as any morphological character. As early as 1898, C. O. Whitman wrote the sentence that marks the birth of comparative ethology. "Instincts and organs are to be studied from the common viewpoint of phyletic descent."

"Comparative anatomy" of behaviour

With the discovery of phyletic homology of behaviour patterns the study of behaviour may be said to have grown out of the purely idiographic stage of its development and to have reached the second systematic stage. This discovery was of the greatest importance in several respects. Not only could innate behaviour patterns be used as very valuable and certainly very welcome new taxonomic characters in ascertaining the phyletic relations between allied forms, but the special form of certain innate behaviour patterns became accessible to a causal explanation based on the understanding of their phyletic origin. Both of these facts, though certainly important enough in themselves, attain an even greater secondary importance by proving the great independence of the behaviour patterns in question as *particulate elements* of behaviour. Thus not only the well-tried method of comparative morphology became applicable to the study of behaviour, but it also became possible to *isolate* a very distinct physiological process as an independent constituent of behaviour and to study it separately, in a legitimate departure from the otherwise obligatory method of correlative analysis on a broad front. Therefore, physiological analysis followed closely on the heels of the comparative and systematic study of the elementary process in question.

Zoological systematics and comparative morphology are, by many, considered tedious and rather mummified branches of research. It is too easy to forget that we owe them the most revolutionary discovery of the last hundred years, the discovery of evolution. Though I emphatically deny the charge that comparative morphology is a spent science which has already borne all its fruits, I should concede that it certainly has borne the most important of them. Now there is some danger that the same might be thought, very prematurely, of comparative ethology. In the study of behaviour the phyletic viewpoint has yielded, as a very valuable result, the possibility of isolating and studying in linear, experimental analysis an important, particulate central nervous function, the endogenous activity. In consequence of this, a second, nearly as independent *perceptual* function, the so-called innate releasing mechanism, has become experimentally accessible. Each of these two functions, as a constituent of behaviour, doubtless plays as important a part as the reflex and the conditioned response, and there is a very great temptation to repeat the error of which we have reproached behaviourists and reflexologists, by believing that we have got hold of "all" the necessary "elements" to explain "everything," and to plunge into linear, experimental analysis without giving any further thought to the necessity of further collecting purely observational facts. Of course, the fulfilling of this duty will not keep us from eagerly following up the leads to linear, experimental analysis which we have discovered. But it is well that we should keep in mind the extreme youth of our branch of science and that we should cherish no illusions about the breadth of the observational basis of induction that has hitherto been attained. A precipitate transition from systematics to nomothesis would be particularly disastrous to our branch of research, and the longer and the more we lay stress on the *descriptive* side of comparative ethology, the better it will be in the long run.

Since the days of Charles Darwin the term "comparative" has assumed a very definite meaning. It indicates a certain rather complicated method of procedure which, by studying the similarities and dissimilarities of homologous characters of allied forms, simultaneously obtains indications as to the phyletic relationships of these forms of life and as to the historical origin of the homologous characters in question. I need not enlarge on the details of this method which is a commonplace to biologists and physiologists. We all know perfectly well what we mean by "comparative" anatomy, morphology, physiology and so on. But it is all the more misleading if psychologists, who evidently are not familiar with what we mean when we speak of the comparative method, apply the same term in a very loose sense to all behaviour studies concerned with different forms of life. I must confess that I strongly resent it, not only from the terminological viewpoint, but also in the interests of the very hard-working and honest craft of really comparative investigators, when an American journal masquerades under the title of "comparative" psychology, although, to the best of my knowledge, no really comparative paper ever has been published in it.

* * *

Historically the first and perhaps still the most thoroughly studied group is the order of pigeons which was investigated by Whitman. Second are the order of Anatidae, as studied by Heinroth, and one family of this order, the Anatinae, more especially studied by Heinroth and by myself. If, as a third group, I add the cichlid fishes, investigated by A. Seitz, G. P. Baerends and Baerends van Roon and A. Steiner, whose paper on dwarf cichlids is, however, not yet published, our list is complete.

* * *

The immense field of observation which is still waiting to be systematically exploited needs whole armies of investigators. These armies are, however, already in existence, at least two of them: the bird-lovers and the fish-fanciers. It is a superlative merit of a book like that of the Rev. E. A. Armstrong that it calls the amateurs' attention to the important details which we need to know. There can never be too many such really scientific and yet generally intelligible books on behaviour. I have already explained why the bird or fish "lover" is in some respects the ideal observer. The task is, to enlist him to help us in concerted effort.

The methodologically first task is, without any doubt, the *intensive* study of all of the innate behaviour patterns of a restricted group of forms. Besides this, it is a legitimate way of proceeding, *extensively* to study throughout the animal kingdom one given homology of behaviour pattern. O. Heinroth (1930), in a paper as interesting as it is short, has given us a survey of the extension and distribution of some exceedingly widely spread and evidently exceedingly old behaviour patterns of vertebrates, such as yawning and scratching the head with the hind leg. Curiously enough, I do not know of another paper proceeding in this way, though this way of pursuing a behaviour pattern as a homologous character throughout a large group is interesting work and gives quick results in comparison with other comparative investigations. At present I am collecting notes on the scratching movements of mammals, particularly of rodents. In this group, and also in primates, the areas scratched with the hind leg and with the forepaw respectively are a taxonomic character whose distribution is most interesting.

Only in two groups (in ducks and cichlid fishes) have our intensive studies of the single species and our extensive studies of the several behaviour patterns progressed far enough to make possible what in comparative morphology is quite a commonplace thing, that is, to represent both lines of investigation in a tabular index. In one of my papers (1947) I have given such an index concerning a number of behaviour patterns of sixteen species of ducks belonging to different families (Anatinae, Cairininae and Casarcinae). This method of letting, metaphorically speaking, two lines of investigation cross each other at right angles, is fundamental to every kind of phyletic investigation of homologous characters. Only this kind of broad survey can give us the possibility of forming a firm opinion about the relative age and phyletic conservativity of any taxonomic character.

In some cases which are particularly favourable to this kind of investigation, it has been possible to form a quite well-founded opinion concerning the lines along which the evolution of certain behaviour patterns has taken place. Always keeping in mind that there are *no* "primitive" forms of life, and that therefore any linear taxonomic arrangement of living animals or plants is necessarily sheer nonsense, it is nevertheless possible and legitimate to attribute the quality of relative primitivity to *single characters*. It is a dangerously loose and misleading expression to call any recent form of life, however many primitive characters it may possess, a "primitive" animal. In regard to all the characters of his mouth and teeth, man is much more primitive than the *Platypus*. What alone may be arranged in a linear sequence are the different forms of actualization which one and the same homologous character shows in related forms of life. The ramifying "line of differentiation" thus obtained does, very probably, correspond to the actual stages of evolutional development of the character in question. A number of indubitable sequences of differentiation could be established among innate behaviour patterns.

In all cases where it has been found possible to do this, the innate behaviour pattern in question was a "releasing ceremony." This is far from being accidental.

* * *

A social releaser is a device—either a property of colour and/or shape, or a special sequence of movements, or, for that matter, of sounds, or a scent—specially differentiated to the function of eliciting a response in a fellow-member of the species. To every releaser, as an organ for sending out sign stimuli, there corresponds a perceptual correlate, an "organ" to receive sign stimuli and to activate the answering reaction. This we call an innate releasing mechanism (*auslösendes Schema*). The function of social releasers and of answering innate releasing mechanisms is very much like that of a human signal code, or of that of human word language. Exactly as the several forms of the single word symbols of our language do not have any direct connexion with their meaning, but are only fixed by the *convention* of one particular sign code, so also is the particular form of any social releaser determined by an intraspecific "convention" between the sender and the receiver of sign stimuli. The important consequence of this is that we may exclude from our consideration the possibility of convergent adaptation when comparing similarity and dissimilarity of releasers. It is immeasurably improbable that the similarity of the words *mother, mutter, mater, μήτηρ* and *māT-b* in English, German, Latin, Greek and Russian, should be due to anything else than to their common derivation from a mutual Indo-European "ancestral form." If, therefore, the comparative philologist is fully justified in neglecting any other possibility, then the very same holds true for any resemblance between releasing "ceremonies" of phyletically allied species. This is why we know more about the origin and the evolution of releasing ceremonies than we do about any other innate behaviour patterns. This is also why, on the basis of releasing activities used as taxonomic characters, it is, in quite a number of cases, possible to ascertain phyletic interrelations of species with a certitude and a wealth of detail, hardly ever attainable on the basis of purely morphological characters.

We know of two phyletically distinct ways, by which non-social, mechanically effective endogenous activities may develop into social releasers: in one case the so-called "intention movement" (*Intentionsbewegung*, Heinroth), in the other so-called "displacement activities." Both types of movement are, in their primary form, mere by-products of action-specific excitation and certainly devoid of any survival value. The intention movement is nothing but a slight hint of a certain innate behaviour pattern, as will occur whenever action-specific excitation only reaches a very low level of intensity. Displacement activities are the result of a very different physiological process. If action-specific excitation is at a comparatively high level of intensity and finds its path to its normal outlet suddenly blocked—for instance, by activation of an antagonistic drive or by sudden disappearance of the adequate object—it "sparks over," as Makkink very expressively describes it, into a nervous path not normally its own and finds its expression in behaviour patterns belonging to a quite different action-specific excitation. Makkink, Tinbergen, Kortlandt and E. A. Armstrong have studied displacement activities very extensively.

Intention movements as well as displacement activities are, to an observer well conversant with the system of actions of a species, very reliable indicators for the present "mood" of an animal, that is, in objectivistic terminology, for its present internal state of specific readiness for certain activities. Now in very many animals an analogous "understanding" of the intention movements and the displacement activities of a fellow-member of the species has evidently been developed on the basis of innate releasing mechanisms. (It seems well worth mentioning that as yet we do not know of a single instance in which a similar "understanding" is brought about by conditioning!) In other words, the characteristic movements in question developed a totally new function as social releasers. From that point onwards both types of releasing movements underwent an exceedingly characteristic process of differentiation. The primarily important, mechanically effective constituent parts of the movement are lessened in strength and amplitude, while all *optically* effective features of the movement are exaggerated and emphasized in a very peculiar and striking way. We know to-day of a very great number of examples of this particular process of evolution; there are all possible gradations between movements, as yet hardly different from their original form and behaviour patterns, whose differentiation along the lines described above has gone so far as to transform them into grotesque "ceremonies" or "dances" whose actual homology to the primary intention or displacement activity can only be ascertained through a thorough study of intermediate forms—if such happen to exist. Releasing ceremonies derived by this over-accentuation on "mimical exaggeration" from intention movements were called "symbolic movements" (*Symbolbewegungen*) by Heinroth. In the case of displacement activities a terminological difference has not yet been made between primary displacement and fixed, ritualized, mimically over-accentuated ceremonies.

In a very great number of known cases, the differentiation of a homologous releasing behaviour pattern has, in allied species, taken place along divergent lines. Very often different morphological devices have been evolved to accentuate the optical effect of a ceremony, thus emphasizing different parts of one homologous behaviour pattern in different species. In surface-feeding ducks there were a good many instances of this. The evolutive process described makes it easily intelligible why in many instances the behaviour pattern should be—as it evidently is—phyletically very much older and much more conservative than the morphological devices, colour and/or structure, that help to accentuate it. Another notable fact is that the sequence of movements, constituting a releasing ceremony, is, in very many cases, much more conservative than its "meaning," than the innate releasing mechanism activated by this particular set of sign stimuli in a fellow-member of the species. The extensive comparative study of one distinct releasing behaviour pattern often leads to amazingly similar results, as the comparative study of a root word in the historical development of language. Just as the meaning of one word symbol may split up into different and even opposite meanings of its derivatives (for instance, in the English and German words *knight* and *Knecht*, which have both been derived from a common root describing an armed follower of a war lord but have assumed the opposite meanings of a lord in English and of a serf in German), so does, to cite only one instance, the behaviour pattern originally expressing inferiority in cichlid fishes assume the exactly opposite meaning of a threatening gesture in the "dwarf cichlids" of the genera *Nannacara* and *Apistogramma*. What is more, we know the phyletic development by which it has done so.

* * *

This kind of "comparative philogeny" of releasing behaviour patterns is not only a fascinating study in itself, but it may, once we know more about its object, develop a very practical value. I would stress the fact that the human expression of emotion is largely built up on exactly the same kind of intention movements as Charles Darwin knew long ago. In man, in whom innate behaviour patterns are, to a great degree rudimentary, the study of these innate expressions of emotion may become a most valuable clue to the human "instincts"!

Approach to physiology

The accumulation of action-specific energy. C. O. Whitman and O. Heinroth were phylogenists and not physiologists. Their chief interest in innate behaviour patterns was of a systematic and taxonomic nature. If they had, as we have reasons to believe, some shrewd suspicions of the physiological singularity of the innate behaviour patterns which they so accurately described, they never said so in so many words and they certainly did not separate them conceptually from other innate behaviour patterns which are of a reflex nature, as taxes, innate releasing mechanisms and others. But while systematically describing and arranging what they called "instincts" and "*arteigene Triebhandlungen*" respectively, they did, maybe quite unconsciously, discriminate a very distinct type of innate behaviour pattern from all others. On the grounds of what we know to-day, it is quite obvious why the endogenous behaviour patterns are much more useful as taxonomic characters than reflex ones. Maybe it was only this that made the pioneers select endogenous activities for their object. But *by* selecting them and grouping them together as something distinct, they brought to light certain striking characteristics which seemed to be clamoring for a causal, physiological explanation. And this is exactly how, in a healthy inductive science, nomothesis will automatically grow out of systematics.

What particularly seemed to demand an explanation was a very unexpected correlation which evidently existed between the *spontaneity* and the *invariability* of the innate behaviour patterns in question. According to the vitalists' opinion, spontaneous activity must automatically be considered as purposive, and must, therefore, show adaptive variability in the pursuit of its end or goal. According to the mechanists' opinion, any innate and individually invariable behaviour pattern must be considered to be a chain of unconditioned reflexes and must, therefore, be totally devoid of spontaneity. In defiance of both these dogmas, just those highly differentiated innate sequences of movements that not only were absolutely invariable individually, but even in phylogeny were rather conservative and slowly changeable characters, proved to possess a very distinct and striking kind of spontaneity.

At the level of superficial observation the innate behaviour patterns in question appeared reflex-like in that they were set off by a sort of "trigger-action" in a very specific stimulus situation. But on closer inspection it became apparent that these activities are, at bottom, to a very high degree independent from external stimulation. Captive animals, deprived of the normal object or releasing situation of an innate behaviour pattern, will persist in discharging the same sequences of movements at a very inadequate substitute object or situation. The longer the normal stimulation is withheld, the less necessary it becomes, in order to set off the reaction, to supply *all* of the stimuli pertaining to it. The longer the reaction does not go off, the finer the trigger that releases it seems to become set. In other words, the threshold of the stimulation necessary to release this type of innate reaction, is not a constant, but is undergoing a continuous process of lowering, going on throughout the time during which the reaction is not released. This gradual lowering of threshold does, in a good many cases, actually reach the theoretically possible limit of zero, that is, the activity in question will finally go off *in vacuo*, with an effect somewhat suggestive of the explosion of a boiler whose safety valve fails to function. This occurrence has been termed "*Leerlaufreaktion*" in German, vacuum reaction and explosion reaction. I would move the general acceptance of Armstrong's term "energy accumulation activity" for reasons discussed later.

The consequences of the "damming up" of a certain innate activity are, however, not confined to the threshold of the mechanism (whatever that mechanism may be) which releases the activity. It is not only a facilitation of the releasing process, not only an increase of passive excitability that takes place, but, quite on the contrary, an active and peculiar excitation. Any one of these particular innate behaviour patterns, however small and unimportant it may seem in itself, develops into an active source of excitation which influences the whole of the organism whenever it finds its outlet blocked. In this case, the undischarged activity becomes a *motive* in the literal and original sense of the word, derivated from *movere*, "to move." In the simplest and most primitive case the organism shows undirected, "random" locomotion, *kineses* as we term it. In more highly differentiated types these kineses are interlaced with taxes orienting the organism's locomotion in space, or even with conditioned responses and all the most complicated and least analysed forms of animal and human behaviour, which, for lack of a better term, we are wont to describe as "intelligent." Though the activities thus elicited comprise the whole range of behaviour, from its simplest to its most complex form, they have one decisive character in common: they are all *purposive* in the sense which E. C. Tolman has given to this term, that is to say, they all tend to bring about, by *variable* movements, an *invariable* end or goal, and they go on until this goal is reached or the animal as a whole is exhausted. The invariable end or goal is represented by the releasing stimulus situation and, therewith, the discharge of the specific behaviour pattern that had been dammed up. The purposive behaviour striving for this discharge was called *appetitive* behavior by Wallace Craig, the behaviour pattern finally discharged was termed *consummatory action*.

I want to emphasize what a great wealth of observational evidence is underlying the facts condensed into the two preceding paragraphs. It took the life's work of at least three excellent observers to gather that evidence: Whitman, Heinroth and Craig. All these facts are strongly suggestive of a very definite assumption, and it is eminently characteristic of these three men that every one of them has conscientiously recorded those facts while none of them put this assumption into words. This is all the more significant of the value they set on observation free of any hypothesis, as they must have found it quite hard to avoid descriptive terms like damming up, discharging, etc., which already imply the assumpton *that some sort of energy, specific to one definite activity, is stored up while this activity remains quiescent, and is consumed in its discharge*.

* * *

The innate releasing mechanism. The conception of a wound-up spring implies the existence of a trigger mechanism, holding it back until the right moment and setting it off when this moment arrives. W. Thorpe says of endogenous activities that "they are set off by a kind of trigger action." J. A. Portielje has, at a very early date, fully recognized this double nature of endogenous activity which always is an "*Aktion-und-Reaktionin-Einem*," an action and a reaction in one, as he expresses it. While it is obviously inadmissible to speak of a re-action to a stimulus not yet received, for instance, when the organism is striving by trial and error to attain a certain releasing stimulus situation (appetitive behaviour in W. Craig's terminology), the organism's answer to this situation, once it is attained, just as obviously *is a re*-action in the truest sense of the word.

Indeed, this reaction has all the earmarks of what I. P. Pavlov would call an unconditioned reflex. With the recognition of the peculiarity and independence of endogenous activity as a distinct physiological process, there evidently arises, as a physiological problem by itself, the question about the "trigger" mechanism releasing the accumulated action-specific energy. The statement that this mechanism is an unconditioned reflex, though indubitably correct, does not help us any further, because the essential part of the problem does not lie in the reflex process as such, but exclusively in its perceptual side. The intriguing problem lies in the *selectivity* of the reaction. The observational facts are suggestive of a mechanism akin to a filter, letting through only certain particular stimuli while strictly excluding others, or of a complicated lock which can only be unlocked by a very specific key. The adaptation of the perceptual organization to certain characteristic key stimuli is very evident.

Through the work of a considerable number of recent investigations some very peculiar facts have become known about the perceptual organization of these innate releasing mechanisms. What acts as a "stimulus" is, in the great majority of cases, not a simple physical influence of *absolute* quality and quantity, but a *relational* property. These *innate* reactions to relational properties differ from reactions to acquired Gestalt perceptions only by their extreme simplicity. In acquired reactions, the Gestalt that elicits the response consists of an indivisible unit of literally hundreds of interdependent relational properties. In innate releasing mechanism we do not yet know a single case in which the innate and immediate reaction is elicited by a relational property comprising the relation between more than *two* elements. The innate releasing and directing mechanisms of the gaping activity of young thrushes (*Turdus*) are a good example. The nestlings will gape at (a) the nearer of two objects, (b) the higher of two objects and (c) the smaller of two objects, the optimum relation of the smaller to the bigger being 30%. All three of these relational properties are characteristic of the normal object of the reaction, the parent's head. Each of them, considered by itself, is a true, if extremely simple, "Gestalt," showing all of Ehrenfels' "Gestalt" criteria, most important of all, transposability. Each of these relational properties represents a key-stimulus setting of the *same* innate releasing mechanism, but each of them does so quite independently from the other. Unlike the immeasurable number of relational properties inextricably woven into the whole of a "Gestalt" perception, *these single relational key stimuli act merely as a sum of non-interdependent elements.* This process of summation of key stimuli, belonging to one and the same innate releasing mechanism, has been termed *Reiz-Summen-Phänomen* by A. Seitz, who was the first to study it intensively. N. Tinbergen translated this into English by "law of heterogeneous summation."

* * *

In his investigation of the gaping reactions of young thrushes, Tinbergen distinguished between releasing and directing mechanisms, according to whether they activate discharge of endogenous activity or a taxis. I think that this terminological differentiation is not advisable. Of course, it is not very elegant to talk of an innate mechanism "releasing" a taxis, or, still worse, a social inhibition. But somehow we ought to express in our terminology the fundamentally important fact that it indubitably is *the same* innate organization of perception that is filtering and selecting the key stimuli activating these very different responses.

The method of dual quantification

A very considerable percentage of all animal activities consists of the typical successive links of appetitive behaviour, attainment of a desired stimulus situation, to which an innate releasing mechanism responds and sets off the discharge of accumulated endogenous action. In the vast majority of cases where we find an organism responding specifically and without previous experience to certain stimulus situations, closer investigation will reveal one or other of the innumerable variations to this theme, always leading up in one way or the other to the final discharge of consummatory actions. What we can objectively observe is exclusively this discharge. But this discharge is dependent upon two absolutely heterogeneous causal factors: (1) the level attained by the accumulated action-specific energy at the moment and (2) the effectiveness of external stimulation. None of these two factors is directly accessible to our observation. Absolutely identical reactions can result, in one case, from an extremely low level of endogenous accumulation and strong stimulation, and, in the other, from a high level of accumulated action-specific energy and a very weak external stimulation, or even, in the case of explosion activity, from internal factors alone, external stimulation not taking any part in the activity at all. This has to be taken into very serious consideration in the experimental study of perceptual organization. If, after the elimination of one or more single stimuli, the activity can still be elicited in its normal form, this does not in any way justify the conclusion that these stimuli are ineffective in activating the releasing mechanism pertaining to the activity in question. If Stone's (1923) rats discharged certain innate behaviour patterns apparently normally, although they were deprived of important sensory organs, this is in no way astonishing. A mother rat who, in spite of the removal of her olfactory bulbs, continues to foster her young, is in no other position than an old virgin Scotch terrier bitch who, with a sufficient lowering of threshold, proceeds to discharge homologous reactions with a cushion for a substitute object.

What is constant is not the effect actually produced by a certain key stimulus at a given moment, but the relation between this effect and the organism's momentary internal state of accumulation of action-specific energy. The constancy of this correlation only becomes apparent when both the internal and the external factors are quantitatively investigated *simultaneously*. The method of doing this truly deserves the name of a *correlative* analysis. In a dabbling and uncertain way I sensed the possibility of this method as early as 1926, when I was experimenting with my first free-flying tame jackdaw. The observations I then made are still very vivid in my memory, and though it has taken me years to see their real purport, they are a good and simple illustration of what I am trying to explain. The following reaction of a hand-raised and fully fledged young jackdaw (*Coloeus monedula*) is elicited by a small number of single stimuli which, though apparently simple, nevertheless imply the necessity of assuming an innate perceptual organization responding to *relational* properties. The experimenter who wants to elicit the reaction which the young birds give to the stimuli emanating from its parents guiding it, must do at least *one* of three definite things. He must either more or less quickly move *away from* the bird, or *upwards* from its position, or he must utter a good imitation of the call-note of the species. (A fourth very effective stimulus, pertaining to the same releasing mechanism, is the sudden unfolding of a pair of black wings, but this is not so easy for the human experimenter to imitate. Its existence is, however, immaterial for our present concern.) When I walked abroad with that young jackdaw, the bird would, at first, follow me very willingly. It would fly up in pursuit even when I walked away from it quite slowly. (Though it was always possible to get away from the bird, without having it following after me, by the method of letting the stimulus "creep in," walking away with extreme slowness.) After a time it became necessary for me to walk away *abruptly*, over-accentuating my movements very much in the same way as a parent cichlid does with his swimming-away movements. After a few more performances I had to *add* another stimulus, that is, I had either to call, or to crouch low to the ground beside the sitting bird and jump up in starting. At last, I had to do *all* of these things simultaneously and with the utmost intensity, crouching close to the jackdaw, jumping up suddenly, running away as quickly as I could and yelling jackdaw-calls all the time, in order to make the bird take to wing at all. In this simple experience two important facts are obtrusively apparent: (a) the activity is *exhaustible* independently from the general state of exhaustion of the organism as a whole, (b) the stimuli which release it are *interchangeable* and the general releasing effect of a situation is dependent on the *sum* of the single releasing stimuli contained in it. Furthermore, another fact mentioned earlier is again illustrated: internal and external motivation may add up to exactly identical results if the weakness of one is compensated by strength of the other.

Whenever we put stimulation in front of an organism and register its answer to this stimulation, we have to deal with an equation containing *two* unknown factors. The quantification of each of these factors is, on principle, impossible without simultaneously ascertaining the quantitative effect of the other. What is visible to us is (a) the stimulus situation we are putting before the animal, and (b) the discharge of a specific activity thereby released. The varying internal state of specific excitability is only accessible to us through the means of these two indicators. If we are looking for an object favourable to an exact solution of this equation containing two unknowns, we must obviously search for an activity, the discharge of which has a number of clearly distinguishable grades of intensity, and the releasing mechanism of which consists of an appreciable number of distinct stimuli, so that by the addition of these stimuli we may obtain a gradation of different releasing values.

A very extensive observational basis is needed to obtain a certain knowledge of all the different forms of movement activated by *one* quality of action-specific excitation. The internal state of heightened action-specific excitability may betray itself by small intention movements, and from the merest hint of these up to the activity of full intensity there are all possible gradations.

* * *

For the purposes of dual quantification it is necessary that not only the intensities of the activity released, but also the releasing effect of external stimulation should be accessible to an exact graduation. This is obviously best in cases in which the innate releasing mechanism is activated by a comparatively great number of independent stimuli, in other words, in the most complex and most highly differentiated releasing mechanisms. These, indubitably, are the ones that correspond to social releasers. In this respect, the fighting reactions of certain fishes again prove to be an admirable object, being released by a number of visual stimuli which not only are easy to reproduce in a dummy, but also very different from each other as to their respective releasing values.

Before describing in detail how we proceed in our simultaneous quantitative study of endogenous generation of action-specific energy and of the respective effectiveness of single releasing stimuli, I want to emphasize what a tremendous amount of purely observational work has to be done, to furnish us with the *measure* with which we quantify these factors.

* * *

The practical way of proceeding in experiments of dual quantification is obvious, after what has already been said. Presenting the animal with a given stimulus situation and recording the intensity of its reaction presents us, as I have explained, with an equation containing two unknowns: we do not know how much of the intensity recorded is due to internal accumulation of action-specific energy and how much is due to external stimulation. The obvious thing to do, is to let *maximal* stimulation impinge upon the organism immediately after the first experiment, in order to see how much specific energy is "left." This already gives us a definite notion about the *relative* effectiveness of the stimulation supplied in the first experiment. What we are doing is best illustrated in a hydro-mechanic model which, in spite of its extreme crudeness and simplicity is able to symbolize a surprising wealth of facts really encountered in the reactions of animals. In Fig. 1 the tap T supplying a constant flow of liquid represents the endogenous production of action-specific energy; the liquid accumulated in the reservoir R represents the amount of this energy which is at the disposal of the organism at a given moment, the elevation attained by its upper level corresponds, at an inverse ratio, to the momentary threshold of the reaction. The cone-valve V represents the releasing mechanism, the inhibitory function of the higher centres being symbolized by the spring S. The

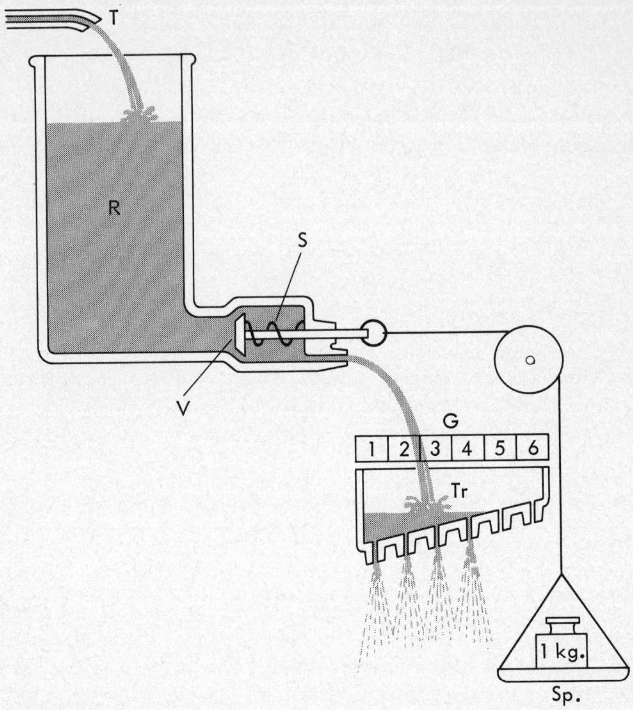

Figure 1

scalepan Sp which is connected with the valve-shaft by a string acting over a pulley represents the perceptual sector of the releasing mechanism, the weight applied corresponds to the impinging stimulation. This arrangement is a good symbol of how the internal accumulation of action-specific energy, and the external stimulation are both acting in the same direction, both tending to open the valve. It can also easily and obviously represent the occurrence of explosion activity. The activity itself is represented by the spout discharged from the jet F. The intensity of reaction is symbolized by the distance to which the jet springs, in other words, by the speed of the outflow. This automatically corresponds to the proven fact that the consumption of action-specific energy in the time unit is in direct proportion to the intensity of the reaction. The intensity of the reaction can be read on the scale G. To this apparatus we can easily attach a gadget exactly symbolizing the way in which a sequence of different movement patterns belonging to one scale of action-specific excitation is activated. A row of little funnels attached below the gradation will meet the case where, with the attaining of a higher level of excitation, the activities corresponding to lower levels are discontinued (as, for instance, in the taking-to-wing ceremony of the greylag goose). It is, however, much more usual that the movements activated at the lowest levels of action-specific excitation are continued unceasingly all the while those corresponding to higher levels are discharged. We can symbolize this by fixing below the scale G an oblong trough Tr which has an oblique bottom perforated by a number of holes. The outflow from these holes then represents the intensity scale of a sequence of different activities, such as fin-spreading, gill-membrane expanding, etc. For reasons subsequently to be expounded, we have arranged the scale tray representing the receiving section of the innate releasing mechanism in such a manner as to let some of the ultimate flowing out of liquid impinge on it in a diffuse way.

This contraption is, of course, still a very crude simplification of the real processes it is symbolizing, but experience has taught us that even the crudest simplisms often prove a valuable stimulus to investigation. As an instrument for the quantification of external and internal stimulation this model has already proved to be of some value. Let me explain its use. Suppose we present an organism with a stimulus of unknown effectiveness. All we can immediately record is, as already explained, the intensity of the reaction. In the terms of our model, we do not know what weight we have applied to the pan. In order to ascertain it, we must try to get some notion of the pressure pushing on the valve from the inside. The simplest way to do this is to open the valve altogether and record the distance which the spout delivered by the jet will now attain. In other words, we shall present our animal with the *normal* object of the reaction which may be roughly (though not theoretically) identified with its optimal object, and record the intensity of which the reaction is capable at the moment. Out of both data the relative effect of the first stimulus can be roughly calculated, as well as the pressure acting from within. In other words, we have now got two equations with two unknowns.

* * *

If a releasing mechanism contains the perceptual correlates to a comparatively large number of key stimuli, it is possible to construe an immense number of combinations and permutations of these stimuli. By this we obtain the further possibility of comparing the stimulating effect of objects whose releasing value is approximately the same, but is brought about by a summation of entirely different constituent key stimuli. By patiently constructing a great number of dummies of approximately equal releasing value, but different elementary stimuli, A. Seitz (1942) was able to show rather conclusively that the releasing value of the single constituent stimulus is absolutely constant in all possible combinations. This is what he termed "*Reiz-Summen-Phänomen*" in German, and what Tinbergen translated into English by "Law of heterogeneous summation." Wolfgang Schmidt (to be published) investigated the human reaction to the visual stimuli of laughing with exactly the same method and obtained strictly analogous results. The effect of heterogeneous summation could be demonstrated even more clearly in this case, because in experimenting with humans it is possible to compare directly the releasing values of two dummies, by applying the choice method. In experimenting with animals, especially with the lower forms, this method has a very serious drawback: it frequently occurs that the stimuli, activating an innate releasing mechanism, are *not identical with those that direct it in space*, and therefore very misleading results can be obtained by letting an organism "choose" between two dummies. As Tinbergen has shown conclusively in the greyling butterfly, *Satyrus semele*, it is quite possible that one object elicits the discharge of endogenous activity, while the second is sending out directing stimulation guiding the organism's reaction to itself.

* * *

When we suddenly deprive an animal of the object of its reaction, the activity never breaks off abruptly but nearly always continues a considerable time *in vacuo*. Doubtless it is a consequence of the same phenomenon that the "momentum" gained by any activity will carry it on for an appreciable time after the moment when its releasing threshold, rising continually throughout the duration of the discharge, has reached the value corresponding to the external stimulation impinging at the moment. This is also the reason why an organism that is left continually in the presence of a releasing object does not continually react to it with a constant intensity, as otherwise would be expected. The inertia of a reaction carries the threshold high above the value corresponding to the ever-present stimulating object. After the reaction has ceased, the gradual lowering of the threshold during its period of quiescence acts exactly as a correspondingly slow increase of stimulation would act upon a constant threshold, that is, it results in the well-known phenomenon of "creeping in" of stimulation. Because of this, the new burst of activity does not occur until the threshold is very appreciably *below* the value corresponding to that of the constant stimulation. The result of this is the usual form of discharge, in rhythmically recurring "bursts" of activity. The heightening of threshold, caused by inertia of reaction, may be so considerable that it can supersede a quite substantial difference between the stimulating effects of two dummies. If we let the animal react to one combination of stimuli until activity ceases and then *fail* to elicit any response by the presentation of another set of stimuli, we are not yet in possession of conclusive proof that the second combination is not quite considerably stronger than the first. For obvious reasons this fact is all the more important, the more quickly the activity in question is exhaustible.

It is probable that the *initial* inertia of endogenous activities and their propensity to *continue* longer than corresponds to present stimulation, are two entirely different phenomena, requiring different physiological explanations. The way in which quiescent activities respond to stimulation is more suggestive of initial *friction*. They "behave" exactly as if the valve releasing their discharge was a bit sticky. Furthermore, there arises the question whether there is not a distinct relationship between creeping in of stimulation and what we call habituation. Contrary to these phenomena of an initial resistance to stimulation, the continuance of activity after cessation of adequate stimulation is easily explainable on the assumption that the activity is *self-stimulating*. A strong argument for this assumption lies in the fact that this type of inertia of an activity is evidently in exact direct proportion to the intensity the discharge has attained. If in our hydrostatical model we arrange the scale-pan receiving the weights representing stimulation in such a way as to let part of the outflow of the activity impinge upon it in a somewhat diffuse way (Fig. 1), we are able to reproduce a very exact replica of the avalanche-like way in which the endogenous activities tend to grow after being released. Also we can account for the fact that the valve stays open much longer than corresponds to the added values of internal pressure and weights on the scale-pan. A very considerable part of all endogenous activities is in this way self-exhausting and self-stimulating at the same time, a fact which has been stressed by Julian Huxley long ago.

Analysis of taxis and kinesis constituents of "instinct"

As long as the whole of innate behaviour was considered a chain reflex, there was no reason to make any particular distinction between the single constituent reflexes which are building it up. The discovery of endogenous generation of action-specific energy has considerably changed this position. I have already shown in what way the recognition of the very peculiar physiological character of these functions forces us to conceive of endogenous activity and of the perceptual organization releasing it as of two very distinct physiological processes which, though they are strictly interdependent constituents of one mutual function, each demand a different method of analytical approach. In very much the same manner as the discovery of endogenous activity had influenced our conception of the releasing mechanism, it also necessitates some corrections of our notions about the mechanisms *directing* activity in space. Correlative analysis is like the solving of a crossword puzzle, in that the correction of one word which had been put in erroneously makes it necessary to revise out notions about all the parts interlocking with it. I need not say anything here about the interdependence of endogenously automatic activities and taxes as constituents of innate behaviour; a much better authority on the subject will do so.

I would, however, say a few words in answer to an evidently widely spread misconception about our conceptional distinction of physiologically independent constituent functions. It is in the very nature of analysis that it leads to a progressive narrowing in of formerly wider and more general conceptions. What formerly was very simply conceived of as "an instinct" is shown by the advance of analysis to be a very complex mechanism of very distinct and very different constituents, such as endogenous automatisms, releasing mechanisms, taxes, kineses and, maybe, quite a number of further as yet unrecognized particulate functions. For this we have been reproached by many psychologists, all of them more or less under the influence of vitalistic and finalistic preconceptions. We were accused of pulling to pieces what really was a whole, and it has been said of us what Goethe said of a certain type of analytical science: *"Sie haben die Teile in der Hand, fehlt, leider, nur das geistige Band."* I would emphatically deny this charge. When we distinguish independent and particulate constituent functions of innate behaviour, we are just as much justified in doing this as the physiological anatomist is justified in distinguishing between the bones, joints, muscles and nerves of, let us say, a human arm or leg. The conceptional distinction of constituent parts and of their particular qualities does not in any way preclude the fullest cognizance of the general mutual interdependence and interaction of these parts. It does not, in any way, prevent us from seeing the organism as a whole or system. This will seem a very commonplace thing to say to physiologists, as, indeed, it really *is*. But it is a deeply regrettable fact that it seems to be exceedingly necessary to say it again and again to some psychologists.

The hierarchy of moods

Doubtless a quite considerable percentage of all the activities of animals and men are compounded of the three successive links described in the classical diagram given by Wallace Craig and further differentiated by Tinbergen and myself: (1) accumulation of action-specific energy giving rise to appetitive behaviour; (2) appetitive behaviour striving for and attaining the stimulus situation activating the innate releasing mechanism; (3) setting off of the releasing mechanism and discharge of endogenous activity in a consummatory action. Doubtless this tripartite organization of behaviour does occur. Pessimists assert that yesterday's truth is to-day's error. In biological research work it is certainly more correct to say that the truth of yesterday is, not the error, but the *special case* of to-day. In biological research, an all-too-cautious abstaining from forming a hypothesis would get us nowhere, and we must have the courage to formulate preliminary hypotheses, though we are well aware that these preliminary formulations are much too simplistic and correspond, at the best, to a particularly simple special case. In fact, the discovery of a natural law has been, in many instances, identical with the discovery of a special case, in which it was actualized in a particularly simple manner. We are therefore neither surprised, nor pained, that the original, tripartite diagram of the organization of innate behaviour has proved a typical example of this kind of simplification. That is to say, it meets a good many cases, but is much too simple for most cases.

It was a most decisive step forward in our understanding of innate behaviour when Baerends (1941) pointed out that appetitive behaviour by no means always leads immediately to the discharge of consummatory action. In the vast majority of cases, the appetitive behaviour with which an activity begins is of a much more *generalized* nature. The releasing situation attained by this first step of appetite and the innate releasing mechanism activated in this situation, do not lead to the discharge of the final consummatory act, but, as the next step, to another form of appetitive behaviour of a distinct and *more specialized* form, striving for another, also more specialized, releasing situation. A very general "mood," in the sense of a readiness to certain activities, as, for instance, the "reproductive mood" of a male stickleback, which comprises the several readinesses to a very considerable but finite number of consummatory acts, is, step by step, narrowed down to the discharge of one of these. There is, in other words, a hierarchical order of wider and narrower readinesses or moods. The action-specific appetite and the discharge of the consummatory action are the *lowest* rung in this ladder of superimposed commanding instances or "centres." Baerends has extensively studied the "Hierarchy of moods" in the digger wasp, *Ammophila campestris*, and Tinbergen in the stickleback. As the latter is going to read a paper on the subject himself, I need not say more here.

From the methodological viewpoint, which is the subject of this paper, it must be emphasized that the conception of a hierarchy of "instincts," at which we have arrived, is something entirely different from what W. McDougall and other purposive psychologists conceive of as chain appetites, first and second order drives and the like. These conceptions have been arrived at from the opposite direction of approach. We have not gone out from finalistically considering what the animal *ought* to do, in order to sustain itself and its species, but from what it actually *does*. We have not started from the hypostatization of a "reproductive instinct," deduced from the fact that sticklebacks reproduce their kind, but from the observation and inventory of the lowest rung of the hierarchical ladder of integration, with the consummatory actions, and worked our way inductively *up* that ladder instead of deductively down. The conceptions of the successive levels of integration are therefore determined *causally* and not teleologically. Therefore Tinbergen's and Bearend's conception of hierarchically superimposed "centres," though determined exclusively from a functional point of view, may prove a most valuable hint to the experimenting physiologist as to the number of localizations for which he has to look.

The psychological aspect of comparative ethology

There is an amazingly widely spread prejudice that a given process of behaviour can only have *either* a physiological *or* a psychological explanation. In F. Hempelmann's otherwise admirable text-book this fundamental error keeps recurring every few pages. Of course there simply is no psychological process which does not have its physiological correlate and which does not demand a physiological explanation as well as a psychological one. The serious consequence of the erroneous "either-or" lies therein *that it implies the superfluity of physiological explanations in all cases where a psychological explanation can be found.* On the other hand, there are a lot of processes comprised in animal behaviour which do not have a psychological explanation. In other words: every psychological process has its physiological side, but not every physiological process is correlated to psychological phenomena.

Though comparative ethology is resolutely and exclusively concerned with an *objectivistic* study of behaviour, as long as it is concerned with animals, we do not, by any means, shut our eyes to one important fact: just those particular physiological processes that are the main object of our investigations, undoubtedly belong to the kind which *does have a correlated psychological side to it.* This is true of all the three most important elementary processes of innate behaviour, of endogenous-automatic activities as well as of innate releasing mechanisms and of taxes.

There cannot be the least doubt that the discharge of accumulated action-specific energy is accompanied by very intense and very specific subjective phenomena. I fully agree with W. McDougall in his fundamental assertion that man has just as many "instincts" as he has qualitatively distinguishable emotions. Jan Verwey, an indubitably objectivistic student of animal behaviour, is evidently of the same opinion when, in his famous paper on the grey heron (1930), he writes: "Where reflexes and instincts can be distinguished from each other at all, there the reflex is functioning mechanically, while instinctive activities are accompanied by subjective phenomena" (translated from the German). Heinroth used to answer to anybody's assertion that animals were reflex machines by saying jocularly: "Quite on the contrary, animals are exceedingly emotional people, possessing very little intellect." The same opinion is evidently underlying Heinroth's conception and treatment of "moods" (*Stimmungen*). Our firm conviction that there exists a subjective, emotional side to "moods," as well as to the discharge of action-specific energy, does not in any way imply our making illegitimate assertions about the nature and quality of an animal's subjective experiences. These are, and ever will be, inaccessible to our knowledge. I would lay great stress upon the fact that Heinroth's conceptions of "moods" can one and all be strictly defined from a purely objectivistic point of view, as specific internal states of readiness to discharge a certain complex of behaviour patterns. Even Heinroth's terms are not derived from human subjective phenomena, but from the objective activity itself, as, for instance, "*Flug-stimmung*," "*Fortpflanzungs-Stimmung*" (flying mood, reproductive mood) and others. We still hold that a real "psychology" of animals is on principle impossible. But in its application to *man*, the fact that the discharge of certain endogenous activities is correlated to certain distinct emotional phenomena may become of tremendous importance.

The psychological aspect of innate releasing mechanisms presents some parallels to that of endogenous activities. We know for a certainty that the attainment of the releasing stimulus situation represents the end or goal to which appetitive behaviour is directed. We can give an exact and purely objective definition of this kind of directedness or purpose. In all human behaviour, which also objectively fits this definition, we furthermore know for certain that the organism, as an experiencing subject, is striving for certain pleasurable subjective phenomena accompanying as well the perception of the releasing stimulus situation as the discharge of the accumulated endogenous activity. I do not think that any observer really familiar with appetitive behaviour in higher animals will ever doubt the fact that the animal as a subject also experiences intense sensual pleasure as the subjective correlate of attaining the releasing stimulus situation. W. Craig gives a convincing description of the behaviour of a young male dove on attaining the specific situation eliciting the nest-calling activity: When "a ready-made nest is put into his cage, the inexperienced dove does not recognize it as a nest, but sooner or later he tries it, as he has tried all other places for nest-calling, and in such trial the nest evidently gives him a strong and satisfying stimulation (the appetitive stimulus) which no other situation has given him. In the nest his attitude becomes extreme; he abandons himself to an orgy of nest-calling (complete consummatory action), turning now this way and now that in the hollow, palpating the straw with his feet, wings, breast, neck and beak, and *rioting in a wealth of new luxurious stimuli*" (italics mine). Again, we cannot say anything about the specific quality of the animal's subjective experiences. But again the laws we have found in animal behaviour find an enormously important application to the special phenomena of human psychology. We do not go far wrong if we suspect the existence of an innate releasing mechanism, wherever we can introspectively ascertain a specific quality of sensual pleasure. To give a good example of this, and quite a proper one at that: It is a distinct and indubitably sensuous pleasure to fondle a nice plump, appetizing human baby. The stimulation emanating from such a baby even elicits one of the very motorically innate behaviour patterns of our species, a quite distinct movement of taking the infant on one arm, holding it with the other. In fondling the baby, we behave to its "sweet" stimulation very much like Craig's pigeon does to the nest. By introspection I can ascertain that the pleasure I feel in fondling and petting the baby is certainly qualitatively different from any other kind of sensual pleasure, especially, whatever psycho-analysts may assert to the contrary, from that of a sexual nature. Indeed, I can assert that my pleasurable sensations in fondling a sweet human child are of the same quality as those I experience in fondling a chow-puppy, or a baby lion. There is not even an appreciable difference of intensity between these reactions; indeed, it is questionable whether a young lion is not, in some respects, what we call a supernormal object to my reaction. But I would seriously warn any psychoanalyst against telling me that I am reacting sexually to that lion! In this case, the existence of a true innate releasing mechanism in man has been clearly proven, the relational properties which the object must possess have been analysed, and the fact that their common effect is obeying the law of heterogeneous summation is quite evident. Also, the objective and subjective reactions activated by the mechanism are clearly distinguishable. A normal man—let alone a woman—will find it exceedingly difficult to leave to its fate even a puppy, after he or she has enjoyed fondling and petting it. A very distinct "mood," a readiness to take care of the object in a specific manner, is brought about with the predictability of an unconditioned response. Quite especially a strong inhibition to hurt or kill the "sweet" baby is activated by the innate releasing mechanism in question. When, much against my emotional inclination, I once forced myself for rational reasons to kill a number of young hooded rats who still were in a "sweet" stage of development, I acquired a very slight, but indubitable neurosis; that is to say, I repeatedly *dreamt* of the occurrence in an emotionally super-accentuated way. This particular innate releasing mechanism of our species is, for obvious reasons, an exceptionally favourable object for the method of approach just described. But there is no doubt that in many more cases introspection, circumspectly applied, can lead to similarly important discoveries.

Perhaps an even greater theoretical interest is due to the psychological aspect of *taxes*. It can be shown irrefutably that it is not possible to draw a sharp distinction between taxes and *insight* either from the objectivistic or from the psychological point of view. Considered from the objectivistic viewpoint, taxes, as constituents of appetitive behaviour, are represented in all possible gradations of complexity, ranging from simple and clearly analysable directing reflexes to those complex processes of spontaneous solution of spatial problems that we are wont to call "insight." The former merge gradually and imperceptibly into the latter, without giving us an opportunity to draw a distinction between both. Neither is it possible to arrive at a clear conceptual distinction between taxis and insight from the psychological, introspective side of the matter. The subjective experience of insight which Karl Bühler so very expressively has termed the "*Aha-Erlebnis*," always characteristically takes place in the exact moment when disorientedness gives way to orientation. But in this also there is no essential difference between the simplest and the most complicated processes of "orientation." The specific experience of the "Aha" crops up in the moment when we attain an "illumination" concerning some very complicated problem, but it also does in the case of a simple and indubitable tropotaxis. When, for instance, our tropotactic orientation to gravity has been removed, let us say by narcosis, or, as I have experienced it twice, in a vasomotoric collapse, then the reorientation sets in again with what I should metaphorically call an audible click, and this click is accompanied by a superlatively strong and qualitatively distinct "Aha-experience"!

In my opinion, every physiological process which can, in the manner just described, be approached simultaneously from the objective and from the subjective side is of a paramount theoretical and practical importance. The cardinal problem of psychology, the question of the interdependence of body and mind, cannot be inductively approached in any other way than by studying those not too common cases, in which the intrinsic unit of one of these highest life processes *can* be studied from the objective, physiological side as well, as from the subjective, psychological one. To refrain from introspection in such cases would mean renouncing a superlatively valuable source of knowledge for purely dogmatic reasons which would be about the worst thing a natural scientist could do. On the other hand, it is high time that social and group psychology began to occupy itself with the physiological side of behaviour and more especially with the innate processes of which I spoke above. It is high time that the collective human intellect got some control on the necessary outlets for certain endogenously generated drives, for instance "aggression," and some knowledge of human innate releasing mechanisms, especially those activating aggression. Hitherto it is only demagogues who seem to have a certain working knowledge of these matters and who, by devising surprisingly simple "dummies," are able to elicit fighting responses in human beings with about the same predictability as Tinbergen does in sticklebacks.

REFERENCES

Allen, Arthur A. (1934), Sex rhythm in the ruffed grouse (*Bonasa umbellus* L.) and other birds, *Auk*, **51**, 180–199.

Antonius, O. (1937–1938), Ueber Herdenbildung und Paarungseigentümlichkeiten der Einhufer, *Z. Tierpsychol.*, **1**, 259; **2**, 115.

Armstrong, E. A. (1942), *Bird Display*. London: Cambridge University Press.

Baerends, G. P. (1941), Fortpflanzungsverhalten und Orientierung der Grabwespe *Ammophila campestris, Tijdschr. Ent.*, **84**, 68–275.

Baerends, G. P. (1950), An introduction to the ethology of cichlid fishes, *Behaviour* (in the Press).

Bierens de Haan, J. A. (1940), *Die tierischen Instinkte und ihr Umbau durch Erfahrung*. Leiden: Brill.

Bühler, Karl (1922), *Handbuch der Psychologie*. Jena: I. Teil.

Craig, W. (1918), Appetites and aversions as constituents of instincts, *Biol. Bull. Woods Hole*, no. 2, 91–107.

Darwin, C. (1874), Der aüsdruck der Gemütsbewegungen (German translation). Stuttgart.

Dewey, John (1918), *Human Nature and Conduct*. New York: Henry Holt and Co.

Eliot, Howard H. (1929), *An Introduction to the Study of Bird Behaviour*. London: Cambridge University Press.

Heinroth, O. (1910), Beiträge zur Biologie, insbesonders Psychologie und Ethologie der Anatiden, *Verh. V. Int. Ornithol. Kongr.*

Heinroth, O. (1930), Ueber bestimmte Bewegungsweisen von Wirbeltieren, *S.B. Ges. Naturf. Fr. Berl.* 333–342.

Heinroth, O. and M. (1924–1928). *Die Vögel Mitteleuropas*. Berlin.

Holst, E. v. (1935), Alles order Nichts, Block, Alternans, Bigemini und verwandte Erscheinungen als Eigenschaften des Rückenmarkes, *Pflüg. Arch. Ges. Physiol.*, **236**, Heft 4, 5 and 6.

Holst, E. v. (1936), Versuche zur relativen Koordination, *Pflüg. Arch. Ges. Physiol.*, **237**, Heft 1.

Holst, E. v. (1936), Vom Dualismus der motorischen und der automatisch-rhythmischen Funktion im Rückenmark und vom Wesen des automatischen Rhythmus, *Pflüg. Arch. Ges. Physiol.*, **237**, Heft 3.

Huxley, J. S. (1914), The courtship habits of the great crested grebe (*Podiceps cristatus*); with an addition on the theory of sexual selection, *Proc. Zool. Soc. Lond.*

Jennings, H. S. *Das Verhalten der neideren Organismen* (German translation by Mangold 1910). Leipzig.

Koehler, Otto (1933), Die Ganzheitsbetrachtung in der modernen Biologie, *Verh. Königsberger Gelehrten Gesellschaft*.

Köhler, Wolfgang (1933), Die physikalische Gestalten in Ruhe und stationärem, *Zustand*. Braunschweig.

Kortlandt, A. (1938), De uitdrukkingsbewegingen en geluiden von *Phalacrocorax sinensis*, Shaw & Nodder, *Ardea*.

McDougall, W. (1933), *An Outline of Psychology*, 6th ed. London.

Makkink, G. F. (1936), An attempt at an ethogramm of the European avocet (*Recurvirostra avocetta* L.) with ethological and psychological remarks, *Ardea*, **25**, 1–60.

Matthaei, R. (1929), *Das Gestaltproblem*. München.

Morgan, C. Lloyd (1909), *Instinkt und Erfahrung* (German translation by M. Semon). Leipzig and Berlin.

Petrunkewitch, A. (1926), The value of instinct as a taxonomic character in spiders, *Biol. Bull. Woods Hole*, **50**, 427–432.

Plath, O. (1938), *Bumble Bees and Their Ways*. Boston.

Portielje, A. F. J. (1938), *Dieren zien en leeren kennen*. Amsterdam.

Seitz, A. (1940), Die Paarbildung bei einigen Zichliden. I. Die Paarbildung bei *Astatotilapia strigigena* Peffer, *Z. Tierpsychol.*, **4**, Heft 1, 40–84.

Seitz, A. (1942), Die Paarbildung bei einigen Zichliden. II. Die Paarbildung bei *Hemichromis bimaculatus* Gill, *Z. Tierpsychol.*, **5**, 74–101.

Seitz, A. (1950), Vergleichende Verhaltensstudien an Buntbarschen (Cichlidae), *Z. Tierpsychol.* (in the Press).

Selous, E. (1905), Observations tending to throw light on the question of sexual selection in birds, including a day-to-day diary on the breeding habits of the ruff (*Machetes pugnax*), *Zoologist*, fourth series, **10**, no. 114.

Stone, C. P. (1922), The congenital sexual behaviour of the male albino rat, *J. Comp. Psychol.*, **2**, 95–153.

Stone, C. P. (1923), Further study of sensory functions in the activation of sexual behaviour in the young male albino rat, *J. Comp. Psychol.*, **3**, 469–473.

Thorpe, W. H. (1948), The modern concept of instinct, *Bull. Anim. Behav.*

Tinbergen, N. (1940), Die Uebersprungbewegung, *Z. Tierpsychol.*, **4**, 1–40.

Tinbergen, N. (1942), An objectivistic study on the innate behaviour of animals, *Bibliotheca. Biotheor.*, **1**, 39–98.

Tinbergen, N. and Kuenen, D. J. (1939), Ueber die auslösenden und richtungsgebenden Reizsituationen der Sperrbewegungen von jungen Drosseln (*Turdus m. merula* L. und *T. e. ericetorum* Turton), *Z. Tierpsychol.*, **3**, 37–60.

Tinbergen, N., Meeuse, B. J. D., Boerema, L. K. and Variosseau, W. W. (1942), Die Balz des Samtfalters, *Eumenis (Satyrus) semele* (L.), *Z. Tierpsychol.*, **5**, 182–226.

Tolman, E. C. (1932), *Purposive Behaviour in Animals and Men.* (Century Psychol. Series.) New York and London.

Uexküll, J. v. (1921), *Umwelt und Innenwelt der tiere.* Berlin.

Verwey, J. (1930), Die Paarungsbiologie des Fischreihers, *Zool. Fb.*, Abt. allgem. Zool., **48**.

Watson, J. B. (1913), Psychology, as the behaviorist views it, *Psychol. Rev.* **20**, 158.

Weiss, P. (1937), Further experimental investigations on the phenomenon of homologous response in transplanted amphibian limbs. *J. Comp. Neurol.*, **66**, 181–209, 481–535, 537–548; **67**, 269–315.

Whitman, C. O. (1899), *Animal Behaviour*. Biol. Lectures of the Marine Biological Laboratory, Woods Hole, Mass.

B. SOME CRITICISMS OF THE VIEWS OF LORENZ AND TINBERGEN

The papers of Lorenz and Tinbergen, such as the preceding article by Lorenz, and particularly Tinbergen's book *The Study of Instinct* published in 1951 created an intellectual stir. Naturally, a close scrutiny of the "instinct theory" followed, and three examples of these criticisms are collected here. Kennedy's trenchant analysis concerns mainly physiological issues of the theory; whereas Lehrman and Schneirla, whose views are similar, give broader scope with emphasis on the role of experience in behavioral development. The four chapters to follow this section single out some critical issues or examples and trace changes in thinking about them. In the last chapter, the essays attempt to determine how much we now know, both about the validity of the original theory and its criticisms collected in this first chapter.

Is Modern Ethology Objective?

J. S. KENNEDY

Ethology is an old word, once used in a wide sense for behaviour in general or even more widely as almost synonymous with "bionomics." Recently it has become associated with one particular theory of behaviour, that due in the first place to Konrad Lorenz (1937a and b, 1950) and amended by N. Tinbergen (1942, 1950, 1951), G. P. Baerends (1941) and others. Behaviour students have now to choose between one or two courses: either attempting to restore the old broad use of the name ethology for the study of behaviour in general (there is no other single word for this); or relinquishing the name to Lorenz and his school. The latter course is chosen here, adding to ethology the qualification "modern" to distinguish the new, restricted meaning of the term from its old broad one. Unless this distinction be made between the study of behaviour in which we are all engaged, and the particular approach to it which has taken over the name of ethology, it will seem that this particular approach has become the agreed basis for work in this field. Yet the view point of modern ethology is not generally accepted by members of the Association, and still less, probably, by potential

---

Reprinted in part from the *British Journal of Animal Behaviour*, volume 2, pages 12–19 (1954).

members. In at least one university the same students have been given lectures both for and against the ethologists' approach to behaviour: It is a matter of lively controversy for which the Association provides the natural forum. The greatest service of ethologists to progress in this field may well turn out to be that they have engendered controversy.

Modern ethologists have carried out stimulating and often beautiful experiments, notably in the analysis of complex stimulus-situations. But there is no doubt that the main attraction of their approach lies in its claim to objectivity in a field, instinct, where objectivity has been conspicuous for its absence in the past. Many of us brought up more or less under the influence of "reflexological" theories of behaviour, with their Loeb-Kuhn-Fraenkel-Gunn elaborations of taxes and kineses, experienced a certain sense of liberation on encountering the ideas of the modern ethologists. On closer examination however, what had seemed at first sight perhaps minor defects in their theory, proved so serious as to invalidate entirely its claim to objectivity.

This may seem a surprising judgment. Tinbergen (1942, 1951) has set forth the scientific objections to teleology, anthropomorphism and subjectivism as cogently as any mechanist could ask. He even coined the word "objectivistic" to distinguish his approach. He deplores the "tenacious hold" of subjectivism on the study of behaviour, with its reliance on analogy and its anti-analytical consequences. What little published criticism there has been of modern ethology points the same way, the misgivings of many physiologists having remained unprinted. Thus Bierens de Haan (1947) took the ethologists to task for being too objective, and Thorpe (1948) expressed some sympathy with his point of view. Others like Schneirla (1952), Hebb (1953) and Armstrong (1950) have criticized particular aspects of ethological theory but have not gone so far as to question its objectivity, being all on the psychological rather than the physiological wing of the subject themselves. There has been one exception in this respect. It was Carthy (1951) who first pointed out that: "The central idea of the motivating force of reaction-specific energy arising within the animal itself . . . is in many ways closely similar to ideas in psycho-analytical theory."

This was not an overstatement. It is true, as Carthy implies there, that the central idea in Lorenz's theory is not reaction-specific energy itself, although contained in it: The idea that the energy is specific is being discarded among ethologists themselves. The central idea is that there exists an endogenous, accumulable, consumable form of nervous energy as a distinct psysiological category. Lorenz (1950) is quite explicit about this, stressing: "The peculiarity and independence of endogenous activity as distinct from what?, is also quite clear: distinct from reflex action." Whitman's and Heinroth's signal contribution was to discover, according to Lorenz, "an independent, particulate function of the central nervous system which . . . is, at the very least, equally as important as the reflex." Internal energy accumulation, and external stimulation, are, he writes, "two absolutely heterogeneous causal factors."

It may be said that these formulations are deliberately extreme and need not be taken too literally. Ethological theory is open to revision and ethologists, by no means agreed on all points, are themselves busy revising it (as Bastock, Morris and Moynihan, 1953). One may perhaps look forward to eventual revision of Lorenz's central thesis itself. But so far his idea of two distinct neurophysiological processes, one internal, and one reflex, continues to provide the agreed basis for the whole school. In fact this idea will need more than revisions if ethology is to make good its claim to objectivity, because it is a subjective one.

ENERGY AND ID

Ethology is subjectivist on its own showing, as a dualist theory. Ethological energy, conceived as absolutely distinct from reflex mechanisms which alone have direct external relations, is quite as subjective a concept as the Freudian id. It matters little whether the prime mover inside the animal is called a "biological need" or "purpose" or by other such explicitly subjective teleological names; or whether it is called "energy" or "potential" or "impulses" which have a neutral and even physiological sound. What matters is that all these are prime movers and inside and quite distinct from reflexes: that is what makes them subjective. That is to say they are all unwarranted analogies from man, in whose behaviour subjective phenomena, appearing to us as prime movers inside, do undoubtedly play some causal role. As Tinbergen (1951) says, even in man subjective phenomena are of very limited use for the causal analysis of behaviour.

The kinship between ethology and Freud which Carthy pointed out lies in their dualism. How closely the two theories approximate should be more generally known, because many zoologists might at first hardly credit the fact that what Lorenz has managed to do is to win wide acceptance among zoologists for the wholly subjective concepts of psychoanalysis. There are of course differences between Lorenz's and Freud's theories, the most important being that Freud admitted only two basic instincts and made no attempts to compromise with physiology. But here are some passages from Freud's final summing up of his theory:

> "The core of our being is formed by the obscure id which has no direct relations with the external world . . . "
>
> "It contains everything that is inherited, that is present at birth, that is fixed in the constitution—above all, therefore, the instincts . . . "
>
> "The power of the id expresses the true purpose of the individual organism's life. This consists in the satisfaction of its innate needs . . . The forces which we assume to exist behind the tensions caused by the needs of the id are called instincts . . . "
>
> "We have found that the instincts can change their aim (by displacement) and also that they can replace one another—the energy of one instinct passing over to another." (Freud, 1949, pp. 2, 5, and 67).

The working ideas of ethology are there in a body: purpose as satisfaction, tensions, fixity, innateness, separation from the external world; even energy, and its displacement between instincts; and even, one cannot but add, to the obscurity at the core. Further, the separate "coat of reflexes"—the ethological term borrowed from von Holst—which regulates the discharge of energy in relation to the external situation, corresponds (despite its physiological flavour) with Freud's ego: "a special organization which acts as an intermediary between the id and the external world" (Freud, 1949). Thus Lorenz and Freud both draw a distinction in principle between the nervous machine and its motor, between reflex action and the energy for action.

This idea was not of course original with Freud, who acknowledged the priority of Empedocles. Empedocles made a separation in principle between the four material Elements (earth, air, fire and water) on the one side, and on the other, the two forces (love and hate) which alone move the material elements. Empedocles's dualism has been abandoned in modern physics and chemistry, along with his short list of four elements. It is only in biology, psychology, philosophy and related fields, that dualism remains a strong current, apparently for social reasons. In general biology such dualism is of course called Vitalism. In the particular field of behaviour, as Lorenz (1950) takes some trouble to point out, vitalism typically appears as subjectivism. But the link between dualism and vitalism he seems to have missed. He seems not to notice that in transferring back to animals the sort of dualism which Descartes restricted to man, he is re-creating a vitalist, subjective theory of animal behaviour (see also Kennedy, 1954).

What is at first sight so confusing about ethology, until one examines its dualist foundation, is that Lorenz places himself firmly on the side of mechanism against vitalism. He regards "the reciprocal errors of the mechanists' reactions as justified in themselves and erring only through exaggeration" (Lorenz, 1950). That is the position of a great many experimental zoologists. It is true, as Thorpe (1953) says, that the idea of instinct has in general been highly repugnant to them. It is also true as Lorenz (1950) says, that this was due partly to the great stress vitalists place upon instinct. The endeavour of mechanists has been therefore not so much to explain, as to explain away instinct. It is the bold bid of ethology to save mechanism from its friends, the earlier mechanists, whose "grotesque simplifications" (Tinbergen) played into the hands of the vitalists, although the latter were "much more fundamentally wrong" (Lorenz). The vitalist essence of ethological theory is thus far from being the most obvious thing about it.

According to the vitalists, Lorenz (1950) writes, spontaneous activity must be purposive and therefore variable; according to the earlier mechanists, such fixed innate behaviour patterns are reflex chains and must therefore be "totally devoid of spontaneity. ... But on closer inspection it became apparent that these activities are, at bottom, to a very high degree independent of external stimulation." Even a very high degree of independence is not of course the same thing as an absolute separation, such as ethologists make on this kind of evidence. Even vacuum activity has not been shown to be absolutely spontaneous no matter how "irrelevant" it may be. An animal achieves absolute independence of external stimulation when it dies. In other contexts ethologists are ready enough to grant the flow of energy from one part of the nervous system to another, in the functioning of hierarchies and displacement reactions. This immediately opens the possibility (to non-ethologists, the certainty) of past and present external stimulation being reflected in a variety of activities according to the circumstances, and not merely in the single one with which the given type of stimulation is most immediately connected—and that only in an excitatory capacity—in the nervous system. The task is to work out the laws governing these complex events. But in this context, ethologists confine our attention to one kind of immediate stimulation, the one typically "releasing" the given response, and, showing that it can be virtually imperceptible, proceed to set up an absolute distinction between external stimulation and endogenous activity.

REFLEXES AND INSTINCTS

The second point in support of the claim that modern ethology is not objective but subjective, is connected with the first, but concerns reflexes. Ethologists have effectively disposed of the old "reflexological," "telephone-exchange" theory that behaviour consists of a sum of fixed reflexes. What they call the "priming" of reactions does occur as well as their immediate elicitation by particular external stimuli. The sharp distinction they make between these two processes of priming and release, fundamental to the theory, rests, however, on a conception of the reflex as rigid and old-fashioned as that of the reflexologists they justly criticise. Instead of going forward from the criticism of reflexology to build upon modern reflex physiology (a very different thing) they curiously ignore key parts of it. They retain the discredited picture of the telephone-exchange reflex, merely limiting its occurrence and making this very limitation the grounds for postulating nervous processes which are not reflex at all.

It is most difficult to see what, other than the subjectivism at the root of ethological theory, can explain this. For in reality there is neither need nor room for a separate category of central nervous process to set against and complement the reflex. Physiologically considered, the change in the excitatory state of a neurone or centre is not something distinct from, but an integral part of every reaction no matter how simple. This is what Creed, Denny-Brown, Eccles, Liddell and Sherrington (1932) have to say about it in their book: "The motor centre even at its spinal simplest is more than a passive relay forwarding impulses on their way out to their muscle." For physiologists, the excitatory state of a centre is determined not by some peculiar independent process inside it, but on the contrary, by outside influence upon it.

> "Under intact natural conditions we have to think of each motoneurone as a convergence-point about which summate not only the excitation processes fed by converging impulses of varying provenance arriving by various routes, but also inhibitory influences of varied provenance and path: and that there at that convergence-point these two opposing processes finally interact. ... Whether the excitatory has the upper hand, or whether the inhibitory, commonly both are at work, and the functional state of the motoneurone indexes the net result from the two," (*ibid*).

There is no room here for any distinction in principle between endogenous and exogenous stimuli in their mode of action at the motor centre. But there is plenty of room in view of the multitudinous connections of every neurone with others, for ordered variation in the effectiveness of any type of peripheral stimulus in bringing about or preventing motor discharge.

It is a not unreasonable complaint of behaviour students that physiology has helped them disappointingly little. But whatever the physiologists have failed so far to do they have as a body provided the firmest possible ground for Pavlov's (1928) contention that "there is no essential difference between the two phenomena designated as reflex and as instinct."

> "When one considers," he went on, "that any given reflex, as a response to certain external conditions, is not only governed and regulated by other simultaneous reflex actions, but also by a multitude of internal reflexes as well as by the presence of many internal stimuli, chemical, thermal, etc., operating in different regions of the central nervous system or even in the executive elements (motor and secretory), then such a conception would include as reflexes the entire complexity of all responsive reactions, and nothing would remain to necessitate the forming of a special group of phenomena known as instincts."

In attributing the crudely "reflexological" view of behaviour to Pavlov, ethologists are very wide of the mark; and what is more important, having justly criticized that erroneous view, they put an even less scientific one in its place. Pavlov preferred the term reflex to instinct, not because he was a "reflexologist" but because reflex had "had from the beginning a purely scientific connotation," embodying as it does "the principle of determinism." Ethologists would readily agree that that is more than can be said for instinct. Pavlov did not of course devote himself to the study of unconditioned reflexes, although some of the laws he established are directly applicable to innate behaviour. He was aware of this omission and said that the study of instincts was the "next important task" (Pavlov, 1928, p. 296). Due credit goes to the ethologists for putting this task again in the foreground for us and for removing some of the theoretical obstacles that have delayed the attack on it. It is certainly very hard, however, to tackle this problem and still keep one's feet firmly on the physiological ground.

By what means did ethologists succeed in directing renewed experimental attention to innate behaviour? They did so by emphasising how much of the variation in behaviour might be due to the working of the innate part of the system without necessarily, the intervention of conditioning processes. This had previously been effectively obscured by views of instinct as either rigid, or if not rigid, then "intelligent" (Thorpe, 1948). The ordered variation of instinctive behaviour is the significant objective point underlying (although misleadingly conveyed by) Wallace Craig's distinction between appetitive behaviour and consummatory act (or fixed pattern), Lorenz's distinction between *Taxiskomponent* and *Erbkoordination*, and Tinbergen's hierarchy of moods.

Having once recognized that ordered variation is a property not only of learned but also of instinctive behaviour, the logical next task is surely to study the interaction of these two kinds of variation in the animal. For comparative ethologists, great interest would attach to the detailed analysis of evolutionary differences in the relative importance of the two—the degree of dominance of conditioned over unconditioned reflexes—as one passes through the animal kingdom, from worms to insects, fishes to birds, mammals to man. It is another point in support of the claim that ethology is still not an objective theory, that this is not yet the sort of programme ethologists set themselves. For they cannot yet consolidate the advance they have made, and even endanger it, because of their tendency to force into the portmanteau of instinct as much of the variation in behaviour as they possibly can. This tendency is the inevitable result of the subjectivist separation of a motivating core from a reflex coat in the causation of behaviour. Given that separation, conditioned reflexes are tacitly excluded from a direct part in changing "motivation." That was the evident logic behind Tinbergen's (1951) programme of studying instinctive behaviour first, and learned behaviour only afterwards. Hebb (1953) has well brought out the impracticability of this. If in fact instincts and simpler reflexes differ only in degree (although Hebb does not make that point), then instinctive and learned behaviour must become an "inextricable tangle" as Hebb put it, and there is plenty of evidence that they do. What this means, unfortunately, is that the whole ethological interpretation of stickleback behaviour, for example, will have to be revised and much experimental work done over again.

Today, it would be most rash to set any *a priori* limits to the extent to which conditioned reflexes intervene in the very processes which ethologists have treated as non-reflex, innate and endogenous. Hebb understated the case when he wrote: "There is no behaviour, beyond the level of the reflex, that is not essentially dependent on learning." No such level can be distinguished physiologically. The reflex cannot be used to mark a level; its value is rather that it characterizes the elementary mode of operation of the nervous system at all levels. The contraction of the human pupil in response to light is not merely a reflex but an autonomic one, yet Hudgins (1933) conditioned it and by stages of conditioning, actually made it voluntary. Vasomotor responses again, have been conditioned (Menzies, 1937; Gottchalk, 1946), although they enter into emotion which Lorenz (1950) correlated solely with endogenous energy discharge. Especially pertinent is the work of Bykov (1938, 1953) and his school on more strictly "internal" factors (ethologically speaking) than these classical reflexes. Among the internal factors cited by ethologists, hormones loom large. Hormone action might seem at any rate a good deal more innate than, say, salivation at the sound of a bell. Nevertheless it and many other visceral functions have been conditioned.

Injections of adrenaline into dogs, in one experiment, were repeatedly accompanied by the sound of a bell, and produced the usual effects of this hormone including rise of blood pressure and pulse rate. After a while the bell alone produced the same effects. Then, even when the sound of the bell was accompanied instead by an injection of acetyl-choline under conditions in which it would normally produce the opposite effects (fall of blood pressure, etc.), the effects of the bell and injection were nevertheless the same as they had been with adrenaline. Thus conditioned reflexes may have more "energy" than unconditioned. The resting metabolic rate, another profoundly "internal" factor which undoubtedly primes many reactions, has also been experimentally conditioned by these workers. As a result of conditioning, the mere sight of particular surroundings where physical work has habitually been done, or merely assuming a posture in which heat loss has habitually had to be compensated, may lead to a rise of metabolic rate. It would be rather surprising if conditioning played no comparable part at all in the territorial behaviour of sticklebacks.

The diuretic activity of the kidneys of dogs has also been conditioned (Bykov and Alexeyev-Berkmann, 1930; Bykov, 1938; Marx, 1931)—an irresistible example to quote, inasmuch as the resultant filling of the bladder is what primes the instinctive act of micturition; Lorenz could have found here a living model, it is seen that the two processes of priming (bladder-filling) and of release (by, say, lamp-post) are neither absolutely nor innately distinct. Both are in principle reflex, and become partially separated only as the animal grows up. Many conditioned reflexes of the reciprocal type have also been elaborated, where some visceral change, as of stomach contents or blood composition, has been used as the conditioning stimulus. Finally it might be mentioned that the alpha rhythm in the cerebral cortex (in any case notoriously sensitive to environmental conditions) is also profoundly affected by conditioning processes.

It seems clear that in an animal like a dog, all reflexes from the simplest up to elaborate instincts are under cortical control, meaning that their performance depends upon the formation and extinction of conditioned reflexes from the internal and external world. It is a very far cry from the old rigid "reflexological" picture of the reflex to the modern picture these workers are slowly constructing. They think in terms of reflex-complexes: on the basis of the instinctive systems of unconditioned reflexes (some of which show some conditioned reflex properties) there are built whole systems of conditioned reflexes which themselves show a certain stereotypy and independence one from another. Here again all workers in the field of behaviour are faced with a choice: given that the old picture of the reflex was a wooden caricature of reality, whether drastically to restrict the term or even drop it entirely, or whether to try and modernize it. Ethologists have chosen to restrict it, Pavlovians to modernize it. Unfortunately, so restricting the term deprives it of physiological meaning, while abandoning it altogether incurs still greater risks. It is extremely hard to break what Tinbergen so justly called the tenacious hold of subjectivism on behaviour. That is why we can hardly afford to follow him in rejecting the "reflex hypothesis" of behaviour. When that goes out of the door, subjectivism flies in at the window.

COST OF SUBJECTIVISM

Only subjectivism, with its pre-occupation with the ends of behaviour sequences, can surely explain, for example, Tinbergen's assumption that migration in the stickleback is motivated by a top reproductive centre. Spurway and Haldane (1953) have recently pointed out that another cause of migration is more probable. Tinbergen's view that the connection between migration and reproduction must be central, with the second internally exciting the first, was an *a priori* one, based on little but the fact that reproduction is normally the ultimate consequence of migration. The advantage of a broadly reflex theory of behaviour here is that it guides the investigator to the efficient causes of sequences or hierarchies of behaviour. In so far as such sequences do lead to a particular, useful end, the causes lie in the occurrence in normal sequence of the environmental conditions to which the animals respond, the responses having become suitably adapted during evolution.

Passing to the other end of the ethological hierarchy, it again seems to require the tenacious hold of subjectivism to explain why ethological workers (Bastock, Morris and Moynihan, 1953) are only now groping toward the physiologically self-evident fact that what causes the sharp change of behaviour after a consummatory act is not the consumption or running out of the impulses (real nerve impulses do not do that sort of thing), but an input of impulses from the periphery. The change in behaviour at this point reflects the input: It is reflex. One other point must suffice to show how little real change has been effected by Tinbergen's replacement of the earlier ethological term energy by a more physiological one, impulse. These ethological ("motivating") impulses, quite unlike physiological ones, are never inhibitory but always excitatory. This discrepancy is so plain that ethological theory will no doubt in due course be amended accordingly. The point is that it need never have occurred if the theory had not been subjectivist; and until the theory ceases to be subjectivist, such amendments will continue to provide an unnecessary labour for ethologists.

The ethological treatment of displacement reactions also suffers from subjectivism. Considered taxonomically, the displacement of a response from one reaction chain to another in the course of evolution is a clear and useful idea, suggesting a change in central nervous connections. Ethological work on this subject including the "ritualized" modification of such elements and their morphological complements in the new context, is a stimulating development (Baerends, 1950). But the concept of displacement is again directly applied, by unwarranted analogy, to the causal explanation of the behaviour. Any response that turns up in the chain but is biologically "irrelevant" or displaced, is by the same token explained as physiologically "displaced." Tinbergen (1951) says it does not "belong to the motor pattern of the instinct activated," on his assumption that "all the activities forming part of a purposive behaviour pattern aimed at a certain goal depend on a common neurophysiological pattern." Evolutionary and physiological causes are confused here in just the way Tinbergen elsewhere says they should not be. The idea that responses properly belong to one or other instinctive reaction chain, depends absolutely on the subjective notion that every chain is internally directed to its one normally-observed end by impulses from a nervous centre or pattern "for" that end. Physiologically considered, displacement appears to be a special case of positive induction. It looks very much like the central excitation of a third reflex as a result of two others which are mutually inhibitory, both being externally excited at the same time: for example, fighting and fleeing. Analysis on such lines must be hampered, however, by a theory which attributes displacement to a conflict between internal forces called drives which are labelled, as if caused, by their ends.

Finally, something should be said of the ethologists' repeated use of subjective terms such as purpose, drive, mood, appetite, consummation, and so on. Ethologists agree that these terms were used impermissibly subjectively before, but now use them freely. It is a matter of importance to enquire whether adequate reasons for doing so have been brought forward; for if not, then the use of these terms is open to all the old objections Tinbergen (1951) enumerates, and is alone strong evidence of the subjectivism of ethology itself.

These terms came to acquire their new respectability for ethologists, originally, on the strength of Wallace Craig's scheme of instinctive behaviour, at a time when energy was supposed to be strictly action-specific. The scheme comprised three parts, set out thus by Lorenz (1950; see also Thorpe, 1948):

1. accumulation of action-specific energy giving rise to appetitive behaviour;
2. appetitive behaviour striving for and attaining the stimulus situation activating the innate releasing mechanism;
3. setting off the releasing mechanism and discharge of endogenous activity in a consummatory action.

It was by treating this sequence as if it were not only a coloured description of the observed sequence of behaviour, but was also, by direct analogy, a description of the underlying physiological process, that the old subjective terms acquired their new objectivist legitimacy. If the centre for an innate behaviour pattern or "endogenous movement"—an act which appeared overtly only at the end of a long behaviour sequence—if that centre were also the source of all the activity preceding the final act, then, it seemed that here was an objective physiological basis for previously subjective concepts like purpose and consummation. As Lorenz (1950) put it, "though the activities thus elicited comprise the whole range of behaviour, they have one decisive characteristic in common: They are all purposive in the sense with E. C. Tolman has given to this term, that is to say, they all tend to bring about, by variable movements, an invariable end or goal."

But did Wallace Craig's tripartite scheme provide an objective basis for the previously subjective terms, or was it the subjective ideas that created the tripartite scheme? It is difficult to accept the claim that the terms ever became objective, when they continue to be used even after their objective basis, the tripartite scheme, has been abandoned as the general case. It is now agreed, thanks to Baerends, by Lorenz and Tinbergen that ends as well as means may be variable. The behaviour leading normally to a certain end-action is not now held to be caused by the centre of that end-action itself, but from a different, higher centre instead. It is now agreed that which one, out of a number of possible consummatory acts, actually occurs, depends upon what external stimuli are received. One might therefore have expected that the idea of behaviour being directed internally, (whether by a stream or "motivating impulses" or otherwise), might have been abandoned too, along with terms based on it like purpose, drive, consummation and so on. That these terms are retained even temporarily, can only be explained by the tenacious hold of subjectivism. It is the basic subjectivist assumption that behaviour sequences must be caused by parallel, counterpart, internal connections that has to be abandoned ultimately. When it is abandoned, it will be seen that the ordered variation both of instinctive behaviour, and that due to conditioning, serve the single function of the nervous system as defined by J. E. Smith (1950): "to take appraisal of environmental changes, and to promote responses which lead to favourable adjustments." It need hardly be added that once ethologists come round to this, their peculiar assets—experimental skill, and especially their determination to tackle the long-neglected higher functions of the intact nervous system—can be turned to enormous advantage.

REFERENCES

Armstrong, E. C. (1950), The nature and function of displacement activities, *Symp. Soc. Exp. Biol.*, **4**, 361–384.

Baerends, G. P. (1941), Fortpflanzungsverhalten und Orientierung der Grabwespe *Ammophila campestris, Tijdschr. Ent.*, **84**, 68–275.

Baerends, G. P. (1950), Specializations in organs and movements with a releasing function, *Symp. Soc. Exp. Biol.*, **4**, 337–360.

Bastock, M., Morris, D. and Moynihan, M. (1953), Some comments on conflict and thwarting in animals, *Behaviour*, **6**, 66–84.

Bierens de Haan, J. A. (1947), Animal psychology and the science of animal behaviour, *Behaviour*, **1**, 71–80.

Bykov, K. M. (1938), The functional connection of the cerebral cortex with the viscera, *Acta Med. U.R.S.S.*, **1**, 178–203.

Bykov, K. M. (1953), New data on the physiology and pathology of the cerebral cortex, Moscow (paper read to 19th Int. Cong. Physiol. Montreal, (1953)).

Bykov, K. M. and Alexeyev-Berkmann, I. A. (1930), Die Ausbildung bedingter Reflexe auf Harnausschneifung, *Pflug. Arch. Ges. Physiol.*, **224**, 710–721.

Carthy, J. D. (1951), Instinct, *New Biol.*, **10**, 95–105.

Creed, R. S., Denny-Brown. D., Eccles, J. C., Liddell, E. G. T. and Sherrington, C. S. (1932), *Reflex Activity of the Spinal Cord.* Oxford.

Freud, S. (1949), *An Outline of Psycho-Analysis.* London.

Gottschalk, L. A. (1946), A study of conditioned vasomotor responses in ten human subjects, *Psychosom. Med.*, **8**, 16–27.

Hebb, D. O. (1953), Heredity and environment in mammalian behaviour, *Brit. J. Anim. Behav.*, **1**, 43–47.

Hudgins, C. V. (1953), Conditioning and voluntary control of the pupillary light reflex, *J. Gen. Psychol.*, **8**, 3–51.

Kennedy, J. S. Modern ethology as a subjectivist theory of animal behaviour, *Quart. J. Exp. Psychol.*, (in preparation).

Lorenz, K. (1937a), Uber den Begriff der Instinkthandlung, *Folia Biotheor.*, **2**, 18–50.

Lorenz, K. (1937b), Uber die Bildung des Instinkbegriffs, *Naturwiss.* **25**, 289–300, 307–318, 324–331.

Lorenz, K. (1950), The comparative method in studying innate behaviour patterns, *Symp. Soc. Exp. Biol.*, **4**, 221–269.

Marx, H. (1931), Diuresis by conditioned reflex, *Amer. J. Physiol.*, **96**, 356–362.

Menzies, R. (1937), Conditioned vasomotor responses in human subjects, *J. Psychol.*, **4**, 75–120.

Pavlov, I. P. (1928), *Lectures on Conditioned Reflexes*. New York.

Schneirla, T. C. (1952), A consideration of some conceptual trends in comparative psychology, *Psychol. Bull.*, **49**, 559–597.

Smith, J. E. (1950), Some observations on the nervous mechanisms underlying the behaviour of starfishes, *Symp. Soc. Exp. Biol.*, **4**, 196–220.

Spurway, H. and Haldane, J. B. S. (1953), The comparative ethology of vertebrate breathing. I. Breathing in newts, with a general survey, *Behaviour*, **6**, 8–34.

Thorpe, W. H. (1948), The modern concept of instinctive behaviour, *Bull. Anim. Behav.* **7**, 12pp.

Thorpe, W. H. (1953), Editorial, *Brit. J. Anim. Behav.*, **1**, 3–4.

Tinbergen, N. (1942), An objectivistic study of the innate behaviour of animals, *Bibliotheca. Biotheor.*, **1**, 40–98.

Tinbergen, N. (1950), The hierarchical organisation of nervous mechanisms underlying instinctive behaviour, *Symp. Soc. Exp. Biol.*, **4**, 305–312.

Tinbergen, N. (1951), *The Study of Instinct*. Oxford.

Interrelationships of the "Innate" and the "Acquired" in Instinctive Behavior

T. C. SCHNEIRLA

INTRODUCTION

The title assigned to me by Dr. Grasse offers an opportunity to question the concepts of innate and acquired as the mental tools with which to investigate instinctive behavior. For too long these concepts have been accepted *a priori*, on the questionable assumption that they represent separable entities in behavior determination. This dichotomy is a heritage from our pre-experimental past, incorporated with too little question in the basic statement of the "instinct problem." But instinct theory, like all other theories, must have concepts substantiated through reliable evidence.

Reprinted from *L'Instinct dans le Comportement des Animaux et de l'Homme*, pages 387–452 (1956). Published by Masson and Cie, Paris.

In this colloquium we have for convenient reference an influential contemporary theory of the nativistic, performistic type, in which the above concepts and their basic distinction form a prominent part of the groundwork. Lorenz (1950) has indicated his point of view in these words:

"We shall call instinctive only a sequence of motions that occurs with an inborn coordination of all its components, independent of training or understanding."

Not only the basic activities, but much of their inner organization, are considered inborn as effected through specific integral neural mechanisms. Such claims require careful examination.

To explain what we mean by challenging these concepts and also what we do not mean, the case of the so-called "Anti-instinctivists" may be mentioned. In the 1920's certain scientists, American psychologists (e.g., Kuo, 1924) in particular, raised objections to nativism and in favor of a more objective attitude. For Watson, a chief figure, this was not merely a reaction to the practices of others. He himself had attempted a classification of animal "instincts" (1914) and in his drive against mentalism he later (1919) defined "instinct" as "an heriditary pattern reaction, the separate elements of which are movements, principally of the striped muscles."

Here we find a suggestion of hereditarily-determined organization and of "the instinctive movement," not unrelated to Lorenz's (1935, 1950) ideas. But the resemblance soon stops, for Watson did not encapsulate the inherited determiners in the nervous system, and mistrusted the Freudian influences to which Lorenz may have been susceptible (Kennedy, 1954). Watson had a conception of hereditary influence that although rather positivity held was not too consistent, for he (1925) denied "it" on the human level.

The anti-instinctivist movement, although criticized as having gone "too far" (Hunter, 1947), made two contributions of value, first an emphasis on objectivity and an insistence on experimentation. In a preceding period of naturalistic investigation, essentially descriptive methods had been followed by Whitman (1899), Mills (1898) and others, and indeed Mills, zealous investigator of behavior and founder of a club at McGill so dedicated, seemed convinced that the "animal mind" was closed to experimentation. Watson in contrast favored experimental study of behavior both in laboratory and field (e.g., Watson and Lashley, 1915), but disengaged it strongly from mentalism. The second contribution of the anti-instinctivists was a stress upon diagnostic, analytical research. This is exemplified in the work of Kuo who asked not only "What happens?" but also "How?". His investigation (Kuo, 1930, 1932) raised serious difficulties for dogmatic distinctions of innate and acquired. In fact Kuo disowned both "instinct" and "learning" as obscure, emphasizing ontogenetic processes in the development of behavior.

The fresh methodological improvements built upon the earlier naturalistic approaches, with their further emphasis on quantitative methods we can accept, and these might be used to better advantage in behavior study today. Through this movement particularly, a healthy skepticism developed to balance overconfident statements such as Kingston's (1928) that "Everybody knows what instinct is." The challenge was directed mainly at the traditional criteria for distinguishing the innate from the acquired, which were: universality of given activities through a species early appearance in the individual, absence of learning, and appearance in isolation. As we shall see, present evidence supports this challenge strongly.

That these objections have survived is indicated by the frequency calls (e.g., Leuba, 1940; Howells, 1945; Anastasi and Foley, 1948) for reconsideration of the "nature-nurture" dichotomy. But the situation is not as unequivocal as might be implied by Beach's (1947) remark in recent symposium on instinct that the "artificiality of the implied dichotomy" (i.e., heredity-environment) "is apparent to everyone." Much uneasiness exists about discarding the distinction, witnessed by the chairman's statement ending the same symposium, that all five contributors had "emphasized the role of heredity in the determination of behavior" (Hunter, 1947).

The need for a contemporary stock-taking is accented by Hebb as follows:

"I would not suggest for a moment that the problems in this area are unreal; I do suggest that they have been poorly stated, inasmuch as we cannot dichotomize behavior into learned and unlearned, environmentally determined and hereditarily determined. I urge that there are not two kinds of control of behavior, and that the term, "instinct," implying a mechanism or neural process independent of environmental factors, and distinct from a neural process into which learning enters, is a completely misleading term and should be abandoned" (1953, p. 46).

As Anastasi and Foley (1948) have said, the term heredity is often defined "indirectly, vaguely or inconsistently, especially when it comes to the domain of behavior phenomena." To replace the unsatisfactory dichotomy, these authors suggest a reconsideration of behavior etymology in terms of structural and functional factors. But progress in biology has emphasized increasingly how intimately functional properties are bound into the principles of morphology, and neither the idea of isolated morphological factors nor that of isolated function, fits the needs of a dynamic behavior theory. As Cobb (1944) has said,

"... no function is possible without an organ that is functioning and therefore no function takes place without structural change."

Consistently we may say that all problems of development are inherently both functional and structural.

Doubtless the traditional criteria have inherent weaknesses. Confidence about excluding learning from any given activity is inversely related to appreciation of the scope of learning and how incompletely this phenomenon is known; early appearance becomes unreliable with demonstrations of conditioning at early stages in insects, birds and mammals; and, as will be shown later, appearance in isolation, however sound at first sight, rests too heavily upon an incomplete understanding of equivalence between environments. Even universality is questionable, first of all because of acts that normally are likely to be learned widely throughout a species (Smith and Guthrie, 1921). Actually there seem to be no hard and fast rules for distinguishing hypothetically innate behavior from other kinds.

Undeniably, the influence of genetic constitution is expressed somehow in the functions and behavior of every animal. Raccoons, for instance, could not readily be brought to peck at their food as do chicks. The "instinct problem" therefore centers around the occurrence of behavior that may be termed species-stereotyped or species-specific, species-characteristic or species-typical. A species, defined dynamically, concerns groups capable of interbreeding under natural conditions, but reproductively isolated from other such groups (Mayr, 1942). Investigations of the mechanisms underlying such isolation in insects (Spieth, 1952) and in fishes (Clark, Aronson and Gordon, 1954) indicate that behavior factors are involved as well as physiological, structural and environmental ones. But what is the causal nexus?

The simplest answer might seem to be that species-typical behavior is inherited, as nativistic theories suggest; yet the unreliability of the criteria for innateness reminds us that the genes do not directly translate themselves into behavior by any means. Between the fertilized ovum and properties of the mature organism lie the complex processes of development. The following preliminary considerations of these matters encourage a broad perspective on the problem of the individual rise of species-typical behavior.

1. Preformism is misleading. "That which is directly inherited . . . is the set of genes, with the accompanying cytoplasm" (Jennings, 1930). This is the initiating cause of development, designated by Johannsen the genotype as distinguished from the phenotype, i.e., from the organism and its functions produced through development. Now, as David and Snyder have said:

"Limitations in the early concept of genes as individual "determiners" of Mendelian characteristics, and of a one-to-one correspondence between gene and characteristic, were exposed by the discovery of genic interactions and of environmentally contingent gene effects. The newer concept of genic balance implies that genetic variability is a function of the genotype as a whole and that isolation of individual gene effects involves an artificial disjunction of gene and total genotype" (1951, p. 54).

Clearly, it is the phenotype that can be studied more or less directly, but from the beginning of development the effects of the genotype can be examined only indirectly through evidence gained thereby. Haldane (1946) thus points out that a given genotype promoting ontogeny can lead in one environment to adaptive results, in a different environment to neutral or maladaptive results. And, as Dobzhansky (1950) has put it:

"The so-called, nature-nurture (genotype-environment) problem is not to distinguish which traits are genotypic and which are environmental, for all traits are genotypic and environmental."

". . . The outcome of development at any stage is a function of both the heredity of the developing individual and the environment in which the process has taken place. The development is apparently epigenetic, not preformistic . . . "

The concept of a direct determination by the genes, a one-to-one relationship with developmental processes, might be considered at least for the molecular plane of development. But modern investigators have their reservations even here.

As Weiss states:

"The genome of the zygote endows all descendant cells with a finite repertoire of modes of reaction. What is commonly called, differentiation potency may be interpreted as a finite assortment of chemical entities. These entities, of course, must not be viewed as direct precursors of any final results, but as a reactive system, the constant interaction of which with systems of the extragenic space will only gradually yield the later specific characteristics of the various cell strains."

"... Since the extragenic space, i.e., the genic environment, is thus undergoing progressive transformation, it is evident that every new reaction must be viewed in terms of the cellular system in its actual condition at that particular stage, molded by the whole antecedent history of transformations and modifications, rather than solely in terms of the unaltered genes at the core. Incidentally, keeping this in mind ought to stop the confusing practice of labelling all intrinsic properties of a cell at an advanced stage as genetic, but those brought out by still later interactions with neighboring cells or diffusable agents as 'environmentally' or 'hormonally' introduced, forgetting that no cell develops independently, but that all of them have gone through a long chain of similar 'environmental' interactions with neighboring cells and the products of distant ones" (1954, pp. 193–194).

Theories concerning how behavior and behavior patterns arise through this mass of finite events must do more than bridge initial and terminal stages with hypothetical shielded intra-organismic determiners.

2. *Genic effects are indirect and mediated.* As Stern (1954) says, we have barely touched the surface of the problem concerning how genic factors actually influence organic development, and to investigators like Dobzhansky (1954) it seems improbable that these factors have anything like a complete determinism at any stage. Systems of intervening variables, which always include the influence of developmental conditions, both intrinsic and extrinsic to cell, tissue, organ, system or organism, mediate the initial genic effects at each successive stage. There is evidence that these variables may range from biochemical (e.g., enzymatic) conditions to the repercussions of action and specific extrinsic effects. Moreover, their effects on development appear to be self-reactive and cumulative. Thus at any stage further development and new organizational gains occur through interrelationships that are only partially the products of genic influence.

3. *Distinctions of the "native" and "acquired" ambiguous.* These considerations suggest the unwisdom of attempts to distinguish what is "innate" from what is "acquired," or to estimate the proportionate effects of these or to judge what kinds of effects they might produce separately. There exist no separate entities of this sort, for conditions at any stage are the complex product of trace effects from previous stages entering into interactions with prevalent extrinsic-intrinsic conditions, themselves composite acquisitions. Learning, the most complex form of acquisition, may have very different forms (Maier and Schneirla, 1935, 1942), and as Hebb (1953)

reminds us, is known so incompletely that its effects may be far from obvious. But indirectly, all learning is somehow influenced by the genes (Howells and Vine, 1945).

4. *The "instinct" problem is one of development, different for each phyletic level.* In each type of organism, the genotype varies characteristically in its range and complexity of effects as evidenced by different types and degrees of sensitivity to surrounding conditions as well as complexity and plasticity of neural and reactive functions. The problem of "instinct" therefore demands careful attention to phyletic differences in those configurations characterized as "levels" (Needham, 1929; Redfield, 1942; Schneirla, 1946; 1949). "Instinct" is not a real and demonstrated agency in the causation of behavior, but a word for the problem of species-typical behavior at all phyletic levels.

Consequently, "instinct" study must examine the ontogeny of behavior in each type of organism, so as not to miss any aspect of the expanding field of relationships in development. Each organism must be studied in its own terms, since when new capacities enter the developmental situation, new configurations must be expected in adaptive behavior as well as variations in patterns more similar to the "phyletically old." Consequently, valuable evidence, both as to the ontogenetic causation and the evolutionary history of behavior patterns, lies in the comparison both of closely related and of distantly related organisms.

THE RELATION OF DEVELOPMENT TO BEHAVIOR AT DIFFERENT PHYLETIC LEVELS

Behavior patterns often reach similar ends in different phyla, as a result of parallel evolutionary processes. To the teleologist, an equivalence of ends signifies equivalence in the organizations attaining these ends. But actually the accomplishment of adaptively comparable results, as through feeding, tells us nothing of the antecedent processes. These may involve complex anticipations, as in a socialized human being, or may be reflex-like and automatic, as a lower invertebrate. Calling two such acts in different phyletic contexts "instinctive" mainly conveys the information that they are both species-typical and nutritively beneficial, but leaves doubt as to what else they may have in common. The implication is of course that both somehow depend upon mechanisms resulting through evolution; however, these may be either homologous or analogous, or both, and if convergent may function to similar ends although very different in internal makeup. Thus, great doubt exists that a common formula can be found for "instinctive behavior" at all phyletic levels.

Feeding in a coelenterate and in a cat would both be called instinctive, since both acts are species-typical products of ontogeny in the normal species habitat. Both are adaptive. Yet the patterns may arise through very different developmental processes with different underlying organizations resulting.

In the typical feeding pattern of the medusoid coelenterate the stimulus, typically a chemotactic effect, if too weak produces only a brief local response of tentacles, if too strong brings a vigorous contraction of the bell which turns the animal away. But adequate stimulation elicits first a contraction of the local marginal tentacles which curl about the object, then a local contraction of the bell margin which, spreading centrally, pulls object and tentacles toward the mouth tube. Meanwhile this tube, the manubrium, has bent toward the stimulated sector and by expanding its opening now engulfs the object.

This response pattern can be obtained normally from an animal of given structure acted upon by stimulation within a given range of intensity. Its respective components depend first of all upon the functional properties of the tentacles, bell and manubrium. In the mature organism the intergration of these into well-timed series depends particularly upon the conductile functions of the nerve net (Bullock, 1943) which tie them into a predictable functional sequence. Their functional patterns are strongly influenced first of all by the morphology of the parts, in locations fixed through development. In one important aspect of control in the act, perhaps the most critical, Bozler's (1926) findings support the hypothesis of Loeb that the diffuse nerve-net impulses first reach and adequately activate the basal part of the mouth-tube on the side nearest the highly aroused marginal sector. Hence typically the manubrium bends directly, "purposively," through functional conditions imposed by its structure and its location in the animal (Maier and Schneirla, 1935).

Pantin's (1943, 1950) studies on hydroid coelenterates show in detail how a patterned action of parts can result from a nerve-net system connecting the components in a given spatial way. Thresholds of receptors and of muscles, together with nerve-net conductile properties of irradiation and summation, are critical for this behavior system. The system is capable of considerable variation according to the conditions of arousal and the current state of the organism. One such variation, a temporary failure of the tentacle responses to food, obviously depends upon intraorganic changes thought by Jennings (1905) to be a "loss of hunger," perhaps grounded in widespread neural changes. But Parker (1919) was able to narrow the cause experimentally to a peripheral adaptation effect which temporarily raised the excitation threshold of the tentacular receptor cells.

Such patterns in lower invertebrates are stereotyped in that they are produced directly through arousal of the functional properties of the species-characteristics tissues. The intervening variables therefore are those of a specific developmental process, and the activity pattern is in this sense "directly determined" without the intervention of special variables such as learning. Yet even these simple organic systems do not arise in a vacuum, for if the individual is to develop in the characteristic form with the typical species activity repertoire, there must be a representative succession of developmental stages, each involving interactions between the respective organic conditions and standard extrinsic conditions.

A marked variability is seen in the activities of these lower invertebrates, but only that possible in a simple organic system of radial symmetry—a variation of multitudinous changes about a fixed center as it were and within set organic limits. Although from time to time behavior can change in somewhat more persistent ways, these seem limited to alterations in sensory adaptation or muscle tonus, peripheral changes which are ephemeral in the sense that the organization of behavior is held to the species norm—the typical individual behavior system that is approximated through life. Despite temporary variations behavior thus is essentially stereotyped, and lacks the plasticity found in higher phyla. In phyletic behavior comparisons, many types of variability and many types of plasticity are to be expected.

Feeding in the domestic cat also is a species-typical pattern which results through development in the standard habitat. But the rise of the chief properties and the organization of feeding in this mammal hardly follows a course duplicating that in coelenterates. For one of the prominent typical patterns of food-acquisition, rodent killing, a study of Kuo's (1930) has revealed that the developmental factors, both intraorganic and extrinsic, are complexly and variably interrelated at all stages. The contributions of bodily growth are essential, but so is a learning process through which the components are integrated into a pattern. These components depend on features of bodily makeup which adapt the cat to being excited by small moving objects, to making swift movements, and to capturing and devouring small animals.

But the normal outcome is by no means inevitable. By appropriately regulating the situations in which kittens encountered rodents, Kuo was able to produce adult cats which attacked and ate rodents and others which obtained their food in other ways—either fearing and avoiding rodents or living peaceably with them according to experience. It is also possible, for example, to train young kittens to eat solid food without using their paws, which are frequently used normally. Still other variations are possible in dependence upon the kind of situation in which the animal develops.

In both coelenterates and birds, the typical feeding pattern emerges in the normal habitat through an organic development basically initiated by the genic constitution. But these ontogenetic processes in the different phyla involve strikingly different intervening variables particularly in neural resources determining plasticity. In both organisms intraorganic relationships introduced through growth influence further stages and the outcome. But in the coelenterate these are held within narrow limits, whereas in the mammal, through wider and more complex intrinsic and extrinsic interrelationships, they become major factors in behavioral development. The eventual mammalian pattern owes its specific organization particularly to the superior capacities of the nervous system for change. In the cat, the function of components such as use of the paws may be eliminated or changed through training; in the coelenterate, comparable changes (e.g., in tentacle responses) are relatively fluctuant and dependent upon peripheral modifications. The developmental system of the mammal therefore may be termed plastic in the sense that under appropriate conditions one or more patterns offering from the species norm may be produced and retained. But in the coelenterate pattern, variability occurs about a fixed axis, as it were, and is apt to be confused with plasticity.

INTERRELATIONSHIPS IN COMPLEX STEREOTYPED BEHAVIOR PATTERNS

The insects are often cited for their "instinctive" behavior. The insect nervous system appears to be first of all a transmitter and summator of impulses from afferent systems. The arrangement of tracts and centers seems particularly to favor direct discharge under the predominant afferent effect, rather than plasticity of organization as in the very different mammalian system (Schneirla, 1953b). Consequently, and with the added important fact of rapid conduction over short arc, we should expect patterns of activities rigidly governed by organic mechanisms and, in the mature individual, readily dominated by abrupt external sensory change. In relation to these conditions, highly sensitive receptors have evolved such as the compound eye, permitting delicate reflex adjustments as to slight movements in the visual field (Autrum, 1952), illustrated by the attack-reaction of the praying mantis.

In a system of this kind, specific organic factors readily translate themselves into behavior. By means of the successive operative removal of head, thorax, legs and various abdominal segments in the mature silkworm moth, McCracken (1907) demonstrated in the egg-laying pattern of this insect an olfactory control as link to the external situation, a reciprocal neuromotor function in the postural execution, and an abdominal process serving as nucleus of the whole pattern. Although the functional systems are typically complex with an organization resisting analysis, Spieth (1952) found the mating patterns of closely related species of *Drosophila* so well differentiated to be taxonomically reliable. And Adriaanse (1947), in a population of campestris, a solitary wasp previously considered a single species, and two distinctive behavior patterns which with previously overlooked morphological differences indicated the existence of two distinct species.

Insect behavior cannot be adequately studied in terms of one functional stage alone, since any stage emerges through relationships effective earlier in development. Critical factors may even be extrinsically introduced by other individuals, as when an ovipositing female incidentally fixes the early environment of her young by laying her eggs in a particular place (e.g., on the species food-plant in a phytophagous insect). Through this act the newly emerged insect is given not merely food and a feeding station (Kennedy, 1953) but also a shelter with predators, diseases and other properties (Dethier, 1954)—a ready-made environmental complex. In this case the main determinative factor through which the next generation is so weightily influenced may be some organic factor affecting adult sensitivity, e.g., in a metabolic influence upon olfaction (Dethier, 1947). Or in certain species, it is even an olfactory habituation established in larval feeding which may persist to the adult stage (Thorpe and Jones, 1937).

To discover the nature of the organization in each type of behavioral pattern and its prerequisite developmental relationships, analytical investigation is essential. Although evidence from this source is not plentiful, some of the salient points may be brought out in the consideration of one such study of a solitary insect and one of a social insect behavior pattern.

The spinning behavior of the Cecropia silkworm was studied by Van de Kloot and Williams (1953). Normally, after a period of wandering, the mature caterpillar settles down at a twig-crotch and spins a double cocoon with a thinner upper end through which the adult eventually escapes. Cessation of feeding is attributable to glandular changes of the period, as is the wandering, which is absent in caterpillars deprived of silk glands. This extirpation also eliminates two movements, the "stretch-bend" and the "swing-swing" basic in normal spinning, which thus depend at least in part upon sensory input from the spinning apparatus itself.

The genesis and temporal order of these movements are attributed largely to stimulation from the changing internal environment, but their execution depends predominantly upon tactual and gravitational cues from without. For example, for the termination of either basic movement, contact of the spinnerets is essential at full flexion or full extension. The pattern of the normal cocoon thus may be considered a compound expressed through a spatial relationship of the body to the external world. So, in a highly uniform environment such as the interior of a balloon, the larva spins a flat layer of silk. This behavior pattern therefore comes about through a complex progressive relationship between changing organic conditions and the sensory input from these states, interacting with that from sets of external circumstances which are altered according to how the act progresses.

In a sense, insect social patterns have evolved as permutations of individual characteristics, since each has arisen in a dynamic environment defined chiefly by complexly interrelated individual properties (Schneirla, 1941, 1946, 1952b). Some of the functional principles involved in one highly specialized pattern of this type may be examined in terms of a case which I have studied (1938, 1944, 1953a) in some detail—the nomadic, predatory system of the terrestrial army ants.

One central feature of army-ant life is the formation and regulation of the temporary nest of bivouac, a complex outcome of worker behavior critical for the functional pattern of the colony. The bivouac is a mass of living workers which not only affords a temporary base of operations and a population reservoir for the colony, but also is an excellent incubator for the great broods that energize the colony (Schneirla and Brown, 1954). Indirectly, as a result of colony behavior, each new bivouac site is opportunely sheltered in the general environment; and may be readjusted to disturbing variations in surrounding atmospheric conditions. A stable microclimate is so regularly established for the successive broods that the periodicity of the developmental stages is highly predictable in brood after brood. Thereby the bivouac, and the worker behavior accounting for its properties, serve as fundamental regulators of colony "drive."

These bivouac properties facilitate the occurrence of a regular cycle of behavior changes in the colony, each successive one caused by a further massive excitation of the adult population by the brood. The shelter properties of the bivouac are greatly enhanced by a diurnal behavior routine in the worker population which, although related only indirectly to environmental atmospheric changes, buffers the brood securely against potentially harmful exposure. Reciprocally, through indirectly contributing to the existence of a rhythmic colony behavior, a normally developing brood makes the bivouacs possible. Through complex interactive organic and behavioral events, adult behavior, brood processes and environmental conditions are interrelated in the functioning of this complex behavior pattern.

An indispensable component in the nomadism and predation of the army ants is their foraging pattern which centers around the making and following of chemical-trail systems. These raiding systems are typical of the species. For example, *Eciton burchelli* may be called a swarm-raider in that it conducts its forays in large masses or swarms connected by columns with the bivouac, *E. hamatum* is in contrast a column-raider in that it operates in systems of branching columns terminating in small raiding groups. Certain characteristics of the adult workers of these species, identifiable in simple tests, seem implicated, no doubt with others less readily identified. Chief among these are differences in olfactory threshold, glandular secretions and their intensity, level and range of excitability. In all of these respects, *E. burchelli* is the more generalized, *E. hamatum* the more specialized. Thus, *E. hamatum* can advance in relatively small numbers and follow its trails readily, with more precise trail-division responses than *E. burchelli*, which requires greater numbers. *E. hamatum* is more specialized in its booty, taking soft-bodied prey evidently found mainly through odor; *E. burchelli* takes a wide variety of booty primarily through response to motion.

Certain basic organic factors thus may be postulated as centrally involved in the species pattern. Another, and a critical one, centers around the fact that the newly emerged worker does not enter at once into an adult function. Instead a few days are required, first within the bivouac, then after a day or two in gradually extending the scope and efficiency of operations on raiding trails. An experience factor is suggested by the fact that callows artificially removed from their cocoons typically are accepted in other colonies of their species and are undisturbed on being introduced, whereas older callows are attacked and are noticeably disturbed on introduction. To account for such results, we may assume a simple habituation to the specific colony odor, perhaps begun in larval feeding, progressing after emergence when the callow feeds voraciously in the presence of this odor (Schneirla, 1941). The young worker therein has the basis for a following reaction to the colony odor trails when she first ventures outside the bivouac. The improvement shown within a few days from the initial clumsy and hesitant condition may be partially due to further maturation, but this cannot be the entire answer. There is also one specialized trail reaction, turning toward the bivouac at trail junctions when booty-laden, which may have its basis in early feeding on the typical booty.

These assumptions find support. Conditioning is a well-known phenomenon in social insects (Schneirla, 1953b), and Thorpe and Jones (1937) appreciably shifted the adult oviposition response of a solitary insect, the ichneumon *Nemeritis*, to an abnormal host upon which experimental subjects had fed as larvae. A conditioning process for the ants is not considered the sole factor, rather as implementing and extending the function of organic factors (e.g., mouth reflexes). That developmental factors place a distinct limit upon such processes would follow from the difficulty of artifically hatching callow workers of *E. burchelli* into *E. hamatum* colonies, and vice versa. Comparable results have been obtained for the highly socialized *camponotine* species (Schneirla, 1952b).

It is suggested that discrete morphological and physiological properties of the individual insect can enter into patterned interrelationships through types of facilitation effective in the group environment. In ants such as *Formica*, it is possible that early discrete mouth-part reflexes may be extended from simple passive feeding, through colony interactions, into active feeding of others and finally into individual foraging (Maier and Schneirla, 1935; Schneirla, 1941, 1952b, 1953b). The conditioning process assumed is doubtless a rudimentary one, with limitations typical for insects, but nonetheless indispensable to the species functional pattern.

In insect colonies, other individuals are prominent environmental agents facilitating such integrations, thereby being somewhat equivalent to the role which the general environment may serve in solitary insects. In social insects, relationships of this kind are indispensable for the functional patterns of individual and group. General colony function in the army ants depends upon adult responsiveness to the brood, which through its physiology and activity furnishes attractive and excitatory tactual and chemical stimulation. The regular rhythmicity which is essential to the species behavior pattern as well as to colony unity depends upon such relationships (Schneirla, 1938, 1952b). When active larvae are present, intensively exciting the worker population, large raids and regular nomadism are the rule; when the brood is quiescent and social stimulation low, raids are small and emigration absent. These findings offer strong support for extending the Wheeler (1928) concept of trophallaxis to all classes of stimulative relationship among individuals in the social group, rather than merely to "exchange of food" (Schneirla; 1946, 1952b).

In the insect social system, each type of individual enters into group relationships more or less indispensable to the species pattern. In the army ants, for example, the prodigious ovulation processes of the queen come into action for short periods of a few days at regular intervals somewhat more than one month apart. This is of great importance for the *Eciton* functional cycle, since it determines the timing of broods which in turn governs timing of the cycle through predictable brood excitatory effects.

At first sight, time relations in the functional pattern might be attributed to an endogenous control in the queen, who becomes physogastric and delivers a new brood approximately midway in each statary phase. But recent findings show that the queen is brought into this condition by a critical extrinsic effect (Schneirla, 1953a). The specific cause arises anew near the end of each active or nomadic phase, when the larval brood then present is nearing maturity. A few days before this phase ends, the larvae begin to undergo a glandular change which underlies the cessation of feeding and the approach of cocoon spinning (Lappano, unpubl.). At this point, although feeding less and less, the brood is constantly more active and capable of stimulating the adult population increasingly. The queen soon begins to feed voraciously, but this is probably not an automatic reaction to the mere presence of abundant food.

More likely, she is started and maintained in the process by an intensified trophallactic stimulation from the excited workers, and augmented social stimulation should increase her capacity to consume food. A recrudescence of the fat bodies soon begins and with a maturation of eggs in the ovarioles causes her gaster to swell. These processes accelerate and reach their peak toward the middle of the statary phase, when the actual egg-laying episode occurs.

This theory finds a natural test. A second physogastric episode begins near the end of each statary phase, when reflex activities of the nearly mature pupal brood stimulate the workers to larger raids, thereby indirectly causing a food surplus (although the brood is still enclosed and not feeding). But soon after the new brood of workers has emerged from cocoons and a new nomadic phase has begun, physogastry in the queen is cut short. Evidently a large-scale feeding by callows, reducing the food, and a diversion of workers to trophallactic relations with callows, account for the abrupt return of the queen to the non-reproductive condition. This condition persists until the current nomadic phase nears its end, when a new episode of egg-production begins and advances to completion.

It seems clear that the normal reproductive episode of the queen is initiated extrinsically and can complete itself only when the essential extrinsic conditions persist. The necessary cause evidently lies in a complex relationship arising periodically between brood and worker population, and interaction indirectly introduced through the queen's own function at an earlier point in the cycle (i.e., when the preceding egg-batch was laid). The cyclic pattern thus is self-rearoused, so to speak, in a feedback fashion. The rearousal is timed with relative precision and is therefore the product of a reciprocal relationship between queen and colony function, not of a timing mechanism endogenous to the queen. Essential to the timing of intervals is the duration of developmental stages in the brood, which owes its regularity particularly to the bivouac as "brood incubator."

The schema of the *Eciton* functional cycle, in Fig. 1, indicates that the interrelationships of the principal contributive factors and their recurrent effects upon the system are numerous and varied. Relationships with the physical conditions of the general environment, under which the pattern has evolved, are crucial. As one example, heterogeneity in the operating terrain is indispensable for execution of the typical raiding-emigration sequence (Schneirla, 1944). There is also the diurnal cycle to which the timing of daily raids and nightly emigrations is geared in the terrestrial species. Light sensitivity, although unimportant for orientation, thus is a key factor in the cyclic pattern of these species. Emigration in the terrestrial species normally occurs in the evening, at the one time of day when atmospheric conditions are most nearly homogeneous throughout the forest and sufficiently near the brood optimum to permit exposing this delicate part of the population during the change of colony base (Schneirla and Brown, 1954).

The complex *Eciton* cyclic pattern therefore arises through the diverse interrelationships of its component processes—the morphological, physiological, behavioral and environmental factors which interact under given conditions. The

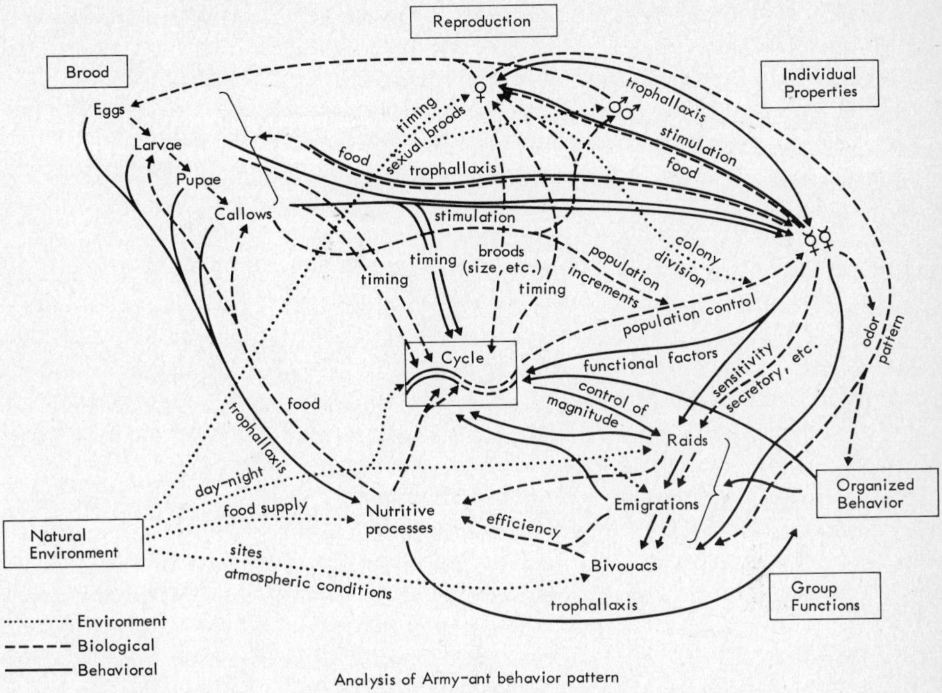

Fig. 1. Schematic representation of the behavior pattern of terrestrial army ants. Major sources of identified influential factors (e.g., natural environment, brood, etc.) are indicated marginally, the principal interrelationships demonstrated are suggested by appropriate arrows indicating the direction of the respective effects. The resulting rhythmic functional pattern, suggested by the "cycle" in the center box, is discussed in the text.

organization does not pre-exist in the heredity of any one type of individual—workers, brood or queen—nor is it additive from these alone. The organic factors basic to the species pattern have evolved in close relationship to the general environment, which therefore supplies key factors essential for their contemporary integration into a functional system.

"MATURATION" AND "EXPERIENCE" IN BASIC VERTEBRATE ONTOGENY

The concept "maturation" in modern morphogenetic theory includes, with growth as such, the excitatory and stimulative effects that arise through growth processes and influence further stages of development. Development is much more than growth therefore. This definition suggests a need to broaden the traditional view of "maturation" as the direct or specific effect of growth processes on development. Their indirect effects, and their interaction with surrounding conditions, can become cumulative. A broader causal pattern is suggested by Holt's (1931) view of self-stimulative relationships in the development of coordinated movements.

The actual breadth of the pattern of relationships governing early behavioral development on different vertebrate levels is not well known. Locomotion is frequently taken as a critical case, and Coghill's (1929) classical experiment with larval salamanders is accepted as demonstrating an innate intraneural determination of coordination. Yet this experiment was not an analysis of function; it was a histological and descriptive study which disclosed a parallel course in the appearance of successive new types of neural connectives and the respective phases of a series of locomotor changes. The functional aspects of the correlation remain unclarified.

The experiment of Carmichael (1926) with drugged salamanders has been widely accepted as eliminating peripheral factors in favor of a neural determination of functional development. When embryos were drugged with chloretone to prevent movement during development, the experiment reported that within 30 minutes for recovery from the drug, adult specimens swam in a manner "indistinguishable" from that of control animals. But Fromme (1941), repeating and extending this experiment with tadpoles, did not altogether confirm this conclusion. In his investigation the recovered experimental subjects differed from normals not only in a slower swimming speed but also in details suggesting deficiencies in coordination, when time-conditioned drug effects were controlled. Particular deficiencies were noted in an experimental group chloretonized only during the advanced stage of "partial movements." Fromme concluded that "without opportunities for the practice of rudimentary movements, the earliest swimming does not compare favorably with that of a control group." It should be noted also that in such experiments, the role of other possible factors such as subtle afferent effects has not been excluded.

Coghill's view that an initial generalized pattern is primary, with partial patterns individuating secondarily from it, is a broad generalization which seems to require secondary qualification. The concept of a pre-set total neurogenic pattern, subsummed under such generalizations, only roughly fits available evidence. Early generalized activities are variable and not too well patterned. On the other hand, discrete activities may appear very early, as Kuo (1932, 1939) found in embryonic chicks, and Carmichael and Smith (1939) in early guinea pig embryos from which they readily elicited local responses with low intensities of tactual stimulations. So many conditions determine the extent or inclusiveness of movements at different stages (Werner, 1940; Carmichael, 1951) that variability and flux deserve equal emphasis with broad patterning in early action. Windle (1940, 1950), in fact, holds that the earliest responses to stimulation are local reflexes, which provide a basis for the integrated patterns of later stages.

The alternatives for the causation of patterning in ontogeny are not simply growth vs. learning. Doubtless, mechanisms like some of the vestibulocular reactions "are not patterned by training but are laid down directly in growth," somehow, as Sperry (1946) concludes from his study of compensatory eye movements appearing in tadpoles without optic nerves. But the "laying down" would (even here) seem to be a matter of development, of intraorganic interrelationships involving other tissues, organs and even systems, rather than just an accretion of tissues, or growth specifically. As a matter of principle, the possibility of even wider relationships cannot be excluded.

Apparent support of a pre-determination of behavior by a neural control from specific centers is found in the investigations of Hess (1949), in which, for example, during electrostimulation of different localities in the cat's hypothalamus, movements suggesting activities such as feeding and sleeping appeared, although unrelated to the present external situation. As Lehrman (1953) points out, Hess himself does not conclude from this that a strict localization of functional control exists in the hypothalamus, and variations in afferent input may account at least in part for the discovered temporal variations in responses to localized electrostimulation. Moreover, as Masserman (1941) found, such "pseudo-affective" reactions obtained by electrostimulation of the hypothalamus differ strikingly from the normal in a number of respects. It should be emphasized that these experiments involved the adult stage only. The role of peripheral loopline processes in normal central coordination may have been underestimated. Thus in the adult cat, Gellhorn, Loella and Ballin (1954) found the responsiveness of cortical auditory and projection areas in lightly anesthetized subjects appreciably increased through peripheral application of nociceptive ("pain") stimuli together with the respective exteroceptive stimuli. Strychnine tests indicated that this is a result of interactions of cortical projection processes with impulses set up in the hypothalamus through nociceptive stimulation.

Adherents of the Lorenz theory often attach considerable weight to certain results from studies on locomotion in fish and amphibians as evidence for an innate central nervous determination of behavior in vertebrates. One such study is that of Weiss (1941), who obtained motor activity in a transplanted salamander limb which had motor connections only with a deplanted section of spinal cord. In the absence of afferent connections, the movements may be attributed to spontaneous neural impulses from the displaced section of cord. But the actual relationship to normal behavior of these effects and of spontaneous discharges such as those recorded by Prosser (1936) from isolated arthropod ganglia has not been established. There also are numerous technical details opposing an unequivocal interpretation of such results (Gray, 1950).

The findings of von Holst (1935, 1950) with the teleost *Tinca vulgaris* and other fishes are often taken to support an innate neural determination of coordinated locomotion. But Lissman (1946), in experiments with dogfish in which all afferent inlets to the central nervous system had been removed, could not obtain such automatic central control. Rather, carrying deafferentation beyond a certain point abolished the rhythm completely. Although positive conclusions await further evidence, it is indicated that the central control in this case depends upon some intact peripheral connections. Since previous experiments have concentrated on mature animals, there is an unfortunate lack of evidence concerning how neural coordination may be established through earlier development in such activities.

Of course nothing is to be gained by minimizing neural capacities for organization appearing through development (Lashley, 1947; Hebb, 1949). To throw needed light on the condition of such functions through ontogeny, let us hope that further studies will involve comparisons of neural control in immature and in adult stages. With respect to hypothalamic functions, it is conceivable, for example, that in the adult, central control may have usurped the role of peripheral functions indispensable in early development. Hence the James-Lange and Cannon theories of emotion may both be valid but for different stages. In regard to the coordination of locomotion, however, Lissman (1950) concludes that ". . . the proprioceptors may well determine the disposition and the state of activity of the central nervous system more permanently than is often assumed."

And this, after all, is only one of the peripheral systems to be considered as influencing central coordination. The complexity of such relationships would appear to underlie findings such as those of Aronson and Noble (1948), who in a systematic operative exploration could not resolve an amphibian behavior pattern, that of mating, into a control of its components within respective neural centers.

There has been a considerable number of observational studies on vertebrate ontogeny, summarized by Carmichael (1946, 1951). These studies, although impressively detailed, exhibit a chiefly descriptive character and a general shortage of systematic analytical procedures which may stem from concepts of an ontogeny dominated by the maturation of predetermined "growth" processes. A concept of maturation thus circumscribed does not seem valid for any stage of development including the earliest and for any vertebrate including the lowest classes. It would seem useful to define the concept more broadly to cover interactions within the developing organism, cumulative from stage to stage, without assuming any sharp cleavage of these processes from changes introduced by extrinsic conditions.

For a discussion of this point at an elementary level, we may refer to some findings of Tracy (1926) in his studies of toadfish ontogeny. The early larva of this fish lies quiescent on the bottom for successive intervals, but intermittently exhibits quick jerking movements, called "spontaneous" because they have no apparent extrinsic cause. According to Tracy's analysis, these movements occur through the presence in the blood of metabolic products, accumulating as growth processes reduce available oxygen during quiescent period. At a critical point in the buildup, an intrinsic excitation somehow occurs. The existing biochemical condition may either cause the direct firing of motor neurones, or may alter thresholds so that sensory impulses from other sources excite motor discharge to muscles. The rhythmicity of the process may be due to the fact that each movement phase alters the embryo's condition so that a further resting phase may ensue, in which metabolites accumulate to the point of setting off another movement, and so on.

These early movements are endogenous in the sense that their basis is mainly within the organism. But for their occurrence, not only must there be interactions of metabolic processes within the embryo, but also interactions of these processes with extrinsic conditions. The point may be stressed by taking up an interesting conclusion of Tracy's, to the effect that if external conditions could be kept constant, the organism's activities would be determined by its own life processes. Such statements can be misleading. What this cannot mean is that extrinsic conditions ever are passive or neutral, much less dispensable, for the developing organism is in constant and progressively changing interaction with its medium. This principle becomes increasingly significant for development as it advances to organizations of processes which include as subordinate the biochemical and other basic functions.

From the initial stages of development, extrinsic conditions are thus implicated in developmental processes. They are not to be included within "maturation" as defined, yet they influence development and activities constantly and vitally. These effects constitute instances of "experience," defined as the effects of extrinsic stimulation upon development and behavior. This concept may be only roughly apposed to "maturation" in its significance for development. It denotes a wide range of phenomena, from the influence of biochemical changes and growth-induced activities to an afferent input variously aroused, with possible trace effects not usually classed as "learning" (e.g., cases of sensory adaptation). It is only at later stages that learning, a higher order of process resulting from "experience" contributes to behavioral development.

No absolute or sharp distinction is intended between these concepts of maturation and experience at their border-line, which may be considered a common zone of overlapping functions. As McGraw (1940) pointed out with respect to attempts to distinguish "maturation" from "learning" through a separation of stimulation from extra- and intraorganismic environments, respectively.

"... the definition is confusing and calls for a clarification of the terms organism and internal and external environments." ... "To what are these environments internal or external?"

Stimuli at first effective externally may, at a later stage, function through their organic traces, as admitted by and in interaction with new orders of function which have entered meanwhile.

Undue stress upon one of these concepts to the exclusion of the other must give a misleading picture of behavioral development. For example, Grohmann (1939) reared pigeons in narrow tubes intended to prevent incipient flight movements, yet in free tests at the age of normal flight these birds were judged equal in their flying to previously unrestrained controls. This case has been cited (e.g., by Tinbergen, 1951) as evidence for an innate control and against experience. But developments contributing to actual flight may still include wing-muscle contractions despite enclosure within a tube, as well as wing movements known to occur in embryonic stages (Kuo, 1932). Also, developmental functions preliminary to flight may be very different in various species of birds. Thus, young turkey vultures kept by Dennis (1941) in restricted quarters until the normal flight time showed poor balance and could not fly on release, without further experience in the open. Was this a case of muscular weakness, or was it lack of experience? What is experience? Tinbergen (1951, p. 132) mentions observations of butterflies and dragonflies which, when disturbed after emergence, made long flights without benefit of incipient flight movements. But previous developmental relationships may actually be relevant in cases of this kind. Also, after such insects emerge, there is a process of wing spreading attended by local actions as the wings dry, which obviously involves a facilitation of respiration to the point necessary for flight. It is not clear how peripheral loop-line relationships and their build-up can be excluded from the development of nervous control in such mechanisms.

The principles of behavioral organization are not to be satisfied by *a priori* postulations of innate organizing centers, but must be worked out in investigations appropriate to each type of behavior. For this purpose, Kuo (1932) directed his attention to the embryonic history of pecking in the domestic fowl. He recorded a time table of events as viewed through a window made in the shell, and in this way could also work out interactions. He traced the development, from indistinct beginnings, of three general kinds of activity significant for the adult pecking pattern. These were, the head lunge, bill-opening and -closing, and swallowing. The head movement first occurs passively as the head, bent down and resting on the thorax, is pushed upward and falls back rhythmically with the heart beat. On the fourth embryonic day a more vigorous nodding movement is observed, interpreted as an active response of the head to touch or to neural irradiation. After the sixth day, flexion of a leg may brush the toes against the head, eliciting a head movement. Such movements are reflex-like in the sense that they are local and automatic, but at the same time they are at first highly variable in form, a feature only partially reduced at hatching. In the course of time, limited integrations appear between the activities of head, bill and throat. At about the eighth day head-nodding may be followed closely by opening and closing of the bill, very possibly through the effect of nervous impulses irradiating from the head action. This movement in time gains increased frequency and amplitude, so that after the ninth day bill-clapping admits fluid into the mouth. Swallowing may then occur, probably as a reflex response to the sensory effect of amniotic fluid drawn into the mouth. A few days before hatching, when the head lifts and thrusts forward, beak-first, opening and clapping of the bill may follow, often with swallowing. All of these movements now are somewhat less stereotyped than before, and would seem to have come predominantly under neural control.

At hatching, the interconnection of these activities is at a relatively low point, with each "component" still variable and remote from its condition in the adult pecking pattern. Then a critical new feature enters:

Visual arousal of the lunge on visual stimulation. As suggested by Maier and Schneirla (1935), this response may be aroused *de novo* by visual effects through a diffusion of impulses from the optic lobe into adjacent midbrain centers, thereby setting off neural mechanisms previously established in terms of a tactual control of the head lunge.

These results seem best accounted for not in terms of an initial central nervous organization, but by an inclusive organization which emerges from the processes and interrelationships of maturation and experience as we have defined them. The eventual components are only vaguely suggested in the unformed, variable and unorganized events of earlier stages. Each component has its own developmental history into which many causes, both directly and indirectly related to organic growth, may enter. Thus tactual effects which at first accompany head movement incidentally, may acquire a direct neural control over this response through the formation of a functional neural pattern—an elementary conditioned response. In each stage, maturation gives rise to effects which influence other activities occurring at the same time and also, through contiguous neural traces, form new integrations influencing further stages. Experience here means the relationship of the organism with its own activities in self-stimulative processes, as well as with its extra-organic situation, at each stage. Typical "experience" in the egg, arising through and related to growth processes, thus seems essential for the normal development of head lunge and other components of the eventual pecking pattern. For the advance of the entire sequence of events, the involvement of feedback processes and their trace effects upon later stages is indispensable.

An efficient pecking response appears after hatching through a discrimination-motor learning in close relation to and in dependence upon further maturation. Studies of this learning process, with due attention to the role of maturation, have been carried out by Shepard and Breed, Bird, Cruze and Padilla (see Maier and Schneirla, 1935). The eventual neural correlation pattern of food-pecking evidently arises as the composite product of an inter, play of forces variously introduced into the typical growth situation. Although the contributions of physical growth are of course indispensable, their role alone is insufficient for the rise of the normal pattern or any other. Also indispensable are the wider reactions into which the growth factors enter as partial contributors. When these wider relationships are changed, the outcome is not the species-typical one. Thus in the experiment of Padilla (1935), a critical change in the post-hatching situation of the chick facilitated the establishment of a pattern significantly different from the typical one.

It would seem to be the prevalence of an intimate, dynamic relationship between the factors of maturation and experience that renders analytical study of behavioral ontogeny so difficult. Methods must be devised appropriate to the complexity and subtlety of these processes and their trace effects influencing later stages. In such work, little may be expected from attempts to estimate the specific or the proportionate contributions of the innate vs. the acquired in ontogeny (Howells, 1945; Anastasi and Foley, 1948). It is another matter to attempt an appraisal of maturational effects as limiting capacity to change through experience (McGraw 1940, 1946). An example is Shirley's (1931) reference to a cephalo-caudad maturational gradient underlying a described ontogenetic sequence of motor changes in the human infant. This is essentially an introductory procedure for appraising how far development has progressed at given stages, and is not the same as attempting to estimate the overall weight of "maturation" as against "learning" in the particular animal. The latter practice can become a dangerous limiting influence in research and theory, especially when one member of the pair is favored unduly. It is thus probable that Gesell (1945, 1950) carried the idea of "maturation" conceived as an innate "growth plan" much too far in his theory of behavioral development in the human infant. For a true picture, "experience" factors must not be held from consideration until advanced ontogenetic stages are reached.

"Experience," as the term is used by Beach and Jaynes (1954), concerns only cases that involve learning and habit acquisition more or less specifically. But the meaning and relevance of the "experience" concept as used here subsumes all types of relationships dependent on extrinsic conditions, including gains through "learning" as one type. Learning concerns only a part of the range of relevant experiences, a chronologically later and more specialized part. For example, indications of conditioning have been reliably identified by Munn (1940) and by Moore and Welch (1940) in the swimming stage of larval amphibia, but we have suggested intrinsic-extrinsic interactions for stages much earlier than this, the organic trace effects of which must be considered as factors in later development. Not the earliest, but an important concurrence of this type, is suggested by the coincidence of a proprioceptive or other organic-arousal process with an exteroceptive effect, as illustrated in Kuo's results. Carmichael (1936) suggested that through such combinations of events, transition to an exteroceptive control might begin and advance, with the stimulus-response pattern thereby assuming a more specific form. Such processes in early stages cannot be expected to fit traditional notions of conditioning.

Prenatal conditioning thus far has been demonstrated only on the bird and human levels. Gos (1935) obtained a specific conditioning of general movement to vibration after the 16th day in the chick embryo; Hunt (1949) after 15 days. The human embryo has been conditioned by Ray (1932) and by Spelt (1948) after $6\frac{1}{2}$ months. The criteria of conditioning appeared broadly satisfied by the results, although the Wickens (1939) rightly questioned whether the ordinary type of conditioning is closely duplicated in the embryonic stage. Significantly, at 10 days in the chick embryo, before a specific conditioning was demonstrable, Gos began to notice signs of habituation to experimental stimulation. Very possibly, with adequate techniques, changes of a simple order may be demonstrated at still earlier stages. Thus far, conditioning in utero seems to be characteristically rather generalized, variable, and unstable in its trace properties. It is likely that, whatever their degree of specialization, processes involving the consolidation of the trace effects of experience must be reckoned with as implementing maturational processes and as indispensably interrelated with them at all stages.

THE "NATIVE" AND THE "ACQUIRED" IN STIMULATION AND PERCEPTION

Objectivity in attempts to appraise an animal's relation to its environment always is influenced by our own perception, which inevitably colors our notions of the adjustment (Schneirla, 1950). From this standpoint Lehrmann (1953) has criticized Lorenz's view of the external stimulative effect, conceived as "releaser" triggering off an innately organized neural center and thereby eliciting the appropriate instinctive act.

The stimulative effect in "instinctive" behavior tends to resist analysis as to its nature, and unfortunately we are too much limited to general impressions which are likely to be unreliable. But the difficulty is only increased by referring to situational adjustments as "perception" in widely different animal types from psychologically lowest to highest (Schneirla, 1948, 1949). The risk is distorting the actual relationship of situation and animal. As an example, Hertz (1931, 1933) found that untrained bees in their initial visual test approached most readily the stimulus cards of greatest contour richness and inner detail (e.g., paddlewheel figures as against crosses). This spontaneous response opposed training, although von Frisch (1923) and others had conditioned bees to brightness, color, and odor differences. Some writers are inclined to interpret the Hertz finding in terms of a fairly complex configurational process comparable to mammalian pattern perception. However, Wolf (1933) found that bees were attracted more readily to rapidly flickering stimulus areas than to slowly flickering areas. He concluded from these and other facts, and in consideration of the poor analyzing power of the bee's eye, that flowers are visually effective for the bee in flight according to the gross effects of intermittent visual change rather than to their unitary figural patterns as such.

The specialized response of the bee is selectively determined first of all by the properties of the receptor and afferent-neural system and not a hypothetical neural releasing center. The afferent mechanism, a product of development, puts the organism directly into relationship with the kinds of external situation under which these mechanisms have evolved. Many other spontaneous responses of insects may depend comparably on receptor properties, for example as responses to air vibrations first depend upon the properties of chordotonal organs or related receptors in insects. The corresponding processes seem to be more adequately characterized as "sensing" than perceiving.

A promising contemporary theory of perception, that of Hebb (1949), describes a process through which cumulative neural organizations arise through further experience in external situations. Perception, the capacity for appropriately organized adjustments to the sensed situations according to their properties, is viewed as additive, starting virtually from zero and progressing to further specializations in dependence upon opportunities for experience and capacity for change through experience. Thus, rats raised in darkness need about six times as many trials as normally reared animals to master a discrimination of vertically from horizontally striped fields (Hebb, 1949). Comparable evidence for other vertebrates will be taken up presently. In the perceptual task itself, the configurational effect evidently is gained essentially through learning, aided in still unidentified ways by early experience.

One difficulty with terms such as releaser is their ambiguity. Even if this term could be limited to the elicitation of spontaneous reactions not dependent upon specific experience with the object involved, the concept of triggering off an innate neural coordination remains unsupported. Nor does "releaser" seem applicable to conditioned stimuli, which must be qualitatively different from those arousing spontaneous responses in their corresponding organic mechanisms, and not merely in the number of items as Baerends (1950) concludes. This view finds support in results to be considered.

Stimuli eliciting "spontaneous" reactions (thus termed for convenience) seem to be simple and generalized, not configurational. The ant guest, for example, does not appear to be initially attracted to the ants as stimulus constellations, but to effects such as odor. The conditions under which impulses through different sensory channels can be equivalent in evoking a common response, as described under the Seitz concept of heterogeneous summation (e.g., Tinbergen, 1951), are not clear. The best examples of this phenomenon seem to exist in the class of postural and locomotory control (e.g., Friedrich, 1932). More specialized types of adjustments often are dominated by specific afferent mechanisms, as in the plant responses of phytophagous insects, frequently odor-specific evidently by virtue of a biochemical aspect of the receptive process (Dethier, 1947).

The influence of afferent equipment in early behavior in higher vertebrates usually depends upon wider properties of the organism. In birds and mammals, the initial "naive" receptor-influenced responses in many cases soon became specialized in species-typical ways. For example, chicks and ducks show differences in their initial encounters with water, due in part, Schooland (1942) thinks, to differences in temperature sensitivity of the feet. Influenced by this afferent property, and also by visually-dependent pecking tendencies, a characteristic water-reaction soon becomes established in the first encounters. In the rise of incubation and nesting in birds (Tucker 1943; Davis 1945), a hormonally produced condition of seasonally increased tactual and temperature sensitivity in specialized ventral areas known as "brood patches" is considered basic. But a considerable amount of variable, random activity is required in the organically disturbed bird before the typical locality adjustments of nesting and incubation can be worked out (Craig, 1918; Lehrman, 1953). Comparably, a hormonally increased temperature sensitivity may be allotted a major role in the "working out" of nest-building activities in parturient rodents (Kinder, 1927).

Doubtless the relation of afferently determined susceptibilities to spontaneous reactions must be established for each type of organism, in terms of prevalent extrinsic conditions. Often the key effect is relatively simple. For example, the nesting herring gull typically pecks at the parent's bill in the feeding situation, and through tests with models Tinbergen and Perdeck (1950) found the critical factor to be a red patch on the adult bill. By varying the position of this spot on the models, these investigators found that the young bird would peck most readily when it was placed near the tip of the bill. The simplest interpretation would seem to be that the spot is effective according to its degree of prominence as a moving object, i.e., on the head it is moving axially and more slowly, near the bill tip circumferentially and faster.

In lower vertebrates, the controlling effect in spontaneous reactions seems more understandable in quantitative than in qualitative terms. Thus the visually-determined "lunge" responses common in fishes, amphibians and reptiles are elicited most effectively by moving stimuli of small area. Toads snap in dependence upon such properties and not the nature of the object (Honigmann, 1945; Freisling, 1948). Critical for the response seems to be energy of the delivered stimulus effect; thus toads snap at small moving objects with the tongue alone, at larger ones with the jaws assisted by the forelegs, but still larger ones bring withdrawal (Eibl-Eibesfeldt, 1952). Specific details may acquire a limited effectiveness as modifiers of the spontaneous response through conditioning (Schaeffer, 1911; Cott, 1936; Eibl-Eibesfeldt, 1952).

Facts such as the above suggest that the mechanism of a spontaneous reaction may depend upon some general organic relationship, such as a critical ratio between afferent, neural and effector thresholds. Deficiency in central nervous correlation capacity seems to be a prime condition, since stereotyped, spontaneous response tendencies are strongest in animals with the lowest supply of internuncial neurones. It may be the resultant dominance of specific peripheral conditions over simplified neural channels that makes spontaneous responses so resistant to reversal through conditioning. Ehrenhardt (1937) found that in its initial visual response, the lizard *Lacerta agilis* snapped most readily at figures with smooth outlines, least readily at those with more varied outlines (e.g., at squares as against circles). In conditioning tests, with stimuli close together in preference, although limited changes were possible, reactions to extremes such as circle vs. cross could not be changed in as many as 850 trials. The neural changes of conditioning may thus be effectively resisted by the wider base of stereotyped response in the organism, fixed through organic development.

When the area of a stimulus model is varied experimentally, and the smallest sizes produce "feeding" responses, intermediate sizes "courting," and still larger sizes "fright" or "attack" (as is typical in fishes and many other lower vertebrates), the underlying response mechanisms would seem to have evolved in relation to the usual energy effects of the respectively different external situations. Thus there tends to be a correspondence in different animals between body size and the magnitude of stimulus thresholds critical for the respective reactions.

The actual effect of a stimulus situation may elude identification through the observer's tendency to offer a diagnosis dependent upon apparent adaptive significance. There is a well-known experiment (Tinbergen, 1948, 1951) in which an airplane-shaped model of a "bird of prey" elicited alarm and escape reactions from young nestling birds when moved over them with the "short-neck" end first, but not when moved "tail-first." This appears sensible since it is the way a man first recognizes a hawk. But the initial effect on young birds may be simple and perceptually non-qualitative, perhaps a shock reaction produced by a sufficiently abrupt stimulus change rather than anything more specific to a "bird of prey flying overhead," a hypothesis of mine discussed by Ginsberg (1952). This idea finds support in a study by McNiven (1954) with nestling young of various birds, which showed disturbance when an airplane-shaped model was moved rapidly overhead, but not when it moved slowly, whichever end was in advance. The artifact method requires careful control against subjective impressions of what common external object or situation the test model may represent to the animal. "Short-neckedness" in the above situation is a cue to the human observer, a perceptual "sign" indicating an innate escape reaction to him but not necessarily to the bird. The animal's actual relation to the situation and its developmental basis may be very different from what the "sign" interpretation suggests.

The "isolation" technique is often considered a critical means of appraising the contributions of the normal environment to the species behavior pattern, and even as the one crucial way of separating "innate" from "acquired" in behavior (e.g., Tinbergen, 1951). But actually, isolation may involve only a relative change from the situation typical for the species (Riess, 1950). The effect of the "normal environment" is not readily diagnosed, and the possibility must be considered that altered situations (e.g., "isolation") may contain extrinsic effects more or less equivalent to those, regarded as species-typical. Isolation merely shows what responses may be obtained in divergent situations. For example, Craig (1914) found that young male doves raised in isolation from species mates, upon sexual maturity would give their mating responses to a miscellany of objects, including the human hand or foot. Such results tell us only that the "abnormal" stimuli are somehow equivalent to the normal one, the species mate. The isolation technique, then, is essentially an introductory means of finding what behavior can appear under different conditions. But what the situation of isolation may have contributed to the observed behavior . . . may not be altogether different from what associations with species mates may contribute normally. In simple logic, the "isolated" animal is not shut off from itself; it is obviously no less a member of its own species, and thus may present to itself, through the processes of its own development and self-stimulative associations, at least some of the influences which normally affect later species associations, as in mating. This must be determined experimentally.

Pattie (1936) reported an "innate gregarious tendency" in chicks raised in isola tion; Howells and Vine (1940) described a limited differential association with species mates in chicks raised with mice; and Schooland (1942) found a similar gregarious tendency in chicks raised with ducklings, interpreted not ". . . in terms of the differentiated environment," but ". . . as rooted in innate constitution." Does "constitution" as used here exclude environmental relationships after hatching? The possibility should be tested that the isolated chick, peeping in situations of organic relief as when fed or warmed, may "develop" a tendency to approach the source of such species-typical stimuli, hence species-mates in the tests. The "tendency" demonstrated in these studies is partial, relative and quantitative. Such conceivable relationships are based on still earlier stages of development. An isolated chick even when satiated will resume pecking when pecking sounds are heard. By the criterion of isolation this response becomes "innate," but what is its actual basis?

Parr (1937) excludes the possibility of self-stimulative gains in an isolated fish, which, without appendages, cannot touch himself and thus become acquainted with his own body. Even occasional natural minor effects are excluded. The argument is that the psychological capacities of these lower vertebrates would not permit sufficient gain from occasional self-stimulative experiences to affect later associations with the animal's own kind. This seems reasonable with respect to total, integrated impressions, but should not be taken to exclude limited experiences through repetitive partial effects as with a self-produced chemical exuded under given conditions, or occasional direct glimpses of a specific part of the body or of the whole as an area. Although psychologically incapable of learning an organized self-reference perception, these lower vertebrates may be prepared through such partial experiences to react to species mates as more than "foreign objects." The value of such relationships through experience presumably is greater for most birds and more contributive still in mammals.

The psychological inferiority of lower vertebrates as compared with man in this respect, although difficult to appraise, is probably not absolute even when judged in terms of the adult stage. But it is great, since man, as Hallowell (1954) concludes:

"... through language and reflective thought is able to integrate perceptions of his own body and his personal experience with a meaningful concept of self that is the common property of other groups of his society."

The difference no doubt is least marked when man is in the prelinguistic stage of his individual perceptual development, before a well-integrated self-perception of the "body-image" type has been established. Then his gains, more like those of his closer mammalian relatives, come through self-stimulative, environmentally linked contiguities experienced in connection with basic adaptive adjustments such as locomotion and feeding.

Developmental relationships are often judged too narrowly as concerns the distinction of maturational and experiential contributions. For instance, conceptions of the ontogeny of bird song have changed since the finding (Scott, 1902, 1904; Conradi, 1905) that singing in many species is influenced by early auditory experience. But other species have been viewed as independent of this factor, as in the roller canaries which Metfessel (1940) found singing recognizable parts of the species song after isolation from hatching. To say that the first type learns its song, the second type has the pattern innately, would be superficial. In Metfessel's isolated birds, organic factors in the development of singing are suggested by the "fatigue curve" trend of individual tours when graphed. The influence of self-hearing experience is suggested by the fact that the order of elements varied and was different from the normal; also, the song tended to be higher in pitch, the song elements were modified when tones were experimentally introduced, and later group experience changed all songs toward the median. The higher pitch may have been due to the fact that isolated birds, more excited and tense than socially-raised birds, produce and hear higher tones than do the latter.

We have postulated one common characteristic among all cases of development:

Namely, they come about through continuous interactions between intraorganic developmental processes (maturation) and the effects of extrinsic conditions (experience). Hypothetically, gains from experience may represent very different types of process, contrasting orders of organization, from direct physiological effects of limited scope to the most advanced patterns of learning. The lowest orders of interaction predominate at early developmental stages, qualitatively more advanced orders (including learning) enter later, based on the accomplishments of preceding stages. The phyla, and the vertebrate classes, differ in their respective capacities for reaching higher orders in maturation, in experience, and in the interrelationships of these (Schneirla, 1949). Therefore, learning cannot be viewed as superimposed upon the effects of maturation and sharply separable from them; rather, these processes occur together in close interaction. But since learning subsumes earlier developmental processes and often therefore represents larger changes, its effects tend to be more discernible than those of simpler orders of experience.

How shall we differentiate gains through learning from other gains through experience as defined above? Attempts to distinguish learning from other processes in development presuppose a more exact knowledge of learning and its organic basis than we now have (Hebb, 1953). Cases of learning, simple or complex, seem all to involve the experiencing of sensory contiguities which modify individual behavior by effecting neural changes. But neither S—R theories with their rigid formula, nor cognition theories with their looser one, has been validated for phyletic comparisons or for individual ontogeny on any phyletic level (Schneirla, 1952a). A theory postulating more than one qualitative process of learning is required. Thus Maier and Schneirla (1935) distinguished conditioned-response patterns from selective (trial-and-error) learning, in that the animal's response in the latter case is fixated in dependence upon changes in the motivational situation. Then, on further evidence, they (1942) distinguished an afferent-afferent type of conditioning from the better known afferent-motor (classical, Pavlovian) pattern, the latter being more wholistic in that a specific change is effected in response control.

In the lowest animal phyla, experience, as we have defined it, is limited to simple, directly effective maturational relationships with learning never involved as a factor. In higher phyla, improved capacities for neural correlation admit learning to varying and increasingly broader roles in behavior development. But attempts to distinguish a stage in ontogeny at which "learning begins," and to separate learning from maturation, are based upon classical notions of learning tailored from the study of adult stages. Simpler forms of change through experience may prevail early in ontogeny, basic to conventionally recognizable signs of learning in later stages. Kuo's pioneer findings indicate how earlier stages may lead into later stages. The earliest changes (as through proprioceptive-tactual contiguities), although variable, imprecise and very slowly cumulative, may nevertheless promote improvements in neural organization essential to the normal succession of stages. Present techniques, although useful to demonstrate what animals can be conditioned in later stages of ontogeny, are insufficient for more than vague glimpses of the form and scope of the process, and admit no view of its antecedent conditions.

Hatching or birth means appearance into a new and complex environment, the occasion for novel behavioral acquisitions arising through a new set of relationships between maturation, growth attainments, and experience. The new environment is a variable, heterogeneous one, only partially equivalent to the former relatively monotonous embryonic setting, and presents new contiguities which promote some striking changes. In many of the changes in birds and mammals, further accessions follow a predictable course in which learning is an essential component. A striking example is the phenomenon which Lorenz (1935) described as "imprinting," after his observation that young goslings, exposed only to the experimenter in the period just after hatching and kept away from species mates, would thereafter follow him but not adult geese. Cushing and Ramsay (1949) were able to form heterospecific groups by placing young birds with a foster mother (e.g., bob-white quail with a Bantam hen) directly after hatching from an incubator. Fabricius (1951), studying imprinting in ducklings of various species, found that the initially effective visual stimulus was not specific, but concerned movement of the attractive object and its parts, and particularly movement-away. A generalized initial effect is indicated by the fact that the size of the "foster parent" could vary within wide limits and that sound (soft, rhythmic effects) was needed in many cases. The imprinted following reaction to the experimenter was often interfered with by a tendency to retreat. This effect was found to increase steadily after hatching, which may account for the fact that the period of effective imprinting is relatively short and typically limited to the first few hours after hatching. In some species such interference operates from the start, as Lorenz found with the incubator hatched young of wading birds such as curlews. Turning-to reactions thus may be considered products of pre-hatching stages which at hatching are elicited by "unconditioned" stimuli of a very generalized nature but with the common property of relatively low intensity. In contrast, stimuli of higher intensity (e.g., "disturbances" such as sudden movements) elicit escape reactions (Schneirla, 1939, 1949). The specific stimuli which may acquire a conditioned control over such reactions through early contiguities may produce fixations for inanimate as well as for living objects (Thorpe, 1944; Fabricius, 1951). As Whitman (1919) and Craig (1914) reported for pigeons, in some species later affiliations (e.g., in mating preferences) depend almost entirely upon the early environment and contiguities it may introduce. This seems to be a relative matter (Cushing and Ramsay, 1941), since in many others barriers exist which Cushing (1941) attributed to "genetically controlled factors." These are developmentally-furnished effects, presumably related first of all to species size and other gross morphological characteristics, to which correspond through evolution the thresholds for arousal of "unconditioned" turning-to or withdrawal reactions, respectively.

Evidence on the effects of early experience in later behavior of birds and mammals is summarized by Beach and Jaynes (1954). It is probable that such lists will be enlarged and their meaning clarified when our conceptions improve as to how still earlier experience antedates and prepares for learning, and how these processes are related to maturation in vertebrate ontogeny. As an example, Tinbergen and Kuenen (1939) concluded that the initial visual arousal of the gaping reaction in nestling thrushes is due to excitation of an innate releasing mechanism, independent of experience. But Lehrman (1953) questions this interpretation from his studies in our laboratory with nestling red-winged blackbirds. In these birds, as in the thrush, the eyes open for an interval during the latter phase of gaping at the stage when this reaction is still initiated only by mechanical stimuli. Visual movement then would augment the response, as would any other stimulus to which the young bird was sensitive. This may be a non-specific neural irradiative effect. But through this summative effect, an object (e.g., the parent) regularly present and visible at such times may, in its specific visual properties, come to initiate and control the response in later stages. This would seem to explain why, when the "visual releaser" stage is reached, the stimuli which Tinbergen and Kuenen consider "innate" have pattern features like those of an adult outlined above the nest edge. For this process which actually seems to be a case of conditioning, it is important that from the start feeding accompanies each repetition of the critical stimulus combination (Craig, 1913).

Gains through early experience are likely to be elusive and only indirectly identifiable. In Siegel's (1953) study carried out in our laboratory with ring doves, the heads of young doves were encased in translucent hoods from 3 days of age, starting before the eyes opened, and the birds were fed by hand. Opportunity for visual form definition was thereby excluded to the time of testing as young adults, although there was normal stimulation by light. Then, in learning to discriminate a circle and triangle by jumping from a perch (a response previously conditioned to contact), the experimental doves were inferior to normal subjects, both in the original learning and in further transfer tests. (Condition of the eye and visual acuity seemed normal in the hood-reared birds, which behaved equivalently to normal birds in optomotor tests). Thus in ring doves, experience somehow provides a perceptual basis essential for form perception and discrimination in the adult.

Nissen, Chow and Semmes (1951) deprived a young chimpanzee of opportunities for tactual and manipulative experience, first by binding the arms, then from the fifth week by encasing all four extremities in tubes. When the impediments were removed at 31 months, the animal although not noticeably deficient in general visual discrimination was very inferior to normal subjects in various tests of tactual-motor coordination and in discriminating touched points on the body. In the following four months these reactions improved, but some deficiencies persisted. Grooming behavior and the sounds normally accompanying it were absent throughout. The chimpanzee's normal tactual adjustments to its own body evidently are established through action in experience, as Werner (1940) finds for the human infant. The absence of grooming would suggest an impaired capacity for social adjustment, based on deficient self-stimulative experience in early life.

Hebb (1949) theorized that early experience benefits later learning in the widest sense, and that the benefit is inversely proportional to age. This idea finds support in numerous studies (Bingham and Griffiths, 1952; Forgays and Forgays, 1952; and Fuller, 1953). Thompson and Heron (1954) raised Scottish terriers under home conditions and later compared them with litter mates moderately or severely restricted in their upbringing, in performance on barrier, delayed-reaction, and other tests. Scores were higher in relation to degree of freedom from restriction in early life. Although the deficit was not specifically defined, the restricted animals seemed handicapped by deficiencies in early perceptual experience rather than by a specific motor impairment. Birch (1945) has shown that chimpanzees lacking the specific earlier experience of "extending" the arm with perceived foreign objects are seriously backward in their ability to use sticks insightfully in later reasoning tests.

The environment does not merely elicit pre-organized mechanisms of behavioral adjustment, but is itself implicated in the development of such mechanisms. Isolation experiments do not tell us what is native in the normal patterns, for if the animal survives, the atypical situation also must have contributed to the development of some adaptive pattern. For each organism, a range of environmental situations exists, variously deviant from the normal or typical, in which may arise functional patterns sufficiently adaptive for survival. Techniques of the "isolation" type thus concern relative-abnormality-of-setting rather than isolation in the full sense, and help tell us how far extrinsic conditions may be changed at a particular developmental stage without preventing or altering further development based on the gains of preceding stages.

The "innate differential" of Howells and Vine (1940) we may paraphrase something like this:

Under given extrinsic conditions, admitting given types of experiences to interact with maturational attainments at any stage, some behavior patterns develop more readily than others in each type of organism. The relative ease or difficulty in obtaining any result is a function both of the species genic constitution and of conditions under which a development may be completed. Since, within limits characteristic of the species, different configurations can develop according to the prevailing extrinsic context, the latter factor must be throughout an indispensable contributor to the rise of a functionally organized system.

Postulation of an original isomorphism, or a functional correspondence of stimulus and reaction, presupposes a preformed organization. But the behavior organization eventually obtained is, from the considerations raised, a developmental product and not one that exists initially as a potential or miniature schema. The eventual relations of stimulation or perception to adaptive behavior can arise only through the conjunction of a whole set of conditions requiring interactions of the endogenous and exogenous at all stages. Instead of expanding from a miniature anlage, a behavior pattern must become synthesized through progressive sets of relationships, and real functional isomorphism is an end product. It is not forecast, but requires the cumulative interactions of many factors in the developmental processes of which the species is capable in the given milieu.

If the developmental capacities of the species permit, one or another neurophysiological organization may arise according to the extrinsic context. Thereby, part-processes such as local reflex-like functions, introduced through development, may be incorporated in wider patterns as nuclear or as peripheral components (Kuo, 1939; Windle, 1940, 1950). The role of implicated factors may vary greatly according to species. Often an afferent factor is crucial, as in spontaneous reactions to visual movement. In the cat, as Kuo (1930) has shown, this factor is nuclear for the rise of the typical reaction to rodents: pouncing and killing. But it is only one component, and the pattern does not arise in its typical form unless circumstances in early infancy admit other factors (as the pounce; claws, jaws and teeth—drawing blood, tasting, biting and eating) for which the kitten is also structurally fitted through normal development. In the typical environs, these combinations are almost certain to rise through experience, permitting the quick formation of an integrated pattern. But, as we have seen, by modifying early experience Kuo was able to produce the patterns of attacking rodents, avoiding them, or living peaceably with them. Are these alternative patterns all to be considered innate for cats? (Crafts, et al., 1938).

PLASTIC DEVELOPMENTAL RELATIONSHIPS IN MAMMALS

On the mammalian level, where successive stages in ontogeny involve the most diversified relationships of maturation and experience in the animal series, variability and plasticity take on new meanings for behavior. Here we find the greatest involvement of feedback processes—a convenient term for relationships in which the functions and activities (i.e., the output) of an organism affect the further course of events in the system that produced them.

There are great differences both in nature and in degree in the capacities of various mammals for such relationships at different ontogenetic stages. Carnivores for instance are doubtless much more limited shortly after birth than ungulates; when the former are still low in sensory, neural correlation, and other capacities, the latter relatively mature with prompt use of visual sensitivity and quadrupedal locomotion (Cruikshank, 1946). Early adjustments in newborn puppies center predominantly around reactions to a "tactile deficit" (James, 1952); kittens improve within the first few hours in finding the mother and their individual feeding stations, utilizing sensory effects gained actively through the individual's own movements (Rosenblatt *et al.*, unpubl.). As these processes vary, we should expect to find important differences in behavior development at later stages (e.g., mating). For cats, as scattered reports indicate for other mammals, an early life apart from species mates accounts for later inhibition of mating reactions, in part through disturbed reactions to "strangeness." Kagan and Beach (1953) found that later sexual responsiveness was significantly limited in male rats held to brief bisexual encounters with species mates in early life, evidently through the disrupting effect of "playful responses."

In mammals, previous experience is closely associated with other age-conditioned variables in affecting developments at later stages (Beach and Jaynes, 1954). There is little doubt that Holt's (1931) conception of self-stimulative relationships in such trace effects will find ample confirmation (Schneirla, 1946, 1949). Many experiences related to this class, influential in later behavioral development, doubtless are gained through the animal's activities in relation to objects and situations apart from its own body. Riess (1950) reports that female rats deprived of early experience with movable objects are deficient maternally, building no definite nests, retrieving young and artifacts seemingly alike, and losing their young through deficient suckling. Carrying of young by the parent, a widespread pattern of behavior in mammals (Causey and Waters, 1936) may depend to an appreciable extent upon previous individual stimulative relationships.

To test the influence of self-stimulative experience in the development of an adaptive pattern normally appearing in maturity, Birch (unpubl.) reared young female rats provided with wide rubber collars worn continually until their first parturition. The rats adapted well to their collars and gained weight normally, although these accessories prevented touching or licking the posterior body including the genitalia, which were licked frequently in the controls. In 13 of 14 observed parturitions with experimental rats, all of the young were lost through cannibalism or neglect. Chewing at unconsumed bodies was common. In the remaining case, 9 of 11 young (rather fortuitously) survived parturition, but most of these were killed by the mother before the 14th day, when the last two died of inanition. These young were at a disadvantage for nursing through being gathered forward under the mother's body instead of being pushed posteriorly in the normal way.

The collars may have deprived these rats of self-stimulative relationships essential for normal parturitive behavior. (Control rats with collars just as heavy but narrower, permitting stimulative access to the posterior body, behaved normally in parturition.) One distorted factor in the experimental rats might be physiological, a deficiency in salt metabolism; another psychological, concerning a self-perception normally gained through adjustments established to the animal's own body in youth. It is possible that normally, a sufficient stimulative equivalence of the newborn pups to the genitalia, which the female has learned earlier to lick but not bite, would prevent cannibalism; but that without this perceptual adjustment the experimental mothers chewed and ate their young. An interesting item is the normal huddling which facilitates nursing. Ordinarily, tactual stimulation from the young causes the mother to maintain and to increase their pressure against her. This reaction is impelled not only through the female's augmented ventral sensitivity, an effect of hormonal processes in pregnancy and parturition, but also through a perceptual adjustment learned in previous grooming and licking of the posterior body and the genital area in particular.

Although no pre-established neural organizer is indicated for parturitive behavior, on the basis of hormonally-induced organic changes through the intervention of learned perceptual adjustments an adaptive pattern may be integrated. The experience factor here is not the specific practice of a pre-existing pattern, since it involves not the young themselves but perceptual effects having an appreciable stimulative equivalence to the young. Uyldert (1946) found that specific experience with young is also beneficial. Virgin females made pseudopregnant with high estrogen were given opportunities to lick, nurse, and tactually encounter young rats. When later tested in their primiparous behavior in comparison with animals lacking previous experience with young, the experimental animals brought through all of their young in 88% of the cases, whereas only 11% of the control animals did so.

Our studies on normal parturitive behavior in rats and domestic cats (Tobach *et al.*, unpubl.) are relevant to this discussion. We are led to characterize the event, for cats, as an interplay or even competition between the stimulative effects of endogenous events (e.g., uterine contractions, emergence of fetus) and the external results of such events (e.g., fluids, neonate). These stimulative by-products of organic processes tend to intrude themselves upon the female's attention in a somewhat variable order, timing and duration. Each one, as it arises, demands a specific perceptual and behavioral adjustment on her part. The parturitive phenomenon here is not a regular patterned flow of events, but a series of rather sporadic organic and behavioral episodes, together with variable activities not specifically parturitive in themselves. Thus the female exhibits, in a not very predictable order, the predictable items on self-licking newborn on floor, eating after-birth, general movement and sitting or lying. Intervals of intense activity indicating a high level of excitement, a condition facilitating delivery operations, are interspersed with intervals of exhaustion and rest, facilitating initiation of nursing and other stimulative relations of mother and newborn.

In these mammals parturitive behavior is indicated as a loose assemblage of functions centering around the stimulative consequences of organic events. A sequence of hormonally-induced endogenous changes sets a loose temporal order for behavioral adjustment, dependent in sequence and timing upon the female's variable attention to competing organic and environmental stimuli. Persistently in evidence is an orientation to the posterior body and particularly the vaginal area, a perceptual set which enormously aids normal parturitive operations. This factor, presumably based upon self-stimulative experience in youth, is very possibly indispensable for an adaptive outcome and survival of the young*.

Reciprocal stimulative relationships between parent and young may provide the unifying basis for a progressive organization of behavioral adjustments in the litter situation. Such relationships initiated in parturition may be sustained both by the physiological condition of the lactating mother and the organically enforced sensory susceptibilities of the young. Within a few days the mother evidences an appreciable discriminative specialization in stimulative interchanges with the young, pointing up her strong attachment to them. And on their side the kittens, as our studies (Rosenblatt *et al.*, unpubl.) indicate, acquire individualized adjustments to the mother (tending, for example, to nurse at specific nipples) which begin within a few hours

*Labriola (1953) confirmed the finding of Weisner and Sheard that female rats display maternal behavior after their young have been delivered by caesarean operation. Nest-building and retrieving can occur, he finds, without the actual process of parturition and associated events such as cleaning young, consuming after-birth, and lactating normally. Since further care of young is not reported, the full effect of the omitted events in the normal process cannot be gauged. Nor has this experiment eliminated the experiential factors indicated by Birch's results as crucial for an adaptive pattern. Labriola's conclusion that the "hormonal changes associated with pregnancy and its termination" can directly and solely in themselves produce maternal behavior, seems premature. Rather, these factors would seem to act as partial contributants to a more inclusive and complex process.

after delivery. There is much to be said for the theory that maternal reactions in mammals develop and become motivated on the basis of stimulative relationships ("Trophallaxis"—Schneirla, 1946, 1949) between parent and young.

Doubtless, afferent susceptibilities initiating and promoting "trophallactic" behavior, together with organic tensions underlying and maintaining the pattern, may be traced back to hormonal processes. But the relationships are more than unidirectional. The existence of feedback relationships in the pattern is indicated by evidence that the endocrine conditions themselves are facilitated and reinforced by stimulation received in routine litter operations, as in nursing (Selye, Collip and Thompson, 1934; Selye and McKeown, 1934; Leblond and Nelson, 1937; Uyldert, 1946; Beach, 1948).

Processes accounting for the usual cessation of maternal behavior at weaning are not well understood. It is possible that age-conditioned behavioral changes in the young so reduce the mother's afferent input from trophallactic processes as to cause a sharp fall in the endocrine secretions maintaining her litter reactions. This change in turn may reduce the female's afferent susceptibilities sufficiently to admit other hormonal processes favoring different behavior (e.g., high estrone promoting heat). Such changes presumably interfere with organic tensions in the mother which evidence (Wiesner and Sheard, 1933; Uyldert, 1946) indicates are normally conditioned to stimulation from the young.

An interesting instance of how the mother's organic processes and behavior may relate to the young is offered by Martins (1949). Some female dogs, in the last stages of lactation and for a few days after weaning, following a meal at a distance from the nest, disgorge regularly in the presence of the pups on returning to the nest. An adaptive relation to weaning is evident, since the pups eat the regurgitated material. Martins suggests an endocrine control for the act; but it is also clear that perception of the young, related to litter experience, also plays a critical part. Regurgitation occurs on the female's return to the nest, but may be inhibited for as long as an hour if the pups are not there, with the mother disgorging promptly when they appear.

Peripheral processes not only contribute to the rise of the components but may also facilitate integration in the "instinctive behavior" of mammals. In sexual behavior, for example, cutaneous stimulation has an important function in promoting activities just before and during coitus (Beach, 1951). In rodents, stroking the rump produces lordosis in hormonally aroused females, and genital stimulation contributes heavily to effective mating behavior in both sexes (Bard, 1939; Beach, 1951). It has been maintained (Lashley, 1938) that the behavior pattern is produced through the priming action of hormones on a surrogate neural mechanism, with peripheral (non-nervous) components playing incidental roles. But the possibility exists and the afferent system is intimately involved in the systemic organization. This is suggested by findings like that of Dusser de Barenne and Koskoff (1934) that male cats with cords sectioned just above the lumbar region responded to stimulation of the penis with gross movements resembling intercourse. Since these movements are not only aroused but maintained for a time, an appreciable measure of interplay is indicated between efferent components, nervous centers (e.g., the sacral autonomic system) and local afferent mechanisms as agents in the production of organized behavior.

The investigations of Kruchinsky (1947) indicate a critical peripheral factor in the pelvic-thrust response of the rabbit's copulatory pattern, involving afferent connections of penis and urethra. In pre-tested male rats castrated and then given regular injections of testosterone in different amounts, Beach and Holz-Tucker (1949) obtained a sex-responsiveness score directly corresponding to the amount of hormone received. The experiment was extended by Beach and Levinson (1950) who found a positive correlation between hormone dosage and the number of receptor papillae in the skin of the penis in experimental subjects. These findings may be taken to suggest hormone effects both in central nervous processes and in peripheral mechanisms relevant to sex behavior.

Although reproductive behavior is grounded on the effects of strong hormonal and related organic processes in development, cortical resources admitting learning as a component increase steadily through the mammals (Beach, 1947, 1951). Even in rodents and lower mammals, this function is indicated. Rosenblatt (1953) found that early experience increases hormonal effects promoting mating in castrated male cats; Warren (1954) reported comparable results with hamsters. Appearance of the mating pattern in previously isolated rodents (Beach 1942) does not rule out experience as a contributor to the normal pattern, on grounds already mentioned. Partial or complete failure of mating in cats visually and tactually isolated from their kind during early life may signify a larger role of learning than in rodents, perhaps magnified by perceptual emotional components. Here also, the learning factor, as distinguished from the influence of developed organic functions, resists disentangling in its specific contributions.

Results reported by Kent and Liberman (1949) seem to offer direct support for a "neural priming" hypothesis. Doses of progesterone too small to produce "psychic estrus" when given subcutaneously to spayed estrogen-primed female hamsters, brought typical mating responses when injected directly into the lateral ventricle of the brain. On this basis the investigators suggest that:

" . . . progesterone may act directly upon one or more neural nuclei, facilitating, in an unknown manner, the manifestation of reflex mating activity in the female hamster, thus effecting the physiological state known as psychic estrus."

But in our laboratory Warren and Rosenblatt (unpubl.), to test the possibility that the Kent-Liberman results may have resulted at least in part from a peripheral or non-nervous differential, injected groups of spayed female hamsters equivalently in the lateral ventricle of the brain, the jugular vein, the subcutaneous tissues. From the results, the factor of differential absorption of hormone into the blood stream (depending on injection site) must be taken into account before results such as those of Kent and Liberman can be interpreted with respect to a hypothetical priming effect on nerve centers.

Available evidence recommends considering the mating pattern as resulting from the interactions of several organic systems. Factors in the excitation and function of specific components and in the organization of the inclusive pattern are not necessarily identical. With respect to the latter, extirpative experiments significantly indicate that in rats the cortex is not directly involved in the integration of mating behavior, but rather facilitates lower centers essential to the responses, in the male although not strongly in the female (Beach, 1952). The possibility is not excluded that cortical facilitation may influence both central and peripheral mechanisms as well as their integration in this behavior.

Another interesting case is the micturition pattern of the dog. This response appears as a squatting posture in young animals and adult females but as elevation of one rear leg in adult males (Berg, 1944). The adult male response is displaced in castrated males by the female-young squatting posture, but testosterone restores it (Martins and Valle, 1948); at discontinuance of injection, however, the squat returns (Berg, 1944). A direct hormonal action on a surrogate neural mechanism might seem indicated; but Clark (1945), in a consideration of the neurological evidence, found conflicting results and appealed for behavioral analysis. Freud and Uyldert (1948) investigated the fact that leg elevation follows object-sniffing, and on the side toward which sniffing occurs. (The urine of females in heat is especially potent for eliciting this response—Beach and Gilmore, 1949). Since leg elevation can occur with or without release of urine, the afferent effect seems important. Osmo-anesthesia abolished leg elevation and produced the squatting pattern in four-fifths of the trials, with the male posture returning on recovery from the anesthetic. From their evidence, the authors regard the micturition pattern as a conditioned response involving both the connections of olfactory receptors and the bladder proprioceptors. Without these afferent limbs the adult male pattern cannot function, and in their absence during ontogeny it is possible that this distinctive pattern could never arise. The action of hormones, in addition to possibly arousing or priming nervous centers involved, thus may extend more widely in the organism to afferent mechanisms essential for such behavior.

Problems of behavioral organization such as these converge in the study of motivation. Simply definable as the impulsion of behavior, motivation in different animals is not soluble in simple, one-process terms. Although the ontogeny of motivation in mammals is still incompletely glimpsed, many of the difficulties doubtless stem from a concentration on adult stages, for at different ages organic tensions may be related very differently to external conditions. Motivational processes certainly are no exception to the rule that plasticity in using organic mechanisms varies with learning capacity. Hebb (1949) postulates proportionately greater gains from early experience than later, but for adequate comparison more evidence is needed. The first step is to analyze the ontogenetic background. For example, food deprivation in youth increases food-hoarding in adult rats (Hunt, 1941), but the nature of the trace effect is still obscure. Perhaps related is the fact that, after regular experiences of thirst and hunger during infancy, rats as adults learn the location of food and water more readily than do normal subjects (Christie, 1952). Related to physiological aspects of the problem, such findings suggest a subtle perceptual factor in the development of motivation, perhaps similar to that already suggested in the case of maternal behavior.

Visceral processes are undoubtedly basic to motivated behavior, yet the relationship has been incompletely explored (Bard, 1939). Emphasis upon the complexity of motivational processes in any mammal suggests a close attention to ontogenetic advances under the influence of contrasting experiences. No visceral mechanism seems immune to a conditioned extrinsic control (Crafts *et al.*, 1938, Chap. 18), although such processes are not as readily modified through conditioning as are so-called voluntary reactions. Trace effects may influence motivation on very different levels of qualitative organization and complexity.

Although visceral processes are considered subsequent energizers of organized motivated behavior, other than in respiration and circulation they are arhythmic at birth, as Carlson and Ginsburg (1916) demonstrated for stomach contractions in the dog. Later, however, visceral rhythms develop in relation to new functions such as feeding, or come under external control (Seward and Seward, 1937). In the neonate as in the embryonic mammal, visceral activities are considered a chief source of the characteristic generalized, diffuse activity then prodominant (Carmichael, 1951). A radical distinction may be necessary in the external relevance of these tensions according to stage. Initially, visceral tensions evidently serve as indiscriminate excitants, deficiency conditions or lacks produced by a shortage of some metabolic ingredient. Only later, when specific relationships of incentive-bound types develop between them and external conditions, may the respectively different visceral deficiency-processes be referred to as "needs." It is probable that a need, or motivating-tension with incentive relevance, cannot exist prior to learning through experience.

The term "striving," which implies an object or goal-relevance in energized behavior, seems applicable therefore only to motivated behavior at later stages of ontogeny. The basic condition is an internal disruption which, while it is in operation, keeps the animal active. In the naive animal that aroused behavior tends to be strikingly variable and persistent, and these characteristics, together with the typical adaptiveness of the outcome, account for the insistence of many authors that all motivated behavior is "purposive." But later behavior can be very different according to how the exciting tensions have been removed or changed on previous occasions and in dependence upon limiting factors peculiar to the ontogenetic stage.

The relation of the neonate mammal to its situation is facilitated in the direction of an adaptive outcome by the proximity and behavior of the mother—whether she bites instead of licking, pushes the newborn forward or backward with respect to her body, and the like. As our studies on the cat indicate, a variety of circumstances in the initial relationship of neonate and mother facilitate the first occurrence of sucking. The newborn is brought into abdominal contact partly through maternal complicity, as in licking, partly by its own crude orienting responses to parentally-furnished tactual and thermal stimulus gradients. A process is thereby initiated whereby the kitten's essentially reflex sucking response becomes the focus around which a more inclusive adaptive pattern can arise. Similar roles often played by specific reflex mechanisms cannot be overlooked (Maier and Schneirla, 1935, Chaps. 6, 11 and 12; Kennedy, 1954). For integration of stimuli and responses into a pattern the crucial occurrence is reduction of the disturbing organic tensions through sucking. The feeding adjustment of the neonate to the mother, initially random and lacking in organization, thus may become specialized as a process of food motivation.

In the neonate, the responses which resolve the drive episode soon became organized thereby into an elementary striving, a crude perceptual adjustment to extrinsic conditions which acquire incentive valence. Repeated combinations of organic tensions and given activities alleviating them increase the directive potency of stimuli recurring at such times. For the kittens, the mother's characteristics, and for the mother, those of the kittens, become strong incentives. In the kitten, an initially random behavior when "hungry" is replaced by efficient responses to incentives leading to resumption of a habituated postural relationship to the mother and consummation at an individually preempted nipple. Thus a randomly disturbing organic "lack" acquires an externalized perceptual relevancy and becomes a motivating "need."

Different principles seem to be involved in behavior changes at later ontogenetic stages. Although the parturient female rodent or cat is not initially capable of patterned adjustments, her behavior is predictably influenced in direction by organic changes, by external circumstances, and by the effects of previous experience. As the Birch "collar" experiment indicates, a pre-existing perceptual factor can influence a trend toward an adaptive combination of these various initially disconnected elements. There are also organic factors which, in their external effects, facilitate the integration of a pattern. One is the licking reaction, which after delivery readily transfers from self-trophallactic reactions to genitalia to amniotic fluids, sac and then to newborn. The equivalence seems purely chemotropic, yet, to an appreciable extent because they bear the fluids, the newborn can become specifically attractive objects, incentives which increase in motivational potency the more they accompany afferent satisfactions and tension-relief.

In the rat's nest-making, as in parturition, a hormonally produced set of organic tensions dominates the animal's activity, and under typical conditions, also can enter into externalized relationships such that an adaptive adjustment to the situation must work itself out. The induced physiological condition and increased sensitivity of ventral surfaces accounts for a temperature reactivity critical for the termination of operations in a situation called a "nest" (Kinder, 1927). In variable activities aroused through heat-loss disturbance, materials are carried and pushed about until an organism-situation relationship exists in which the animal is quieted through sufficient reduction of the disturbing afferent condition (Lashley, 1949). The behavior is not altogether random; for one thing because, as Reiss' results indicate, in the absence of earlier object-experience the materials might be handled differently with maladaptive results. Dependence of environmental susceptibilities upon organic condition is emphasized by Richter's (1942) experiment in which the critical afferent condition was introduced hormonally through removing the thyroid or pituitary gland, thereby bringing about nest-building in rats of either sex.

Presumably each type of motivation may be considered a pattern that develops through progressive interrelationships between organism and environment. Hunger is an example of one complex system. Although initially and basically it depends upon gastric contractions (Wada, 1922; Patterson, 1933), food-striving may later exist independently of stomach contractions (Bash, 1939; Morgan and Stellar, 1950). For some time it has been known that, as in the dog (Carlson and Ginsburg, 1916), hunger contractions of the empty stomach decrease with age. This change with age may parallel increases in the process Anderson (1941) terms "drive externalization," in which incentives can arouse the drive independently of internal facilitation. The relationship of organic tensions to activities and their various stimulative agents must normally become specialized in mammals according to individual experience with responses changing or removing tensions under specific stimulus conditions. Thus Harris *et al.* (1933) found that rats deficient in vitamin B1 could learn a preference for food rich in this vitamin, provided that such food was distinctively flavored in discrimination-learning. The cocoa flavor used then became a controlling property of the incentive.

The complexity of motivation is seen when the act occurs in a situation that seems biologically inappropriate for it, as in "displacement activity," (Armstrong, 1950) when feeding activity occurs in a presumed "mating" situation, or the reverse. This reminds us that presumably different "consummatory" reactions may be closely related basically. Similar autonomic functions may underlie differently motivated acts, as for example stroking a satiated cat may initiate cranio-sacral activity capable of supporting either a rearousal of feeding or a rubbing against the hand. Thus under appropriate conditions the special stimuli for different responses based upon a given autonomic function may exhibit an appreciable equivalence. The explanation, it would seem, lies in the ontogeny of motivational patterns.

The pattern of motivation characteristic of a species must develop through progressive functional relationships between organism and environment. The adult patterns are most complex and plastic in mammals, yet even here, although experience is a factor of importance, the function of learning is not readily separable. Most accessible to description are specific conditioning episodes, in which a new or conditioned stimulus can be discerned in temporal contiguity with an unconditioned stimulus. In such cases, the latter is called "unlearned" in the sense that it produces the critical response (e.g., feeding) somehow as an outcome of earlier development. Even less distinguishable are the old and the new in trial-and-error or selective learning, and particularly at early ages, as in the shifting of appetites at weaning. The solution seems to lie in closer study of ontogeny, with less arbitrary separation of innate "from acquired" influences.

The traditional instinct problem has much in common with that of motivation, in that both require a theory which can account wholistically for behavioral organization through development. The former emphasizes genetic constitution, but this is also central to motivation study, in which the role of organic factors is apparent at all stages. Although Hebb (1949) mistrusts configurational theories because they have neglected the experience factor, this is by no means an essential gap in them, but one that can be repaired by a systematic attention to ontogeny (Werner, 1940; Birch, 1945).

The analysis of behavior patterns in relation to hereditary background, or "psychogenetics," although potentially valuable, is "as yet more a promise than an actuality" (Hall, 1951). Useful beginnings have been made in the description of correspondences between phenotypic behavioral traits and genes (e.g., Keeler and King, 1942; Fuller, 1953). But such facts emphasize the need for analytical studies of ontogeny to avoid an over-simplification of both genetic mechanisms and developmental processes through which these exert their effects. The gaps between genes and somatic characters are great; those between genes and behavior, which are greater still, can be bridged only by studies directed at understanding the intervening variables and their interrelationships. Earlier notions of a one-to-one relationship between these agencies and their effects have fallen away before advances in physiological genetics.

As an example, the general characteristic "wildness" broadly differentiates many animal species. Each species has its characteristic "flight distance" (Hediger, 1950), only relatively reducible through taming. Strain differences may be demonstrated in rats, with "wild" dominant over "tame" (Yerkes, 1913; Dawson, 1932; Keeler, 1942). Other studies (Donaldson, 1928; Yeakel and Rhoades, 1941; Richter, 1952) point to the adrenal gland as foremost among somatic factors basic to this behavior character. This discovery improves upon the statement that the behavior trait is hereditary, yet merely points up the need for analysis of the developmental processes producing the gland in the milieu of early ontogeny, on the one hand, and of its functional relationships in behavior on the other. To this system there are contributants of many kinds, including even learning, as indicated by the fact that experience in being reared with a mother or other species mates which are relatively "wild" promotes wildness in young rats (Rasmussen, 1939).

Genetic influences in the development of behavior characters can be exerted only indirectly, as we have seen for various phyletic levels, through the mediation of organismic processes requiring the interaction of given intrinsic and extrinsic conditions. The traditional dichotomy of heredity and environment promotes a fundamental misstatement of these matters. Rather, the picture is one of a typical species ontogeny in which a flux of relationships exists between the developing organism and the species environment, differing with the conditions of each developmental stage. Aspects of behavior may be termed "genetic" only in the sense that they come about through the ontogenetic processes of the different species or strains under their respective typical developmental conditions.

Two points will bear final emphasis. First, the tendency to oversimplify statements of genetic-behavioral correspondences on the strength of their heuristic value, as for taxonomy, may lead to serious misrepresentation. Thus with respect to one conception of reproductive isolation, Clark, Aronson and Gordon (1954) point out that this condition is not a simple lock-and-key mechanism controlled by a few genes, merely because a striking character of structure or behavior may often seem to be the critical factor isolating two species. Actually, there are complex patterns composed of many small characteristic differences. Many inherited characters, most of which are probably overlooked, are undoubtedly involved. Genotypically, these accumulations of small differences, which probably underlie determinations such as behavioral isolating mechanisms (Müller, 1940), despite their multiple and complex nature may be easier to build up than a single-factor basis (Mayr, 1948). And on the phenotypic side, only confusion is added and problems obscured when a single alleged crucial intervening variable is postulated, such as a pre-organized neural mechanism, a function not sufficiently known to have theoretical validity.

The second point concerns the validity of inter-phyletic extrapolations of common mechanisms based upon broad adaptive similarities in behavior. It is probable, for example, that major qualitative differences exist between the "motivation" of animals which operate on a physiological or biosocial level in their interindividual behavior, and those animals which operate on an involved psychosocial level (Schneirla, 1946; David and Snyder, 1951). The complexity of intelligent behavior in mammals and its resistance to psychogenetic analysis may be illustrated by studies of rats in the Tryon "bright" and "dull" strains, progeny of selective breeding for highest and lowest performance, respectively, in a multiple-T maze. One important difference is sensory, since Krechevsky (1933) found the brights more responsive tactually, the dulls more visual in discrimination situations. Still other factors are indicated in Searles' (1949) finding that the brights are higher in food-getting and lower in water-avoidance motivation, also more timid in open spaces although less disturbed by mechanical maze sounds than the dulls. On a lower phyletic level, as in insects, the development of comparable adaptive behavior might well involve lesser capacities, differently organized than in mammals (Schneirla, 1949).

"Instinct" is thus a problem concerning not only the lower animal phyla with their more stereotyped behavior, but developmental relationships producing species-typical behavior throughout the animal series. To the extent that it has gradually revealed this fact, and thereby has stimulated the rise of appropriate methods of comparative behavior study, the traditional instinct controversy may have made its best contribution to science.

CONCLUSIONS

1. In the development of any animal, systems of intervening variables mediate between genic influences and processes in ontogeny through which adaptive behavior appears. The range of these variables, from biochemical and physical conditions to the effects of experience, depends basically upon limitations imposed by the genic constitution of the species.
2. On all phyletic levels, behavioral organization arises in development through interrelationships of intrinsic and extrinsic factors influencing growth and differentia-

tion. Only in a misleading sense can the genic constitution be said to determine the organization of behavior patterns, even in phyla characterized by the most rigid lines of development.

3. In a strict theoretical sense the terms "innate" and "acquired" cannot therefore be applied validly to behavior or to the organization of behavior. Accordingly, it is suggested that the term "instinct" be retired from scientific usage, except to designate a developmental process resulting in species-typical behavior.

4. "Maturation," as conceptualized for behavioral development, subsumes not only growth of the morphological basis but also the excitatory and stimulative effects arising through growth processes in their influence upon further stages of development. "Experience" is defined as the effects of extrinsic stimulation upon development and behavior. The effects of maturation and experience tend to overlap and become integrated through their interactions and organic trace effects. These concepts involve fewer trammeling theoretical assumptions than do "innate" and "acquired," and should replace the latter in theoretical usage.

5. The role of experience may have narrow or wide limits in behavioral development and may range into simple or complex learning functions according to phyletic properties.

6. Intervening variables in the development of behavior differ characteristically on each phyletic level as to their relative weight and patterns of relationships. It is therefore extremely doubtful that a single formula of "instinctive" behavior can be devised for the animal series in general.

7. There are characteristic phyletic differences in the degree of relationship between stages in behavior development, later stages being based upon earlier ones and modifiable under conditions of maturation and experience admitted according to phyletic capacities.

8. In the more advanced invertebrates and the vertebrates, functions of a disparate, local character appearing in early ontogenetic stages frequently serve as foci or nuclei in the later integration of more complex patterns.

9. In behavior development on various phyletic levels, "feedback" relationships involving interactions between the progressing system, its component processes and its "output," frequently play an essential role in the rise of integration and in advances from stage to stage.

REFERENCES

Adriaanse, M. S. (1947), *Ammophila campestris* Latr. und *Ammophila adriaansei* Wilcke. Ein Beitrag zur vergleichenden Verhaltensforschung, *Behaviour,* **1**, 1–34.

Anastasi, A. and Foley, J. P. (1948), A proposed reorientation in the heredity environment controversy, *Psychol. Rev.*, **55**, 239–249.

Anderson, E. C. (1941), The externalization of drive. I. Theoretical considerations, *Psychol. Rev.*, **48**, 204–224.

Armstrong, E. A. (1950), The nature and function of displacement activities. Chap. in *Sympos. Soc. Exper. Biol.*, **IV**, 361–384. New-York: Academic Press.

Aronson, L. R. and Noble, G. K. (1948), The sexual behavior of Anura. 2. Neural mechanisms controlling mating in the male leopard frog, *Rana pipiens, Bull. Amer. Mus. Nat. Hist.*, **86**, 87–139.

Autrum, H. (1952), Uber zeitliches Auflosungsvermogen und Primarvorgange im Insektenauge, *Naturwiss.*, **39**, *Jg.*, **13**, 290–297.

Baerends, G. P. (1950), Specializations in organs and movements with a releasing function, *Sympos. Soc. Exper. Biol.*, **IV**, 337–360, New-York: Academic Press.

Bard, P. (1939), *The Hypothalamus and Central Levels of Autonomic Function.* Baltimore: Williams and Wilkins.

Bash, K. W. (1939), An investigation into the possible organic basis for the hunger drive, *J. Comp. Psychol.*, **28**, 109–135.

Bastock M., Morris, D, and Moyhihan, M. (1953), Some comments on conflict and thwarting in animals, *Behaviour*, **6**, 66–84.

Beach, F. A. (1940), Effects of cortical lesions upon the copulatory behavior of male rats, *J. Comp. Psychol.*, **29**, 193–239.

Beach, F. A. (1942), Analysis of the stimuli adequate to elicit mating behavior in the sexually-inexperienced male rat, *J. Comp. Psychol.*, **33**, 163–207.

Beach, F. A. (1947), Evolutionary changes in the physiological control of mating behavior in mammals, *Psychol. Rev.*, **54**, 297–315.

Beach, F. A. (1948), *Hormones and Behavior.* New-York: Hoeber.

Beach, F. A. (1951), Instinctive behavior: reproductive activities. In S. S. Stevens (Ed.) *Handbook of Experimental Psychology.* New-York: Wiley, Chap. 12, 387–434.

Beach, F. A. and Holz-Tucker, A. Marie (1949), Mating behavior in male rats castrated at various ages and injected with androgen, *J. Comp. Physiol. Psychol.*, **12**, 433–453.

Beach, F. A. and Jaynes, J. (1954), Effects of early experience upon the behavior of animals, *Psychol. Bull.*, **51**, 240–263.

Beach, F. A. and Levinson, G (1950), Effects of androgen on the glans penis and mating behavior of castrated male rats, *J. Exper. Zool.*, **114**, 159–171.

Berg, I. (1944), Development of behavior: the micturition pattern in the dog, *J. Exper. Psychol.*, **34**, 343–367.

Bingham, W. E. and Griffiths, W. J., Jr. (1952), The effect of different environments during infancy on adult behavior in the rat, *J. Comp. Physiol. Psychol.*, **45**, 307–312.

Birch, H. G. (1945), The relation of previous experience to insightful problem-solving, *J. Comp. Psychol.*, **38**, 367–383.

Birch, H. G. (1945), The relation of factors involved in early experience to maternal behavior in the rat. (unpubl.).

Bozler, E. (1926), Sinnes-und Nervenphysiologische Untersunchugen an Scyphomedusen, *Z. Vergl. Physiol.*, **4**, 37–80, 797–817.

Bullock, T. H. (1943), Neuromuscular facilitation in *Scyphomedusae, J. Cell. Comp. Physiol.*, **22**, 251–272.

Carlson, A. J. and Ginsburg, H. (1916), Contributions to the physiology of the stomach XXX, *Am. J. Physiol.*, **39**, 310–312.

Carmichael, L. (1926-1927-1928), The development of behavior in vertebrates experimentally removed from the influence of external stimulation, *Psychol. Rev.*, **33**, 51–58; **34**, 34–47; **35**, 253–360.

Carmichael, L. (1936), A re-evaluation of the concepts of maturation and learning as applied to the early development of behavior, *Psychol. Rev.*, **43**, 450–470.

Carmichael, L. (1946), The onset and early development of behavior. In *Manual of Child Psychology*. New York: Wiley, 43–166.

Carmichael, L. (1951), Ontogenetic behavior, in S. S. Stevens (Ed.) *Handbook of Experimental Psychology*. New York: Wiley, Chap. 8, 281–303.

Carmichael, L. and Smith, M. R. (1939), Quantified pressure stimulation and the specificity and generality of response in fetal life, *J. Genet. Psychol.*, **54**, 425–434.

Causey, D. C. and Waters, R. H. (1936), Parental care in mammals, *J. Comp. Psychol.*, **22**, 241–254.

Christie, R. (1952), The effect of some early experiences in the latent learning of rats, *J. Exper. Psychol.*, **43**, 281–288.

Clark, Eugenie, Aronson, I. R. and Gordon, M. (1954), Mating behavior patterns in sexual isolation, *Bull. Am. Mus. Nat. Hist.*, **103**, 139–335.

Clark, G. (1945), The central control of micturition, *Urol. Cutan. Rev.*, **49**, 612–617.

Cobb, S. (1944), *Borderlands of Psychiatry*. Cambridge: Harvard Univ. Press.

Coghill, G. E. (1929), *Anatomy and the Problem of Behavior*. New York: Macmillan.

Conradi, E. (1905), Songs and call notes of English sparrows when reared by canaries, *Am. J. Psychol.*, **16**, 190–198.

Cott, H. B. (1936), The effectiveness of protective adaptations in the hive bee, illustrated by experiments on the feeding reactions, habit formations, and memory of the common toad (*Bufo bufo bufo*), *Proc. Zool. Soc.*

Crafts, L. C., Schneirla, T. C., Robinson, Elsa N., and Gilbert, R. W. (1938), *Recent Experiments in Psychology*. New York: McGraw-Hill. (Rev. Ed., 1950).

Craig, W. (1914), Male doves reared in isolation, *J. Anim. Behav.*, **4**, 121–133.

Craig, W. (1918), Appetites and aversions as constituents of instinct, *Biol. Bull.*, **34**, 91–107.

Cruikshank, Ruth M. (1946), Animal infancy, in L. Carmichael (Ed.) *Manual of Child Psychology*. New York: Wiley, Chapter 3.

Cruze, W. W. (1938), Maturation and learning in chicks, *J. Comp. Psychol.*, **19**, 371–409.

Cushing, J. E. (1941), Non-genetic mating preference as a factor in evolution, *Condor*, **43**, 233–236.

Cushing, J. E. and Ramsey, A. O. (1949), The non-heritable aspects of family unity in birds, *Condor*, **51**, 82–87.

David, P. R. and Snyder, L. H. (1951), Genetic variability and human behavior, in J. H. Rohrer and M. Sherif (Eds.) *Social Psychology at the Crossroads*. New York: Harper.

Davis, D. E. (1945), the occurrence of the incubation patch in some Brazilian birds, *Wilson Bull.*, **47**, 188–190.

Dawson, W. M. (1932), Inheritance of wildness and tameness in mice, *Genetics*, **17**, 296–326.

Dennis, W. (1941), Spalding's experiment on the flight of birds repeated with another species, *J. Comp. Psychol.*, **31**, 337–348.

Dethier, V. G. (1947), *Chemical Insect Attractants and Repellents.* Philadelphia: Blakiston.

Dethier, V. G. (1954), Evolution of feeding preferences in phytophagous insects. *Evol.*, **8**, 33–54.

Dobzhansky, T. (1950), Heredity, environment, and evolution, *Science*, **111**, 161–166.

Donaldson, J. (1928), Adrenal gland in wild gray and albino rat: corticome-dullary relations, *Proc. Soc. Exper. Biol. Med.*, **25**, 300–301.

Dusser De Barenne, J. G. (1934), Further observations on the flexor rigidity in the hind legs of spinal cat, *Amer. J. Physiol.*, **107**, 441–446.

Ehrenhardt, H. (1937), Formensehen and Sehscharfebesrimmungen bei Eidechsen. *z. Vergl. Physiol.*, **24**, 258–304.

Eibl-Eibesfeldt, Irenaus (1952) Nahrungserwerb und Beuteschema der Erdkrote, (*Bufo bufo* L.) *Behaviour.*, **4**, 1–35.

Fabricius, E. (1951), Zur Ethologie junger Anatiden, *Acta Zool.* Fennica, **68**, 1–175.

Forgays, D. G. and Forgays, Janet W. (1952), The nature of the effect of free-environmental experience in the rat, *J. Comp. Phys. Psychol.*, **45**, 322–328.

Freisling, J . (1948), Studien zur Biologie und Psychologie der Wechselkrote (*Bufo viridis* Laur.). *Ostl. Zool. Zsch.*, **1**, 383–440.

Freud, J. and Uyldert, Ina E. (1948), Micturition and copulation behavior patterns in dogs, *Acta Brev. Neerl.*, **16**, 49–53.

Friedrich, H. (1932), Studien uber die Gleichgewichtserhaltung und Bewegungsphysiologie bei Pterotrachea, *Z. Physiol.*, **16**, 345–361.

Frisch, K. V., (1923), Uber die Sprache der Bienen, *Zool. Jahrb., Zool. Physiol.*, **20**, 1–186.

Fromme, A. (1911), An experimental study of the factors of maturation and practice in the behavioral development of the embryo of the form, *Rana pipiens. Gent. Psychol. Monogr.*, **24** 219–256.

Fuller, J. L. (1953), Cross-sectional and longitudinal studies of adjustive behavior in dogs, *Ann. New York. Acad. Sci.*, **56**, 214–224.

Fuller, J. L. and Scott, J . P. (1954) Genetic factors affecting intelligence, *Eugen. Quart.*, **1**, 23–43.

Gellhorn, E., Koella, W., and Ballin, H. M. (1954), Interaction on cerebral cortex of acoustic or optic with nociceptive impulses: the problem of conciousness, *J. Neurophysiol.*, **17**, 14–21.

Gesell, A. (1954), *The Embryology of Behavior: The Beginnings of the Human Mind.* New York: Harpers.

Gesell, A. (1950), Human infancy and the ontogenesis of behavior, *Amer. Sci.*, 529–553.

Ginsberg, A. (1952), A reconstructive analysis of the concept, "Instinct" *J. Psychol.*, **33**, 235–277.

Gos, M. (1935), Less reflexes conditionnels chez l'embryon d'Oiseau, *Bull. Soc. Sci. Liege*, **4–5**, 194–199; **6–7**, 246–250.

Gray, J. (1950), The role of peripheral sense organs during locomotion in the vertebrates, *Sympos. Soc. Exper. Biol.*, **4**, 112–126.

Grohmann, J. (1939), Modifikation oder Funktionsrelfung? *Z. Tierpsychol.*, **2**, 132–144.

Haldane, J. B. S. (1946), The interaction of nature and nurture, *Ann. Eugen.*, **13**, 197–205.

Hall, C. S. (1951), The genetics of behavior, in S. S. Stevens (Ed.) *Handbook of Experimental Psychology*. New York: Wiley, Chapt 9, 304–329.

Hallowell, A. I. (1954), The self and its behavioural environment, in E. S. Carpenter (Ed.) *Explorations Two*. Toronto: Univ. of Toronto.

Harris, L. J., Clay, J., Hargreaves, J., and Ward, A. (1933), Appetite and choice of diet. The ability of the vitamin B deficient rat to discriminate between diets containing and lacking the vitamin, *Proc. Roy. Soc.*, Ser. B., **113**, 161–190.

Hebb, D. O. (1949), *The Organization of Behavior*. New York: Wiley.

Hebb, D. O. (1953), Heredity and environment in mammalian behavior, *Brit. J. Anim. Behav.*, **1**, 43–47.

Hediger, H. (1950). *Wild Animals in Captivity*. London: Butterworths.

Hertz, Mathilde (1931), Die Organisation des optischen Feldes bei der Biene, *Zeit. Vergl. Physiol.*, **14**, 629–674.

Hertz, Mathidle (1933), Ueber figurale Intensitaten und Qualitaten in der optischen Wahrnehmung der Biene, *Biol. Zbl.*, **53**, 10–40.

Hess, W. R. (1949) *Das Zwischenhirn—Syndrome, Lokalisalionen, Funktionen*. Basel: Schwabe.

Hingston, R. W. (1929), *Problems of Instinct and Intelligence*. New York: Macmillan.

Holst, E. V. (1935), Erregungsbildung und Erregungsletung im Fischrukenmark, *Pflüg. Arch.*, **235**, 345–359.

Holst, E. V. (1935), Uber den Prozess der zentralnevosen Koordination, *Pflug. Arch.*, **236**, 149–158.

Holst, E. V. (1950), Quantitative Messung von Stimmungen im Verhalten der Fisch, *Sympos. Soc. Exper. Biol.*, IV, New York: Academic Press 143–174.

Holt, E. B. (1931), *Animal Drive and the Learning Process*. London: Norgate.

Honigmann, H. (1945), The visual perception of movement by toads, *Proc. Roy. Zool. Soc.* Ser. B, **132**, 291–307.

Howells, T. H. (1945), The obsolete dogmas of heredity, *Psychol. Rev.*, **52**, 23–34.

Howells, T. A., and Vine, D. O. (1940), The innate differential in social learning, *J. Abnorm. Soc. Psychol.*, **35**, 537–548.

Hunt, E. L. (1949), Establishment of conditioned responses in chicken embryos, *J. Comp. Physiol. Psychol.*, **42**, 107–117.

Hunt, J. McV. (1941), The effects of infant feeding-frustration upon adult hoarding in albino rat, *J. Abnor. Soc. Psychol.*, **36**, 338–360.

Hunter, W. S. (1947), Summary comments on the heredity-environment symposium, *Psychol. Rev.*, **54**, 348–352.

James, W. T. (1941), Morphological form and its relation to behavior, in *The Genetic and Endocrinic Basis for Differences in Form and Behavior*. Am. Anat. Mem., **19**, Sect. VI, 525–643.

James, W. T. (1952), Observations on the behavior of newborn puppies. I. Method of measurement and types of behavior involved, *J. Genet. Psychol.*, **80**, 65–73.

James, W. T. (1952), Observations on the behavior of newborn puppies. II. Summary of movements involved in group orientation, *J. Comp. Psychol.*, **45**, 329–335.

Jennings, H. S. (1911), (1905), Modifiability in behavior. I. Behavior of sea anemones, *J. Exper. Zool.*, **2**, 447–472.

Jennings, H. S. (1930), *The Biological Basis of Human Nature*. New York: Norton.

Kagan, J. and Beach, F. A. (1953), Effects of early experience on mating behavior in male rats, *J. Comp. Physiol. Psychol.*, **46**, 204–208.

Keeler, C. E. (1942), The association of the black (non-agouti) gene with behavior in the Norway rat, *J. Hered.*, **33**, 371–384.

Keeler, C. E. and King, H. D. (1942), Multiple effects of coat color genes in the Norway rat, with special reference to temperament and domestication, *J. Comp. Psychol.*, **34**, 241–250.

Kennedy, J. S. (1953), Host plant selection in Aphididae. *Tr. IXth Int. Cong. Entom. Amsterdam*, 106–113.

Kennedy, J. S. (1954), Is modern ethology objective?, *Brit. J. Anim. Behav.*, **2**, 12–19.

Kent, G. C., Jr. and Liberman, M. J. (1919), Induction of psychic estrus in the hamster with progesterone administered via the lateral brain ventricle, *Endocrin.*, **45**, 29–32.

Kinder, Elaine F. (1927), A study of the nest-building activity of the albino rat, *J. Exper. Zool.*, **47**, 117–161.

Krechevsky, I. (1933), Hereditary nature of a hypotheses, *J. Comp. Psychol.*, **16**, 99–116.

Kruchinsky, L. (1947), The role played by peripheral impulses in the sexual form of behaviour of males. C. Rend. (Doklady) *AC. Sci. U. S. S. R.*, **55**, 461–463.

Kuo, Z. Y. (1924), A psychology without heredity, *Psychol. Rev.*, **31**, 427–448.

Kuo, Z. Y. (1930), The genesis of the cat's response to the rat, *J. Comp. Psychol.*, **11**, 1–30.

Kuo, Z. Y. (1932a), Ontogeny of embryonic behavior in Aves. I. The chronology and general nature of the behavior of the chich embryo, *J. Exper. Zool.*, **61**, 395–430.

Kuo, Z. Y. (1932b), Ontogeny of embryonic behavior in Aves. II. The mechanical factors in the various stages leading to hatching, *Ibid.*, **62**, 453–489.

Kuo, Z. Y. (1932c), Ontogeny of embryonic behavior in Aves III. The structure and environmental factors in embryonic behavior, *J. Comp. Psychol.*, **13**, 245–272.

Kuo, Z. Y. (1932d), Ontogeny of embryonic behavior in Aves. IV. The influence of embryonic movements upon the behavior after hatching, *ibid.*, **14**, 109–122.

Kuo, Z. Y. (1939), Total pattern or local reflexes? *Psychol. Rev.*, **46**, 93–122.

Labriola, J. (1953), Effects of caesarean delivery upon maternal behavior in rats, *Proc. Soc. Exper. Biol. Med.*, **83**, 556–557.

Lashley, K. S. (1938), Experimental analysis of instinctive behavior, *Psychol. Rev.*, **45**, 445–471.

Lashley, K. S. (1947), Structural variation in the nervous system in relation to behavior, *Ibid.*, 54, 325–334.

Lashley, K. S. (1949), Persistent problems in the evolution of mind, *Quart. Rev. Biol.*, 28–42.

Leblond C. P. and Nelson, W. P. (1937), Maternal behavior in hypophysectomized male and female mice, *Amer. J. Physiol.*, **120**, 167–172.

Lehrman, D. S. (1953), A critique of Lorenz's objectivistic theory of animal behavior, *Q. Rev. Biol.*, **28**, 337–363.

Leuba, C. (1940), The need for a systematic study of innate nature, *Psychol. Rev.*, **47**, 486–490.

Lissman, H. W. (1946), The neurological basis of the locomotory rhythm in the spinal dogfish (*Scyllium canicula, Acanthias vulgaris*). II. The effect of de-afferentation, *J. Exper. Biol.*, **23**, 162–176.

Lissman, H. W. (1950), Proprioceptors, in *Sympos. Soc. Exper. Biol.*, IV, New York: Academic Press, 34–59.

Lorenz, K. (1935), Der Kumpan in der Umwelt des Vogels, *J. Ornithol.*, **82**, 137–213, 289–413.

Lorenz, K. (1950), The comparative method in studying innate behaviour patterns, in *Sympos. Soc. Exper. Biol.*, IV. New York: Academic Press, 221–268.

Maier, N. R. F. and Schneirla, T. C. (1935), *Principles of Animal Behavior*. New York: McGraw-Hill.

Maier, N. R. F. and Schneirla, T. C. (1942), Mechanisms in conditioning, *Psychol. Rev.*, **49**, 117–134.

Martins, T. (1949), Disgorging of food to the puppies by the lactating dog, *Phys. Zool.*, **22**, 169–172.

Martins, T. and Valle, J. R. (1948), Hormonal regulation of the micturition behavior of the dog, *J. Comp. Physiol.*, **41**, 301–311.

Masserman, J. (1941), Is the hypothalamus a center of emotion?, *P. Med.*, **3**, 3–25.

Mayer, E. (1942), *Systematics and the Origin of Species*. New York: Columbia Univ. Press.

Mayer, E. (1948), The bearing of the new systematics on genetical problems, in Demerec, M. (Ed.), *Advances in Genetics*. New York: Academic Press, **2**, 205–237.

McCracken, Isabel (1907), The egg-laying apparatus in the silkworm (*Bombyx mori*) as a reflex apparatus, *J. Comp. Neurol. Psychol.*, **17**, 262–285.

McGraw, Myrtle (1940), Basic concepts and procedures in a study of behavior development, *Psychol., Rev.*, **47**, 79–89.

McGraw, Myrtle (1946), Maturation of behavior, in L. Carmichael (Ed.) *Manual of Child Psychology*. New York: Wiley, Chapt 7, 332–369.

McNiven, M. A. (0000), Responses of the chicken duck and pheasant to a hawk and goose silhouette. A controlled replication on Tinbergen's study. (Unpubl.).

Metfessel, M. (1940), Relationships of heredity and environment in behavior, *J. Psychol.*, **10**, 177–198.

Mills, W. (1898), *The Nature and Development of Animal Intelligence*. London: Unwin.

Moore, A. R. and Welch, J. C. (1940), Associative hysteresis in larval Amblystoma, *J. Comp. Psychol.*, **29**, 283–292.

Morgan, C. T. and Stellar E. (1950), *Physiological Psychology*. (2nd Ed.) New York: McGraw-Hill.

Muller, H. J. (1940), Bearings of the *Drosophila* work on systematics, in J. Huxley (Ed.) *The New Systematics*. Oxford: Clarendon.

Munn N. L. (1940), Learning experiments with larval frogs, *J. Comp. Psychol.*, **20**, 97–103.

Needham, J. (1929), *The skeptical Biologist.* London: Chatto.

Nissen, H., Chow, K. L., and Semmes, Josephine (1951), Effects of restricted opportunity for tactual, kinesthetic, and manipulative experience on the behavior of a chimpanzee, *Amer. J. Psychol.*, **64**, 485–507.

Padilla, S. G. (1935), Further studies on the delayed pecking of chicks, *J. Comp. Psychol.*, **20**, 413–443.

Pantin, C. F. A., and Pantin A. M. P. (1943), The stimulus to feeding in *Anemonia sulcata, J. Exper. Biol.*, **20**, 6–13.

Pantin, C. F. A., and Pantin, A. M. P. (1950), Behavior patterns in lower invertebrates, in *Sympos. Soc. Exper. Biol.* IV, New York: Academic Press, 75–195.

Parker, G. H. (1919), The Elementary Nervous System. Philadelphia: Lippincott.

Parr, A. E. (1937), On self-recognition and social reaction in relation to biomechanics, with a note on terminology, *Ecol.*, **18**, 321–323.

Patterson, T. L. (1933), Comparative physiology of the gastric hunger mechanism, *Ann. New York Acad. Sci.*, **34**, 55–272.

Pattie, F. A., Jr. (1936), The gregarious behavior of normal chicks and chicks hatched in isolation, *J. Comp. Psychol.*, **21**, 161–178.

Prosser, C. L. (1936), Rhythmic activity in isolated nerve centers, *Cold Spring Harbor Sympos. Quant. Biol.*, **4**, 339–346.

Rasmussen, E. W. (1939), Wildness in rats, *Acta Psychol.* (Hague), **4**, 295–304.

Ray, W. S. (1932), A preliminary study of fetal conditioning, *Ch. Devel.*, **3**, 173–177.

Redfield, R. (Ed.) (1942), *Levels of Integration in Biological and Social Systems.* Lancaster: Cattell Press.

Richter, C. P. (1942–1943), Total self-regulatory functions in animals and human beings, *Harvey Lect.*, **38**, 63–103.

Ricter, C. P. (1952), The effect of domestication on the steroids of animals and man, in *Ciba Found. Colloq. Endocrin.*, III. London: Churchill.

Riess, B. F. (1950), The isolation of factors of learning and native behavior in field and laboratory studies, *Ann. New York Acad. Sci.*, **51**, 1093–1102.

Rosenblatt, J. (1953), Mating behavior of the male cat. The role of sexual experience and social adjustment, Phd. D. Thesis: Library, New York Univ.

Rosenblatt, J., Wodinsky, J, Failla, Marie Lou, Cohn, Ruth, and Frank, A. Analytical studies on maternal behavior and litter relations in the domestic cat. II. From birth to weaning. (Unpubl.).

Schaeffer, A. S. (1911), Habit formation in frogs, *J. Anim. Behav.*, **1**, 309–335.

Schneirla, T. C. (1938), A theory of army-any behavior based upon the analysis of activities in a representative species, *J. Comp. Psychol.*, **25**, 51–99.

Schneirla, T. C. (1939), A theoretical consideration of the basis for approach-withdrawal adjustments in behavior, *Psychol. Bull.*, **37**, 501–502.

Schneirla, T. C. (1941), Social organization in insects, as related to individual function, *Psycho. Rev.*, **48**, 465–486.

Schneirla, T. C. (1944), A unique case of circular milling in ants, considered in relation to trail following and the general problem of orientation, *Amer. Mus. Nov.*, No. 1253, 1–26.

Schneirla, T. C. (1946), Problems in the biopsychology of social organization, *J. Abnorm. Soc. Psychol.*, **41**, 385–402.

Schneirla, T. C. Psychology, Comparative. Encyclop. Brit. (1948 Ed.) **18**, 690–760.

Schneirla, T. C. (1949), Levels in the psychological capacities of animals in R. W. Sellars *et al.* (Eds.) *Philosophy for the Future.* New York: Macmillan, 243–286.

Schneirla, T. C. (1950), The relationship between observation and experimentation in the field study of behavior, *Ann. New York Acad. Sci.*, **51**, 1022–1044.

Schneirla, T. C. (1952a), A consideration of some conceptual trends in comparative psychology, *Psychol. Bull.*, **49**, 559–597.

Schneirla, T. C. (1952b), Basic correlations and coordinations in insect societies with special reference to ants, *Coll. Int. Cent. Nat. Rech. Sci.*, 247–269.

Schneirla, T. C. (1953a), The army-any queen: keystone in a social system, *Éull. Un. Int. Etude Ins. Soc.*, **1**, 29–41.

Schneirla, T. C. (1953b), In K. Roeder (Ed.) *Insect Physiology.* New York: Wiley. Basic problems in the nature of insect behavior. Chapt. 25. Insect behavior in relation to its setting. *Ibid.*, Chapt. 26. Modifiability in insect behavior. *Ibid.*, Chapter 27. Collective activities and social patterns among insects. *Ibid.*, Chap. 28.

Schneirla, T. C., Brown, R. Z. and Brown, Frances (1954), The bivouac or temporary nest as an adaptive factor in certain terrestrial species of army ants, *Ecol. Monogr.*, **24**, 269–296.

Schooland, J. B. (1942), Are there any innate behavior tendencies? *Genet, Psychol Monogr.*, **25**, 219–287.

Scott, W. E. D. (1902), Data on songbirds, *Science*, **15**, 178–181.

Scott, W. E. D. (1904), The inheritance of song, *Science*, **19**, 957–959.

Searle, L. V. (1949), The organization of hereditary maze-brightness and maze-dullness, *Genet. Psychol. Monogr.*, **39**, 279–325.

Selye, H., Collip, J. B. and Thompson, D. L. (1934), Nervous and hormonal factors in lactation, *Endocrin.*, **18**, 237–248.

Selye, H. and McKeown, T. (1934), Further studies on the influence of suckling, *Anat. Rec.*, **60**, 323–332.

Seward, Georgene H. and Seward, J. P. (1937), Internal and external determinants of drive, *Psycho. Rev.*, **44**, 349–363.

Shepard, J. F. and Breed, F. S. (1913), Maturation and use in the development of an instinct, *J. Anim. Behav.*, **3**, 274–285.

Shirley, Mary (1931), The sequential method for the study of maturing behavior patterns, *Psychol. Rev.*, **38**, 507–528.

Siegel, A. I. (1953), Deprivation of visual form, definition in the ring dove. I. Discriminatory learning, *J. Comp. Physio. Psychol.*, **46**, 115–119.

Siegel, A. I. (1953), Deprivation of visual form definition in the ring dove II. Perceptual-motor transfer, *Ibid.*, **46**, 249–252.

Smith, S. and Guthrie, E. R. (1921), *General Psychology in Terms of Behavior.* New York: Appleton.

Snyder, L. H. (1950), *The Principles of Heredity* (4th Ed.). Boston: Heath.

Spelt, D. K. (1948), The conditioning of the human fetus in utero, *J. Exper. Psychol.*, **37**, 338–346.

Sperry, R. W. (1946), Ontogenetic development and maintenance of compensatory eye movements in complete absence of the optic nerve, *J. Comp. Psychol.*, **39**, 321–330.

Spieth, H. T. (1952), Mating behavior within the genus *Drosophila (Diptera)*, *Bull. Amer. Mus. Nat. Hist.*, **99**, 401–474.

Thompson, W. R. and Heron, W. (1954), The effects of restricting early experience on the problem-solving capacity of dogs, *Canad. J. Psychol.*, **8**, 17–31.

Thorpe, W. H., and F. G. W. (1937), Olfactory conditioning in a parasitic insect and its relation to the problem of host selection, *Proc. Roy. Soc. Lond.*, Ser. B, **124**, 56–81.

Tinbergen, N. and Kuenen, D. J. (1939), Uber die aulösenden und die richtunggebenden Reizsituationen der Sperrbewegung von jungen Drossln (*Turdus M. Merula* l. und *T. e. erictorum Turton*), *Z. Tierpsychol*, **3**, 37–60.

Tinbergen, N. (1948), Social releasers and the experimental method required for their study, *Wilson Bull.*, **60**, 6–52.

Tinbergen, N. (1951), *The Study of Instinct.* New York: Oxford Univ. Press.

Tinbergen, N. and Perdeck, A. C. (1951), On the stimulus situation releasing the begging response in the newly hatched herring gull chick (*Larus argentatus argentatus* Pont.), *Behaviour*, **3**, 1–39.

Tobach, Ethel, Failla, Marie Lou and Schneirla, T. C. (0000), Analytical studies on maternal behavior and litter relations in the domestic cat. I. Parturition. (Unpubl.).

Tracy, H. C. (1926), The development of motility and behavior reactions in the toad fish (*Opsanus tau.*), *J. Comp. Neurol.*, **40**, 253–369.

Tucker, B. S. (1943), Brood-patches and the physiology of incubation, *Brit. Birds*, **37**, 22–28.

Uyldert, Ina E. (1946), A conditioned reflex as a factor influencing the lactation of rats, *Acta Brev. Neerl.*, **14**, 86–86.

Van Der Kloot, W. G. and Williams, C. M. (1953a), Cocoon construction by the *Cecropia* silkworm. I. The role of the external environment, *Behaviour*, **5**, 141–156.

Van Der Kloot, W. G. and Williams, C. M. (1953b), Cocoon construction by the *Cecropia* silkworm II. The role of the internal environment, *Ibid.*, **5**, 157–174.

Wada, Tomi (1922) An experimental study of hunger in its relation to activity, *Arch. Psychol.*, **57**, 1–65.

Warren, Roslyn P. (1954), The sexual behavior before and after castration of the adrenalectomized male golden hamster treated with DCA. Ph. D. Thesis: Library, New York Univ.

Warren, Roslyn P. and Rosenblatt, J. (0000), Does progesterone induce estrus behavior by action on the hypothalamus? (Unpubl.).

Watson, J. B. (1914), *Behavior: An Introduction to Comparative Psychology.* New York: Holt.

Watson, J. B. (1919), *Psychology from the Standpoint of a Behaviorist.* Philadelphia: Lippincott.

Watson, J. B. (1925), *Behaviorism.* New York: Norton.

Watson and Lashley, K. S. (1915), Homing and related activities in birds, (Papers Dept. Mar. Biol., 7). *Carn. Inst. Publ.*, No. 211, 9–60

Weiss, P. (1914), Autonomous versus reflexogenous activity of the central nervous system, *Proc. Amer. Phil. Soc.*, **84**, 53–64.

Weiss, P. (1954), Some introductory remarks on the cullular basis of differentiation, *J. Embryol. Exper. Morphol.*, **1**, 181–211.

Werner, H. (1940), Comparative Psychology of Mental Development. New York: Harpers.

Wheeler, W. M. (1928), *The Social Insects.* New York: Harcourt, Brace.

Whitman, C. O. (1899), Animal behavior. Biological lectures, 1898, *Marine Biol. Lab., Wood's Hole, Mass.*, 285–338.

Whitman, C. O. (1919), Behavior of pigeons, (v. 3, post. works; H. A. Carr, Ed.), *Carn. Inst. Publ.* No. 257, 1–161.

Wickens, D. D. and Wickens, Carol (1939), A study of conditioning in the neonate, *Psychool. Bull.*, **36**, 599.

Wiesner, B. P. and Sheard, Norah M. (1933), *Maternal Behavior in the Rat.* Edinburgh: Oliver and Boyd.

Windle, W. F. (1940), *Physiology of the Fetus: Origin and Exient of Function in Prenatal Life.* Philadelphia: Saunders.

Windle, W. F. (1950), Reflexes of mammalian embryos and fetuses, in P. Weiss (Ed.) *Genetic Neurology.* Chicago: Univ. Chicago Press.

Wolf, E. (1933), Das Verhalten der Bienen gegenuber flimmernden Feldern und bewegten Objekten, *Zsch. Vergl. Physiol.*, **20**, 151–161.

Yeaker, E. H. and Rhoades, R. P. (1941), A comparison of the body and endocrine gland (adrenal, thyroid and pituitary) weights of emotional and non-emotional rats, *Enocrinol.*, **28**, 337–340.

Yerkes, R. M. (1913), The heredity of savageness and wildness in rats, *J. Anim. Behav.*, **3**, 236–296.

1. HOW IS BEHAVIOR MAINTAINED IN A POPULATION?

THE BIOLOGICAL VIEW OF FUNCTION

The first question asked by ethologist and layman alike concerning a given behavior pattern is usually "What is its function?" Evolutionary studies suggest that behavioral as well as morphological characters are not maintained in a population unless they are functional, which is to say unless they contribute to the longevity or reproductive proclivity of the possessing organism. The exceptional cases appear minor: Some characters, morphological and behavioral, may be maintained only indirectly by natural selection. Such characters may be invariably inherited along with other, avowedly functional characters, and thus be maintained because of natural selection for the other characters (the phenomenon of pleiotropism). Or, in very small populations of animals, accidental deaths of animals could drastically lower the incidence of perfectly functional characters; this and other aspects of accidental events that affect the genetic composition of small populations are grouped under the term "genetic drift." As a rule, though, we believe behavior patterns to be functional: They lengthen the life or increase the reproductive potential of the possessor.

Sometimes the function of a behavior pattern presents no problem to the student of animal behavior: An animal must eat to live, must escape predators, must reproduce if its genes are to be part of the next generation. However, there are many variants on these simple processes, and the reason for such variation is yet far from clear. There are even more puzzling behavior patterns. Why do micturation patterns of male and female dogs differ so obviously? Why do parent birds remove the eggshells of their newly-hatched offspring? Why do Kittiwakes build their nests of mud when almost all other gulls use vegetation?

The study of functional aspects of behavior is in its infancy, perhaps because this kind of question appears so difficult to deal with experimentally. However, recent studies have begun to sketch valid experimental methods in this area of investigation, as the brief selection of papers to follow will illustrate. The first collection explores several different kinds of evidence that can be gathered about function, while the second collection illustrates the comparative approach toward understanding of whole complexes of interrelated behavioral characters. (For a fuller discussion and entrance into the literature on behavioral function, the reader is referred to Chapter 9 of *An Introduction to Animal Behavior* by Klopfer and Hailman.)

A. METHODS FOR ASSESSING FUNCTION

The first paper included is a straightforward attempt to discover one possible function of the spines on some small fish. The spines were known to be used in communicative displays to conspecific fish, but were also thought to be used in behavioral defense against predators. The paper gives experiments comparing the survival of prey fish, with and without spines, when these fish are pursued by various larger predatory fish. The method illustrated, then, is one of direct comparison between species possessing a behavior pattern (and morphological character, in this case) with species not having these characters.

There are obvious dangers and limitations in comparing different species, since they will always differ in more ways than the behavior pattern of interest. A far stronger comparison involves behavioral variants in a population of a single species. The second paper illustrates such a comparison with regard to colonial nesting behavior in birds.

In order to be termed "colonial," birds must nest in the same place at the same time. Therefore, Patterson has assessed the breeding success of variants that are less colonial, i.e., those that nest on the fringes of the colonial and those that nest before or after the main population. These deviants leave fewer offspring than their conspecifics, and so the colonial nesting habit is maintained directly by the forces of natural selection.

The last paper in this series illustrates a third method for assessing function. Often, there are no behavioral deviants of which we can compare the survival and reproductive output with normal animals. Therefore, one can sometimes experimentally create such deviants. This last example is particularly interesting, because the behavior pattern to be studied is seemingly trivial. It takes but a few seconds each year for a parent bird to remove the eggshells from the nest containing its newly-hatched offspring; yet the parents are extremely tenacious in performing this ritual. The question is "Why?" A partial answer was achieved by creating nests where the eggshells remained behind, and then noting the survival of the young and unhatched eggs in the nest.

*The Spines of Sticklebacks (*Gasterosteus *and* Pygosteus*) as Means of Defence Against Predators (*Perca *and* Esox*)*

R. HOOGLAND, D. MORRIS AND N. TINBERGEN

Feeding experiments with Pike (*Esox lucius* L.) and Perch (*Perca fluviatilis* L.) as predators, and Three-spined Sticklebacks (*Gasterosteus aculeatus* L.), Ten-spined Sticklebacks (*Pygosteus pungitius* L.), Minnows (*Phoxinus phoxinus* L.), Roach (*Rutilus rutilus* L.), Rudd (*Scardinius erythrophthalmus* L.), and Crucian Carp (*Carassius carassius* L.) as prey species show that both species of stickleback enjoy a demonstrable degree of protection from these predators. This protection is much better for *Gasterosteus* than for *Pygosteus*; tests with de-spined sticklebacks show that it is mainly due to the spines.

A description of the predatory behaviour of Perch and Pike is given; analysis of the experiments shows that (1) sticklebacks are rejected when, after being snapped up, their spines hurt the predator's mouth; (2) after a very few experiences both Perch and Pike become negatively conditioned to the sight of sticklebacks and avoid them before they have made contact. As a result of this conditioning, the last links of the predator's feeding chain drop out; in Pike, mere fixation with two eyes was often sufficient to recognise the prey as unwanted.

The possession of few and large spines in *Gasterosteus* must be regarded as more specialised than that of many small spines as found in various other species of stickleback. There is a correlation between the extreme development of this anti-predator device in *Gasterosteus* and (1) its boldness, (2) its tendency to select a more open nesting habitat, (3) the schooling and wandering tendencies of the females in the spawning season, and (4) its relatively conspicuous nuptial colors; these correlations suggest an interrelationship between anti-predator devices and reproductive behavior and structures; the evolution of these two systems must have occurred in conjunction with each other.

INTRODUCTION

Many species of fish have evolved sharp spines which are often alleged to serve as defensive weapons against predators. Most frequently these spines are modified fin-rays. The sticklebacks possess this type, and it was the purpose of this study to analyse the effect of these weapons upon two predators, the Perch (*Perca fluviatilis* L.) and the Pike (*Esox lucius* L.). The species of stickleback studied were the Three-spined Stickleback (*Gasterosteus aculeatus* L.) and the Ten-spined Stickleback (*Pygosteus pungitius* L.), the only two members of the family which occur in European fresh-waters. The reactions of the predators to these fish were compared with those to various non-spined freshwater fish, *viz.*, the Minnow (*Phoxinus phoxinus* L.), the Roach (*Rutilus rutilus* L.), the Rudd (*Scardinius erythrophthalmus* L.), and the Crucian Carp (*Carassius carassius* L.).

These two predators and five prey species are all known to occur together in the wild, and Hartley (1947) has found sticklebacks, Minnows and Roach amongst the stomach contents of the predators.

Reprinted from *Behaviour*, volume 10, pages 205–236 (1957). Original numbering of figures is retained.

Some preliminary tests by Tinbergen in 1939 appeared to show that young Pike of approximately 15 cm long had considerable difficulty in swallowing adult Three-spined Sticklebacks, and also that they soon learned to leave sticklebacks alone, while continuing to eat non-spined fish of the same and even considerably larger size. Since sticklebacks, when seized by a Pike, respond by raising all their spines, and keeping them rigidly erected as long as they are in the Pike's mouth, it was naturally supposed that the conditioning process in the Pike had something to do with the presence of the spines. This problem was tackled experimentally by Hoogland in 1948. Later, in 1952, Morris, engaged in a study of the behaviour of *Pygosteus*, extended Hoogland's work studying Perch as well as Pike, and *Pygosteus* as well as *Gasterosteus*. The observations done so far throw light not only on the function of the sticklebacks' spines but enabled us at the same time to analyse the predator's behaviour to a certain extent. Also, striking differences were found between the two species of sticklebacks, and these could be correlated with differences in their reproductive behaviour and other aspects of their ecology generally.

A film of the predatory behaviour of Pike towards sticklebacks and non-spined fish was taken by Tinbergen, and copies are in the Department of Zoology at Leiden and in the Department of Zoology and Comparative Anatomy at Oxford.

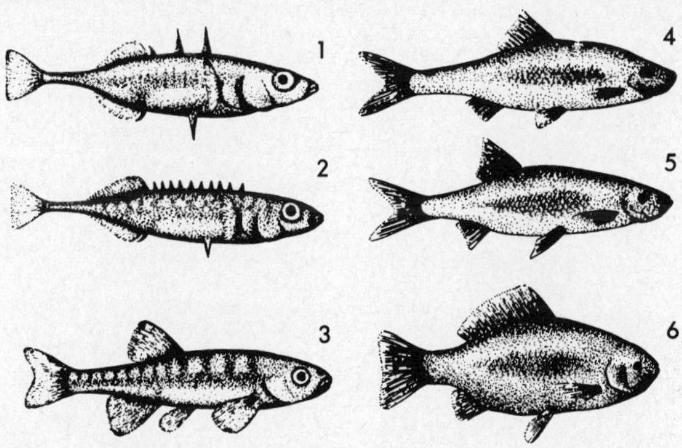

Fig. 1. The prey species used. 1. Three-spined Stickleback; 2. Ten-spined Stickleback; 3. Minnow; 4. Roach; 5. Rudd; 6. Crucian Carp.

MATERIAL AND METHODS

Five Pike and several hundreds of *Gasterosteus*, Rudd, Roach and Crucian Carp were collected near Leiden in Holland; three Pike and five Perch, as well as all sticklebacks and Minnows used in Morris' experiments were collected from ponds and streams in Wiltshire and Oxfordshire in England. The prey specimens offered were adult sticklebacks of both species and non-spined fish of approximately the same size as adult Three-spined Sticklebacks, or slightly larger (Fig. 1). The Pike and Perch varied from 10 to 25 cm; naturally, their size increased during the course of the tests.

All tests were made in the laboratory in aquarium tanks, varying, unless stated otherwise, from 60 × 35 × 35 cm to 125 × 50 × 35 cm. Hoogland's experiments were generally made in sparsely planted tanks; most of Morris' tests were done in bare tanks. In some tests the tanks were screened from their neighbours. In almost all tests the predators were kept singly. Experimental procedure was different from one test to another; details will be given with the individual tests.

As will be seen below, the details noted differed considerably from one test to another. In all tests information was obtained whether a prey fish was eaten or not, and these data were considered sufficient for the purpose of our primary ecological problem. In some tests, particularly those of Hoogland, more detailed observations were made on the Pikes' behaviour, and these throw some light on the nature of the modifications in the Pikes' behaviour in the course of the tests.

Hoogland's Pike showed signs of positive conditioning to the experimenter's movements prior to the introduction of the fish. Morris' predators showed escape tendencies at the beginning of some tests. Since in both cases the disturbing factor influences experiments and controls in the same way, no attempts were made to eliminate them entirely. In some of Morris' tests, where a large number of prey fish of various kinds were left with the predator for weeks on end, neither of these disturbances played a part.

THE REACTIONS OF STICKLEBACKS TO PIKE AND PERCH

Neither species of stickleback shows any tendency to escape from Perch or Pike as long as the predators stay motionless or move only very slowly. Quick movements of either Pike or Perch, particularly when directed towards the sticklebacks, evoke immediate flight, and subsequent hiding in cover. If a Pike really dashes towards a stickleback in a serious attempt to capture it, the stickleback's chances of escape are small; it is never able to approach, let alone to exceed the speed of an attacking Pike. A Perch is much less superior in speed, and both sticklebacks have a chance to escape even from a hunting Perch, as long as the latter is alone. However, in the limited space of our tanks, even single Perch could easily corner and capture any stickleback. Sticklebacks may manage, by a quick avoiding movement, to make even a Pike miss them, and they may escape into cover. In our tests, even in well-planted aquaria, such a frightened stickleback never stayed in cover for long, and if a Pike attempted to catch it, it always got hold of it eventually. Negative results indicate almost always unwillingness of the predator, and in only very few cases (which will be indicated in the protocols) was the escape of a stickleback due to its keeping completely motionless.

The escape by a stickleback can be released by visual stimuli alone, for we have seen sticklebacks fleeing from a Pike in a neighbouring tank. In spite of several attempts, following the method used by von Frisch (1941), we could never see any evidence of a chemical response to a substance released by the damaged skin of another stickleback.

When a stickleback is seized by a Pike, it reacts by erecting all its spines, keeping quite motionless except for breathing. Hoogland (1951) has shown how *Gasterosteus* can lock its spines in this position, which enables it to keep them raised for hours on end without muscular exertion.

Gasterosteus aculeatus possesses three unpaired dorsal spines in front of the dorsal fin, one small spine in front of the anal fin, and a pair of large ventral spines. The dorsal spines are modified parts of the dorsal fin. The two anterior ones are quite large, about 6 mm in an adult fish of 65 mm; the third one rarely exceeds 1 mm in length. The unpaired anal spine, which is a modified part of the anal fin, measures also about 1 mm. The paired ventral spines, which are modified pelvic fins, are, in animals of the same size, approximately 9 mm in length. When fully raised they stand out in line with each other at right angles to the body axis; together they form one strong, rigid and very sharply pointed cross-bar of approximately 20 mm. The two foremost dorsal spines, and to a lesser extent the ventral spines, usually bear a row of small teeth along their edge.

The response of keeping motionless and raising all spines can be elicited by strong mechanical stimuli. When water is sucked from the tank through a rubber hose and a stickleback happens to be firmly caught in the hose's mouth, it will immediately show this response. When the suction is only weak, it will struggle and usually get free.

Ten-spined sticklebacks respond in the same way as Three-spined. Their spines however are all much smaller (Fig. 1).

THE PREDATORY BEHAVIOUR OF PERCH

The hunting behaviour of the Perch has been described in some detail by Deelder (1951). He reports that in nature it is necessary for Perch to cooperate with one another if prey fish are to be caught. However, under the experimental conditions used in the present study, involving unplanted aquaria of moderate size, hunting was reduced to a minimum and the prey had no chance of escaping unless the predator rejected or avoided them. The biting of the prey by the Perch is reported by Deelder as usually being aimed at the head, and this we can confirm.

A hungry Perch suddenly seeing a prey fish may, however, take it tail-first on occasions. If it is a Minnow, it may be swallowed tail-first, but if it is a stickleback, it will almost always be spat out again and then, usually immediately, re-bitten head-first and perhaps swallowed. The Perch has no other, more specialized, method of turning its prey round without letting go of it. In this respect it contrasts with the Pike.

When it has cornered a prey fish, the Perch may attempt to edge round it towards the head. The prey usually responds to these movements by turning also, so that its tail points towards the Perch, and these manoeuvres may continue for some seconds. Sticklebacks under such circumstances have all their spines fully erected. Sometimes they may tilt their ventral surfaces towards the hesitating Perch.

A Minnow which had been bitten head-first usually vanished except for its tail. This was then frequently seen to disappear slowly and smoothly into the now closed mouth of the Perch. A stickleback which had been bitten and rejected a number of times before being eaten or abandoned was sometimes held in the mouth for several minutes and sometimes released as soon as the spines touched the inside of the predator's mouth. Often, after an instantaneous rejection, the stickleback was quickly bitten once more and this was frequently seen to occur over and over again. It seemed as if the Perch could not resist the visual stimulus offered by the stickleback outside its mouth, despite the fact that the mechanical stimulus of the spines always resulted in a rejection.

On a number of occasions a completely swallowed stickleback was regurgitated alive after some minutes. The Perch appears to damage its prey very little, if at all, when biting it. This may well be adaptive, as certain prey species give off a chemical substance when damaged, the presence of which in the water warns nearby members of the same species of danger (von Frisch, 1941).

After eating, the Perch frequently gulps, gapes, or belches, and we have even seen the whole body vibrate with rapid jerks after a stickleback had been swallowed. The gaping movement is seen in a very exaggerated form usually just before a regurgitation. Sometimes a half-swallowed stickleback that has become lodged in the throat of the Perch may be dislodged by a combination of gaping movements and vigorous sideways shaking of the head. In one instance, when a *Gasterosteus* had been half-swallowed in a single gulp by a hungry Perch, blood began to pour from the latter's gills. It swam slowly around the tank leaving a trail of blood behind it until all the tank-water was cloudy. But despite this it was the Perch and not the stickleback that survived in this case.

THE PREDATORY BEHAVIOR OF PIKE

Apart from a few occasional observations in the wild, and from some observations on large Pike in captivity, most of our knowledge of predatory behaviour of Pike is based on laboratory observations on our smaller Pike.

Locomotion in the Pike is carried out by fin movements, by movements of the trunk and the tail, or by movements of both fins and trunk. Before describing these locomotory patterns, the observed fin movements will be described separately.

The pectorals usually perform undulatory movements, in which the fin rays move back and forth in alternation, with a slight phase difference between the neighboring rays. The wave runs from the frontal-median ray towards the caudal-distal edge. The forward movement of each ray may be as strong as the opposite movement. In this case the fins are kept in a plane at right angles to the body axis, and the result is a water current directed downward (Fig. 2A). The fins may also be directed forward, and then the forward movement of each ray is stronger than the backward motion, and the resulting water current is directed forward (Fig. 2B). When the fin is turned back, the backward motion of the rays is stronger than the opposite

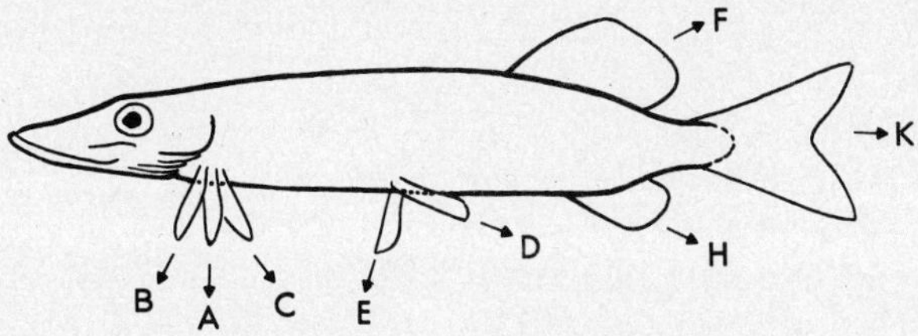

Fig. 2. Types of in movement in Pike; explanation in text.

movements, and the water current is directed backward (Fig. 2C). These currents of course have roughly the opposite effect on the fish's movements.

The pelvic fins are usually directed backward. They undulate also, and effect a downward and backward current (Fig. 2D). They can be switched forward, and when undulating in this position effect a downward and forward current (Fig. 2E).

The dorsal fin, particularly its caudal part, can undulate regularly, the waves usually travelling backward. The result is an upward, backward current (Fig. 2F). A synchronous sideways movement of all rays of the caudal part of the dorsal fin is used in turning. The same movements can be carried out by the anal fin (Fig. 2H); its movements are usually less pronounced than those of the dorsal fin.

Finally, the caudal fin, if moving, also undulates (Fig. 2K); the amplitude of the fin rays' pendulum movements decreases from dorsal to ventral. The rays can also be moved synchronously.

* * *

Summarising, the prey-catching behaviour of a Pike is a sequence of activities which could be named as follows:

1. Eye movements towards the prey
2. Turning towards the prey
3. Stalking
4. Leaping
5. Snapping
6. (Turning the prey head-forward)
7. Swallowing

Leaping and snapping are distinguished because, as will be seen, a Pike may leap at a stickleback without attempting to seize it.

THE RESPONSES OF PERCH TO STICKLEBACKS AND MINNOWS

Two simple tests were carried out to ascertain the reaction of two Perch to a mixed group of prey species. A larger tank (210 × 36 × 36 cm) was used for this purpose.

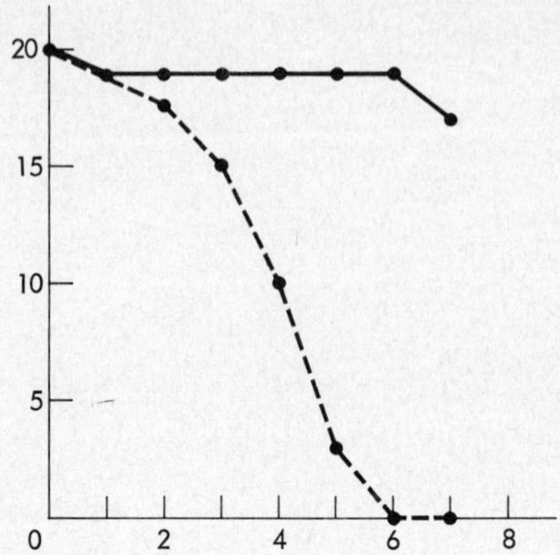

Fig. 4. The effect of predation by Perch on a mixed shoal of 20 *Gasterosteus* (solid line) and 20 Minnows (broken line).
Ordinate: numbers of prey present. Absicssa: time in days.

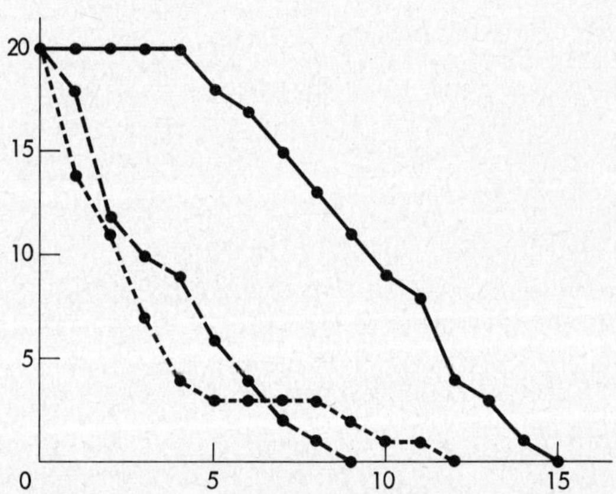

Fig. 5. The effect of predation by Perch on a mixed shoal of 20 *Gasterosteus* (solid line), 20 *Pygosteus* (broken line), and 20 Minnows (dotted line).
Ordinate: numbers of prey present. Abscissa: time in days.

It was unplanted, and contained two predators. A known number of prey was introduced into this tank, and from then on, the number of prey surviving was scored at twenty-four hour intervals. One test was taken with 20 *Gasterosteus* and 20 Minnows. The results are given in Fig. 4. It is seen that *Gasterosteus* was not eaten after one on the first day, and then until all Minnows had gone. In another test, 20 *Gasterosteus*, 20 *Pygosteus* and 20 Minnows were supplied at the start. Fig. 5 shows the result. *Gasterosteus* was left alone until more than half of the other two

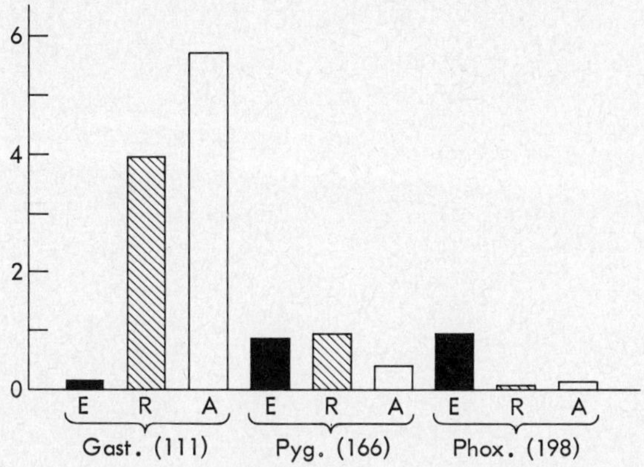

Fig. 6. Results of 475 single presentations of 111 *Gasterosteus*, 166 *Pygosteus*, 198 Minnows to 5 perch.
Ordinate: average number of responses per presentation. Abscissa: responses classified for prey species and for eating (E), rejecting (R) and avoiding (A) by Perch.

species had disappeared. Until the 7th day the Minnows were preyed on slightly more than *Pygosteus*; after that, *Pygosteus* was taken mainly. The difference between *Gasterosteus* and Minnows is striking in both cases.

The vast majority of observations concern the responses of a single Perch to a single prey fish. The various species of prey fish were given in different sequences. Before giving the results obtained from using special sequences, the total results from all these tests, taking each presentation singly, is recorded in Fig. 6. In all, 475 such presentations were made to 5 Perch. At the outset it was hoped that the tests could be standardized in respect of the length of time during which the prey was available to the Perch, but this desirable method was found to be unsatisfactory, because the time lag until the occurrence of the first response varied erratically, and this could not be correlated with the nature of the response that followed it. However, since the prey was always either eaten very soon after its introduction or, if not eaten, was completely ignored after a period of 10—20 minutes, tests were concluded either when the prey had been eaten or when the Perch had lost interest in it.

In Fig. 6 the average number of reactions per test is given for each of the three species when presented to Perch. The predator's reactions have here been classed as Eating, Rejecting, and Avoiding. It ought to be stressed that, although a prey fish can only be eaten once per presentation, it can be rejected a number of times. A Perch may bite and then spit out a fish repeatedly and each time it is spat out it is recorded as one rejection. Thus, although the ordinate value for eating can never rise above one, it can do so for rejecting. This is also true of avoiding. An avoidance was scored each time a Perch made an intention movement of feeding, but stopped before actually making contact.

Figure 6 shows clearly that the Minnow is more frequently eaten and less frequently rejected, or avoided, than *Pygosteus*. Further, *Pygosteus* is more frequently eaten and less often rejected or avoided than *Gasterosteus*. Also, the difference between *Gasterosteus* and *Pygosteus* is much more striking than that between *Pygosteus* and the Minnow.

Part of the above results were obtained by giving the prey fish in the following sequence: 1. *Gasterosteus*, 2. *Pygosteus*, and 3. Minnow. As soon as the predator had lost interest in, or had eaten, the *Gasterosteus*, this prey was, if necessary, removed, and a *Pygosteus* was introduced. Similarly, when the predator had eaten (or lost interest in) this, it was removed if necessary, and a Minnow introduced.

In all, 64 such test sequences were carried out. There were four possible ways in which the sticklebacks could be treated in each series, if the predator's response was simply recorded as Eating (with or without preceding rejections) or Not Eating. The frequency in which these four sequences occurred were:

Gasterosteus not eaten,		*Gasterosteus* eaten,	
Pygosteus eaten	56	*Pygosteus* not eaten	1
Gasterosteus not eaten,		*Gasterosteus* eaten,	
Pygosteus not eaten	7	*Pygosteus* eaten	0

It should be noted that in all these 64 cases the Minnow, presented afterwards, was eaten.

THE RESPONSES OF PIKE TO STICKLEBACKS AND OTHER FISH

The observations were made mainly on eight Pike. Five were used in Hoogland's tests; they were kept separately in tanks with open vegetation. Morris used three Pike which were kept in tanks bare of vegetation. Hoogland usually watched the Pike's behaviour until the prey offered was eaten, or, if it was not eaten, until 30 minutes after it had been presented. Morris broke off his observations after the prey had been either eaten, or when the Pike lost interest.

All Pike gave chase to a stickleback when it was first presented. In these first tests the stickleback was always caught as promptly as any other fish of the same size. Usually it was rejected immediately after it had been snapped up, and the Pike would make violent coiling movements, and would "cough" intensely several times. The stickleback would swim away into cover, but it soon calmed down, and began to move about again. This might cause the Pike to catch it again, and it might then try to swallow it. In this it rarely succeeded, and if so, it took a long time. The stickleback, keeping entirely motionless with spines raised, was maneuvered into various positions by violent head movements of the Pike; sometimes it would almost disappear from sight only to be shifted back forward again, in any position. The drawings of Fig. 10 were made from Tinbergen's film.

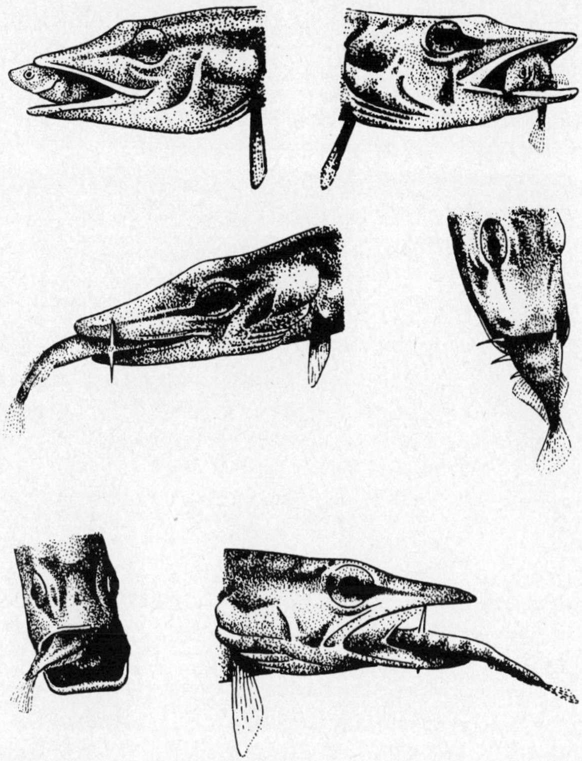

Fig. 10. Drawings taken from 16 mm film showing the variety of positions of *Gasterosteus* in the mouth of Pike.

This might go on for over an hour without the stickleback being either swallowed or even killed. In most cases it was finally rejected, sometimes dead, but more often alive. In other cases it was eventually eaten.

As with the Perch, some tests were done with mixed groups of prey species. In one experiment, 2 Pike were offered 12 *Gasterosteus* and 12 Minnows. As shown in Fig. 11 the Minnows were eaten first, and no *Gasterosteus* was eaten until all Minnows had gone. Again, the sequence was not recorded in detail, so that it is uncertain whether the first stickleback was eaten after all Minnows had gone, or a little earlier on that same day.

Figure 12 shows the results of a test in which 12 *Gasterosteus*, 12 *Pygosteus* and 12 Minnows were offered simultaneously. Apart from one *Gasterosteus* and one *Pygosteus* eaten on the first day, both species were left alone until, on the 5th day, all Minnows had been eaten. The Pike then attacked both species of sticklebacks, but *Pygosteus* was slightly preferred.

In other tests, various fish were presented singly to a single Pike, and the Pike's behaviour observed. In Fig. 13 the results are given as averages of the three types of response scored: Eating, Rejecting, and Avoiding. Again, as in the tests with

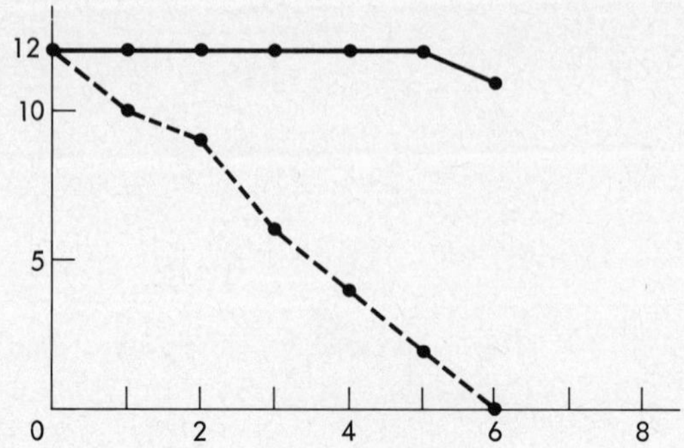

Fig. 11. The effect of predation by Pike on a mixed shoal of 12 *Gasterosteus* (solid line) and 12 Minnows (broken line).
Ordinate: number of prey present. Abscissa: time in days.

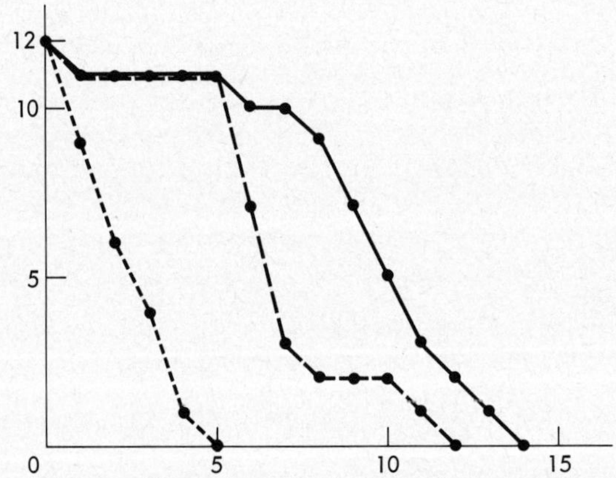

Fig. 12. The effect of predation by Pike on a mixed shoal of 12 *Gasterosteus* (solid line), 12 *Pygosteus* (broken line) and 12 Minnows (dotted line).
Ordinate: number of prey present. Abscissa: time in days.

Perch, Rejecting and Avoiding might be observed more than once during a test, whereas Eating could only be scored once.

The figures for Rejecting were on the whole higher in Perch than in Pike, because the Perch spits out a fish in order to turn it round, and this was recorded as a reject. The Pike turns a prey fish without allowing it to leave its mouth. Avoidance is also much more easily seen in Perch than in Pike, because a Perch will often swim some distance towards a prey before giving up, while a Pike may merely turn its eyes

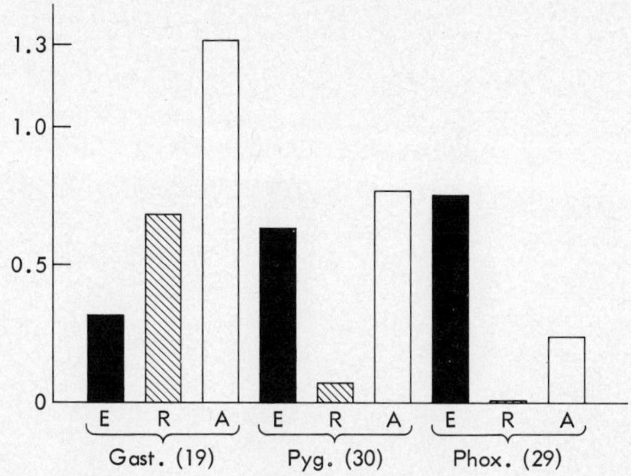

Fig. 13. Results of 78 single presentations of 19 *Gasterosteus*, 30 *Pygosteus* and 29 Minnows to 3 Pike, scored as in Fig. 6.

towards a prey without even turning its body, a response that we may have missed in some of the series. The figures for Perch and Pike are, therefore, not entirely comparable, but both are consistent in themselves, and are reliable as evidence on our main problem.

Figure 13 summarises Morris' results obtained with 3 single Pike, and *Gasterosteus, Pygosteus,* and Minnow as prey species (78 presentations in all). . . .

Pike therefore respond in roughly the same way as Perch: they reject and avoid *Pygosteus* more than Minnows or other non-spined fish, and *Gasterosteus* more than *Pygosteus* and much more than non-spined species.

* * *

THE RESPONSES OF PERCH AND PIKE TO DE-SPINED STICKLEBACKS

In a number of tests, sticklebacks were offered to Perch and Pike after their spines had been cut off. In order not to injure the sticklebacks, the bases of the spines were left intact, and the result was that "de-spined" fish were not as smooth as the non-spined fish used; yet the main obstacles preventing the predators from swallowing sticklebacks were gone, and the remaining stumps were less sharp than the intact spines. In short series with Perch and Pike, involving about 150 *Gasterosteus*, such de-spined fish were almost invariably eaten. The following figures apply to two series done with Pike A and C and de-spined *Gasterosteus*. Of 131 L. offered, 121 were eaten and 5 refused. Of 63 de-spined *Gasterosteus* offered, 55 were eaten and 8 refused. The slightly larger proportion of *Gasterosteus* refused may be partly due to their being still coarser than the non-spined species, partly to the fact that the sticklebacks were always offered after the L. Even with this handicap, it is obvious from a comparison of these series with the foregoing ones that the protection enjoyed by sticklebacks is mainly due to their spines.

DISCUSSION

It seems clear that both *Gasterosteus* and *Pygosteus* are, to Perch and Pike of the sizes used in these tests (10–25 cm) less attractive prey than four non-spined species of fish, and also, that the larger-spined *Gasterosteus* is much less vulnerable to such Perch and Pike than *Pygosteus*; the difference between *Pygosteus* and non-spined fish is however still demonstrable.

The series with the de-spined sticklebacks show that it is actually the spines that afford this protection. The sticklebacks' behaviour, which makes them raise their spines exactly when needed, and the structure of the spines and their erection mechanisms (studied by Hoogland, 1951) are both beautifully adapted to this function.

The rejection of sticklebacks by these predators is really based on two distinct processes. Many of the rejections were due to a non-conditioned response to the strong mechanical stimuli received after the predator had seized the stickleback. This response alone accounts for a sufficient number of rejections to prove the survival value of the spines, particularly since the overwhelming majority of sticklebacks survived after having been snapped up, and many of them survived even after very severe mauling. In addition to this, however, both Perch and Pike learn to respond to visual stimuli by avoiding sticklebacks before they have even touched them. All records scored as "Avoidance" in Figs. 6 and 13, refer to this conditioned response.

The break in the reaction chain may occur either at the end of stalking, or already much earlier. Of the total of 162 observations recorded in series E in which the chain was interrupted before contact was made, 14 were broken off after the leap, 18 after stalking, 81 after turning the body towards the stickleback, and 49 already after the Pike had merely turned its eyes towards the stickleback. It would certainly be interesting to know which the visual stimulus situation is to which the Pike's avoidance becomes conditioned. From some preliminary tests it seems that the main characteristic is the way a stickleback moves; which of course is quite characteristic.

An ecological interpretation of our findings meets with some unexpected difficulties. We saw already that a starved Pike, instead of learning to avoid sticklebacks, learns to eat them. We have no indications however that such a Pike would actually prefer sticklebacks to non-spined fish, and in fact this seems unlikely.

A more serious problem is whether Pike and Perch of larger size would avoid sticklebacks. Our tests, all done with predators less than 25 cm long, naturally cannot answer this. However Frost (1954) has made a thorough study of the stomach contents of Pike of various size caught in Windermere, and her results seem in some respects contradictory to ours. Frost reports that Minnows are much more abundant in the lake than *Gasterosteus*; that Pike up to about 40 cm eat more Minnows than *Gasterosteus*; and that larger Pike eat more *Gasterosteus* than Minnows. The percentage of Minnows declines with increasing size of the pike; this may be explained by assuming that the optimum size of prey grows with increasing length of the Pike. The percentage of *Gasterosteus* increases with increasing size of the Pike up to the size-class 50—59 cm; this would mean that as the Pike grows the spines bother it less. After that there is a drop, which again may have to do with the shift in optimum prey-size. The puzzling feature is however that the percentage of *Gasterosteus* actually mounts above that of the Minnows.

The only conclusion we can draw is that in Windermere *Gasterosteus* is on the average less well protected from Pike than the Minnow. In this connection it may be worth pointing out that, although we used Minnows as one of four non-spined fish, our conclusion is not that sticklebacks are generally better protected than Minnows, but that they are better protected than de-spined sticklebacks. The Minnow may have anti-predator adaptations which work even better than spines, at least against large Pike.

In this context it may also be worth speculating whether the conditions under which *Gasterosteus* live in Windermere are typical of the species. Since Frost reports that "sticklebacks, although at no time of great importance as food, are eaten mainly in May and June, at which time they congregate for breeding" (p. 351), her findings may mean that the stickleback's habit of migrating to shallow waters to breed has evolved as an adaptation which enables it to get away from large predators. Perhaps the Windermere population has too few shallow streams at its disposal and thus is subject to more severe predation in spring than other populations.

Whatever the solution may be, we believe that we are entitled to conclude that 1) sticklebacks derive a certain protection from their spines, and 2) *Gasterosteus* is much better protected than *Pygosteus*. The second point is worth elaborating. Morris (1954) has shown that there are considerable differences in behaviour between the two species, and we suggest that our present study enables us to relate some of these differences with those in the effectiveness of the spines as anti-predator devices. *Pygosteus* is much more timid than *Gasterosteus*, and it is partly due to *Gasterosteus'* boldness that it is so well suited for laboratory studies of its behaviour. Any *Gasterosteus* brought into the laboratory in spring can be observed without special precautions within a day, even in tanks with very little cover. *Pygosteus* requires much cover, and even so remains shy for weeks on end, and often retires to the least accessible corners for nest building, so that mirrors, observation screens, etc., have often to be used in order to see its complete behaviour. It would seem as if *Gasterosteus* can afford to be "tame" (and thus waste less time by fleeing) because of the protection offered by its spines.

Another difference between the two is in the type of nesting habitat they select. *Gasterosteus*, while preferring some vegetation, chooses a relatively open type of habitat; and often a few vertical weeds (offered in or outside the tank), or even crude visual imitation of weeds will induce it to settle and build. *Pygosteus* builds in dense weeds, and will only accept an open habitat after long habituation to the observer.

The difference in behaviour of the females is equally striking. Morris (*loc. cit.*) has shown that female *Pygosteus* do not school in spring, but settle on territories of their own round those of the males, where they stay in cover most of the time. None of us has ever seen large schools of female *Pygosteus* roaming about in the spawning season, whereas the females of *Gasterosteus* are always easily seen, travelling about in schools up to 59 individuals strong, out of which the males pick those that respond to their courtship.

Finally, it seems not impossible that the difference in the nuptial colors of the males has something to do with the different need for procryptic coloration. Both species are usually camouflaged, showing general color resemblance with the background, countershading, and disruptive crossbarring on the back. In the breeding season, *Gasterosteus* males turn a vivid red underneath; *Pygosteus* males become jet black. There seems little doubt that the red of *Gasterosteus* makes it much more conspicuous than the black *Pygosteus* males ever are. Moreover, during the actual sexual phase of the cycle, the back of *Gasterosteus* becomes a bluish white, which reverses the usual countershading, and makes the fish very conspicuous indeed, both when seen from above and when seen from the side. *Pygosteus* males also reverse their countershading, but the effect is far less spectacular.

All this leads us to conclude that *Pygosteus* has so to speak compensated for its inferiority as regards spine effectiveness by means of other anti-predator devices such as better procryptic colouration, greater timidity, and a stronger tendency to keep in cover.

A number of authors have argued that *Gasterosteus* must have evolved from a *Pygosteus*-like ancestor. Bertin (1925) gives anatomical evidence in support of this. A comparative study of nest building behaviour tends to support this conclusion, for *Gasterosteus* is the only species of stickleback that does not build its nest suspended in water weeds. In view of the evidence given in this paper it seems reasonable therefore to consider *Gasterosteus* as a species which has specialized in reduction of number and increase of size of spines, which has enabled it to reduce the time to be spent in fleeing and hiding in cover. This has resulted in various peculiarities of its reproductive behaviour, such as the selection of more open habitats, and the schooling and wandering behaviour of the females.

We believe that this is an example of the way in which diverse and seemingly independent systems such as anti-predator defence (involving morphological and physiological-cum-ethological components) and the reproductive system are in fact interrelated, and do not evolve quite independently. The two species, derived from common stock, have specialized in different types of anti-predator defence, and this has had repercussions in a wide field of other characteristics.

REFERENCES

Bertin, L. (1925), Recherches bionomiques, biomètriques et systèmatiques sur les Èpinoches (Gastèrostèidès), *Ann. Inst. Ocean. Monaco*, **2**, 1–204.

Breder, C. M. (1926), The locomotion of fishes, *Zoologica* (N.Y.), **4**, 159–297.

Deelder, C. L. (1951), A contribution to the knowledge of the stunted growth of Perch (*Perca fluviatilis*) in Holland, *Hydrobiol.*, **3**, 357–378.

Frisch, K. v. (1941), Uber einen Schreckstoff der Fischhaut und seine biologische Bedeutung, *Z. Vergl. Physiol.*, **29**, 46–145.

Frost, W. E. (1954), The food of the Pike, *Esox lucius* L., in Windermere, *J. Anim. Ecol.*, **23**, 339–360.

Hartley, P. H. T. (1947), The natural history of some British fresh-water fishes, *Proc. Zool. Soc.*, **117**, 129–206.

Höller, P. (1935), Funktionelle Analyse des Hechtäschadels, *Morphol. Jahrb.*, **76**, 279–320.

Hoogland, R. D. (1951), On the fixing-mechanism in the spines of *Gasterosteus aculeatus* L., *Konink. Nederl. Akad. van Wetensch. Proc.*, series C, **54**, 171–180.

Morris, D. (1954), The reproductive behaviour of the Ten-spined Stickleback (*Pygosteus pungitius* L.), Doctor Thesis, Oxford University; also *Behaviour*, Supplement 6, in press.

Wunder, W. (1927), Sinnesphysiologische Untersuchungen über die Nahrungsaufnahme verschiedener Knochenfischarten, *Z. Vergl. Physiol.*, **6**, 67–98.

Timing and Spacing of Broods in the Black-headed Gull Larus ridibundus

I. J. PATTERSON

The nests of the Black-headed Gull *Larus ridibundus* are closely aggregated into dense colonies and their use synchronized, these two phenomena together tending to produce a maximal clumping effect. Within such a colony however, nests were found to be spaced out to produce a non-random uniform distribution. The commonest distance between neighbouring nests was found to be about one meter, in contrast to related species. This study was concerned with two aspects of this distribution pattern: its survival value and its behavioural causation.

It was found that pairs nesting just outside the colony had a much lower breeding success than those nesting in the colony and that nests on the colony fringe had a slightly lower success than those in the center. Pairs laying during the peak laying period had a higher breeding success than pairs laying either earlier or later in the season. Since by far the most important mortality agent was predation, it seems likely that both clustering and synchronisation of nesting function as anti-predator systems and arguments in favour of this are discussed.

Variations in nest-spacing within the colony were not correlated with variations in breeding success.

In the causation of the spacing between nests, territorial aggression was demonstrated to be an effective dispersion mechanism and the way in which this mechanism works was investigated in detail.

This spacing mechanism was not sufficient by itself to explain the observed densities, which were higher than one would expect from the aggression alone; there was also some tendency for birds establishing a new nest-site to cluster close to others. The interaction between this, the territorial aggression of the residents and the subsequent avoidance responses of the settling birds, can explain the nest spacing pattern and probably also the observed densities.

Reprinted in part from *Ibis*, volume 107, pages 433–459 (1965). Original numbering of figures and tables is retained.

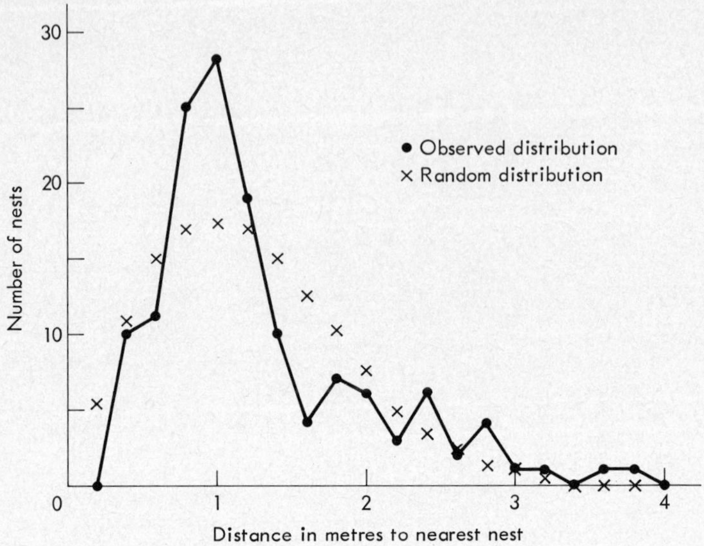

Fig. 1. Distribution of inter-nest distances in the Black-headed Gull.

INTRODUCTION

Many characteristics of a species are taken for granted, sometimes because they have been interpreted intuitively, but perhaps more often merely because we are used to them. Among such characteristics are the species-specific patterns of nest distribution in time and space. Yet these patterns are of great importance in the behaviour and ecology of a species and usually pose considerable problems of function and causation. I have tried to investigate some of these problems in the colonial nesting Black-headed Gull *Larus ridibundus*, whose nests are closely clustered in a relatively small part of the total area occupied by the birds in the breeding season. This distribution contrasts with that of most small passerines, whose nests tend to be more evenly distributed over the area occupied by the population. In addition to this spatial clustering, there is also clustering in time, there being considerable synchronisation of broods (see Fig. 3) in Black-headed Gull colonies.

However, clustering is not maximal. Within the gull colony, nests are not crowded side by side but are spaced apart to some extent. For a sample of nests the distribution of the center to center distance between each nest and its nearest neighbour is shown in Fig. 1 along with the distribution which would be expected if the nests were distributed entirely randomly with respect to one another. (The method of obtaining this random curve is given in Appendix 3.) A comparison of the two curves shows that the nests in the sample were distributed more uniformly than random. Figure 1 also shows that the commonest distance between nests was about one meter. This degree of spacing, though of course variable within the species,

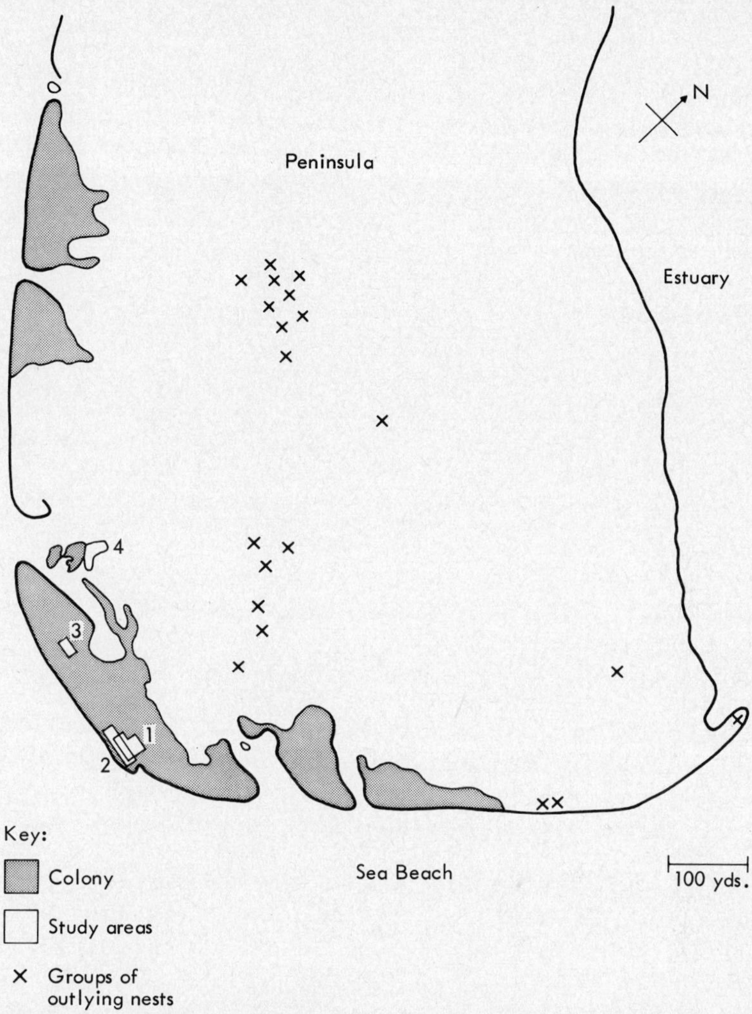

Fig. 2. Map of the Black-headed Gull colony at Ravenglass and the study areas. Area 1 was studied in 1962; areas 2, 3, and 4 in 1963.

seems to differ between colonial species. Thus Sandwich Terns *Sterna sandvicensis* tend to nest closer together than Black-headed Gulls, and Herring Gull *Larus argentatus* much further apart.

There thus seems to be a number of points of interest in the distribution of broods in the Black-headed Gull. Nests are clustered and synchronised and yet spaced apart in a non-random fashion within the clusters. The degree of spacing differs quantitatively between the Black-headed Gull and other species with a similar general pattern. Of the various questions provoked by these features, I have been primarily interested in two groups:

a) The functional questions. What is the survival value of each feature of the distribution?
b) The causal questions. What behaviour patterns lead to this distribution pattern? I have not attempted to answer the question of the problem of the causation of synchrony.

In the breeding seasons of 1962, 1963 and 1964 I worked at the large colony (8,000 pairs) at Ravenglass (Cumberland) in the north of England, where a group of studies is being carried out on the Black-headed Gull under the direction of Dr. N. Tinbergen, F.R.S. The experience of previous members of this group has of course been of the greatest help in the planning and execution of the present study.

At Ravenglass the gulls occupy a small area at the southeast tip of a sand-dune peninsula (Fig. 2). Most of the gulls now breed along the seaward edge of the dunes, although Marchant (1952) showed that the colony previously occupied a much wider area. Roughly half of the birds nest in Marram Grass *Ammophila arenaria* which grows on the exposed sea-facing slopes, and the remainder in Nettle *Urtica dioica* which grows in the more sheltered areas. In the areas deserted by the gulls, Nettles quickly die out.

PROBLEMS OF FUNCTION

There seem to be two possible functions of the nest distribution pattern: it might aid survival of the adults or it might ensure their maximum breeding success, or be a compromise between these two. The present study was concerned only with testing the influence of timing and spacing on breeding success.

Since it was impossible to compare colonies which had differing distribution patterns and very difficult experimentally to alter the pattern, my general approach was to exploit the naturally occurring variations in nest distribution within the Ravenglass colony. I have made the assumption that if the existing nest distribution pattern has survival value, then any birds which deviate from it should breed less successfully than the remainder.

The method used was to follow the case-histories of some 800 marked nests over the breeding seasons of 1962 and 1963. Groups of these, differing in their spacing or timing, were compared for breeding success, measured as the number of young reared per 100 eggs laid. (The details of the method used in measuring breeding success are described in Appendix 1.)

Colony-nesting

A minority of Black-headed Gull pairs each season nested outside the main closely clustered colony, scattered solitarily and in small groups (Fig. 2). Most of these nests were over 100 yards from the nearest colony, so that the birds concerned were deviating considerably from the distribution pattern of the majority. Since they were probably not a random sample of the population it was necessary to show that they did not differ from the main population in ways which might affect breeding success. It seemed possible that such individuals might either be less aggressive birds unable to secure a place in the colony or else young birds. Both these differences could produce less successful breeders, since less aggressive birds might well be more susceptible to predation and young birds of many species are known to breed less successfully than adults.

Outlying birds did show differences in behaviour towards a person walking among their nests. They flew up at a greater distance, stayed further away and attacked less. However this was also seen in the less dense parts of the main colonies and also in normally dense parts on days when few birds were in the colony; so such behaviour may be an effect of isolation rather than a characteristic of the birds. A rough check of the proportion of immature birds showed no noticeable difference between outlying and colony birds.

Most of these scattered nests occur on parts of the peninsula previously occupied by dense colonies (map by Marchant 1952; U. Weidmann, pers. comm.). It thus seems likely that these birds are ones which previously bred in these areas and as such, are comparable with those breeding in the present main colony.

Previous workers at Ravenglass (N. Tinbergen; H. Kruuk, pers. comm.) have noticed that these outlying pairs did not rear any young, at least in the two breeding seasons prior to this study. I checked this in 1962 and 1963 by comparing 125 outlying nests with 590 nests from the four separate study areas within the main colony (Fig. 2). The method used to assess breeding success was the same for both groups, except that the amount of disturbance was in general much less for the outlying nests since there were fewer of them and checking took a shorter time. The results are shown in Table 1. Outlying nests failed to rear any young; in fact few of the eggs survived more than 7–10 days and none hatched.

TABLE 1. Breeding success of outlying Black-headed Gull nests compared with that of colony nests.

	Number of eggs laid	Number of young fledged
Colony nests	1703	157
Outlying nests	358	0

Note: $x^2 = 34.4; p < 0.001$.

In view of this result it seemed interesting to compare the different study areas in 1963. Area 4 was on the edge of the colony nearest to the outlying nests and might be expected to have a lower breeding success than the more central nests in area 3, while area 2 was near the opposite (seaward) edge (Fig. 2). The results are shown

TABLE 2. Black-headed Gull breeding success at the different study areas in 1963

Study area	Number of eggs laid	Number of young fledged
2	197	18 (9.2%)
3	255	33 (13.0%)
4	358	31 (8.7%)

Note: x^2 (2–3) = 0.13, x^2 (3–4) = 2.47, both non-significant.

in Table 2. The breeding success of both area 2 and area 4 was lower than that of area 3, although these differences were not statistically significant.

Within area 4, egg losses were earliest and most frequent at the edge furthest from the main colony, decreasing towards the centre. Young were reared only from nests in the part of the study area closest to the main colony. The same phenomenon was noticed in other marginal parts of the colony, though it seemed to affect only inland margins, not areas on the seaward edges.

Synchronisation of broods

In spite of the marked synchrony in egg laying there also was considerable variation. In particular many birds were still laying late in the season. Thus it was again possible to test survival value by using naturally occurring variation to see whether variations in laying date were correlated with variations in breeding success. It had been noticed in previous seasons that late birds were less successful than the peak ones (N. Tinbergen, pers. comm.), supporting the idea that those pairs laying on the commonest date were the most successful.

In the main study areas nests were grouped by laying date and these date groups compared for breeding success. A 5-day grouping was used in 1962 and a 3-day one with the larger data of 1963. Laying was more protracted in 1963, allowing a better comparison of early with peak nests. Very late nests were studied in 1962 only.

The strict comparability of the different groups had first to be established. In the Kittiwake *Rissa tridactyla* Coulson and White (1958) have shown that young females have a lower breeding success and breed later in the season than older females. However, this factor was unlikely to be important in the present study since birds breeding for the first time made up only about 6% of the population in 1964 (Appendix 4). Another possibility was that many of the later nests were second clutches of birds which had lost their first clutch. Such birds might be expected to be less successful. The interval between losing a clutch and relaying is about ten days (Weidmann 1956). In three out of my four study areas there was a small secondary peak of laying in mid-May which was preceded by a peak of egg losses 9–12 days earlier. Since few nests had been lost up to the end of April in both years, few of the nests

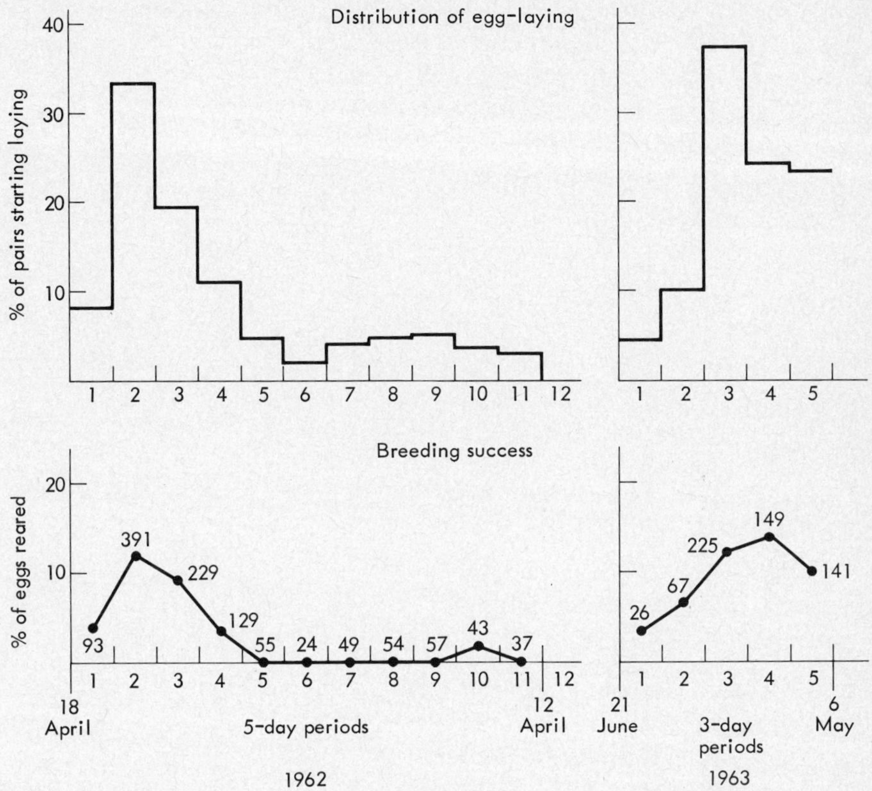

Fig. 3. Changes in breeding success of Black-headed Gulls with laying date, compared with the distribution of laying. The number of eggs on which each percentage is calculated is written beside the point. The histograms are based on 406 nests (1962) and 208 nests (1963).

before about 10 May were likely to be repeat clutches. Therefore only the first five date-groups in both years were used in the subsequent statistical analyses, to reduce the possibility of this source of bias.

The results are shown in Fig. 3. In both years there was a strong correlation between the percentage of birds breeding in a given time interval and the subsequent breeding success of that date group. (Kendal Rank Correlation Coefficient $\tau = +0.95$, $p = 0.010$, in 1962 and $\tau = +0.80$, $p = 0.025$, in 1963.)

It is necessary, before accepting these results to allow for any variations in density through the season, which might partially or wholly explain the differences in breeding success. This can be measured using a Kendal Partial Rank Correlation analysis together with data on the correlation between breeding success and density (see next section) and the correlation between laying date and density. (This latter was $\tau = 0$ in 1962 and $\tau = 0.80$, $p = 0.025$ in 1963.) This analysis reduces the correlation between synchronisation and breeding success slightly (+0.80 to +0.63) in 1963 but leaves the 1962 results unchanged. Thus variations in density between date-groups had only a relatively slight effect, leaving a high correlation between breeding success and the degree of nest synchronisation.

Spacing of nests

Although the spacing between nests in the colony was more uniform than random there was a considerable variation in density, which could be made use of to find the relationship between density and breeding success. Those pairs breeding at the commonest density would be expected to be successful and both higher- and lower-density groups to be less successful.

Each nest in the main study areas was given a density rating, the nests were grouped by density and then the groups compared for breeding success. The density rating of a nest was the number of contemporary nests within a radius of two meters. The results are shown in Table 3. In 1962 there was no correlation between density and breeding success, yet in 1963 there seemed to be a positive correlation between density and success. However if the correlation between density and laying date is allowed for, using a partial correlation analysis as before, the correlation between density and success is reduced to $\tau = +0.83$. The apparent positive correlation between density and breeding success in 1963 was thus entirely due to the correlation in that year between laying date and density. Thus in both years there was no correlation between nest-density and breeding success.

TABLE 3. Relation between Black-headed Gull nest-density and breeding success

	Density; number of nests within 2 meters								
	0	1	2	3	4	5	6	7	8
1962									
Number of eggs laid	51	110	94	98	152	110	101	75	102
% of eggs fledging	8	9	8	11	6	7	10	5	11
1963									
Number of eggs laid	63	85	89	60	73	54	34	23	0
% of eggs fledging	6	2	11	18	14	13	26	22	—

Note: Kendal Rank Correlation Coefficient, $\tau = -0.056$ in 1962 and $= +0.643$ in 1963.

DISCUSSION

Certain conclusions can be drawn from these results giving some information on the survival value of the nest distribution pattern.

The minority of pairs, those which nested outside the main colony group, were strikingly less successful than the birds inside the colony. In addition, within the colony itself, birds nesting along the edges were slightly less successful than those in the center. The results on nest-timing show a similar picture; the birds which laid during the peak laying period were more successful than both earlier and especially later birds. Thus in both cases, the results show that there is a selection pressure against birds deviating from the normal pattern of distribution and hence that selection tends to maintain that pattern. Although this strongly suggests that the present clumped, synchronized pattern is the most successful, the method used cannot strictly show that an entirely different distribution, perhaps dispersed and non-synchronous, would necessarily be less successful. However, a consideration of some of the mechanisms involved gives at least a speculative answer to this problem.

The data on mortality agents (Appendix 2) suggest that predation was by far the most important cause of chick and egg losses. It thus seems probable that both colonial nesting and synchronisation of egg-laying have an anti-predator function. The mechanisms through which these two phenomena could reduce predation on the population have been discussed by Kruuk (1964). He suggests that since the total number of predators is probably limited by the amount of food available during the rest of the year and by intra-specific aggression, concentration of the Black-headed Gull nests into the shortest possible time will reduce the total losses over the season; and that clustering of nests closely together allows a number of behavioural defence mechanisms, such as the flight response to alarm calls and the mass attacks on a predator by numbers of gulls together. Kruuk has shown experimentally that such attacks greatly reduce predation by Herring Gulls and crows *Corvus* spp. In addition to these mechanisms, Cullen (1960) has suggested that close clustering may reduce the chance of a predator finding the group since the area occupied by it will be reduced. This mechanism however may be to some extent counteracted by the tendency of some predators to return to where they have already found one nest and this will endanger others if nests are close together (Kruuk 1964).

These mechanisms certainly suggest that it would be most advantageous for the Black-headed Gull to nest in its present clumped and synchronous pattern. The advantages of a dispersed pattern do not seem as great. Any advantage gained in the wide spacing of the rather cryptic nests would tend to be offset by the conspicuousness of the adults. However, since other types of distribution did not exist and so could not be tested, the general conclusion on the survival value of the present pattern must be regarded as tentative.

In other species also having a fairly vulnerable brood, synchronisation of egg-laying probably functions to reduce total predation, as Darling (1938) for the Herring Gull and Ashmole (1961) for the Wideawake *Sterna fuscata* have already concluded. However the synchrony found in yet other species has undoubtedly quite different functions. Thus, in a number of passerine species such as the Great Tit *Parus major* (Perrins 1963), the peak of laying varies from season to season and seems related to the variations in date of the peak of the nestling food supply. It is possible that a similar function also exists to some extent in the Black-headed Gull although an important objection is that at Ravenglass the breeding season hardly varies, the date on which the first egg was found having only varied between 12 and 15 April over the last eight years (N. Tinbergen, pers. comm.), while the Great Tits showed rather large seasonal variation. There is, of course, the possibility that the period of maximum abundance of earthworms (mainly *Lumbricus* spp.), which form the main food of the Black-headed Gull at Ravenglass does not vary in timing between seasons, a question which might well repay further study.

Colony-nesting appears to be rather widespread among small- to medium-sized bird-species, such as most of the gulls and terns. Their color pattern may very well have considerable survival value; for example white underparts make sea-birds less visible to fish (Phillips 1962), although it will usually make the birds more conspicuous on the nest. It would be interesting to know whether, as in the Black-headed Gull, colony-nesting in these other species has an important anti-predator function.

The survival value of the spacing of nests within the colony differed rather markedly from the results expected. Tinbergen (1956) has stressed the importance of dispersion in cryptic prey so as to minimize the formation of search images in their predators. Since the eggs of the Black-headed Gull are rather cryptic it would seem advantageous to have them spaced apart to some extent. The lower limit of density would however be influenced by the need for clustering already dealt with, and the upper density limit influenced by the possibility of increasing intra-specific predation since the Black-headed Gull is known to take eggs and small chicks of its own species. Since most nests were found to be about one meter from their nearest neighbour, it was expected that pairs breeding at this density would be most successful. Since this obviously was not so, there being no change in breeding success over the fairly wide range of density studied and hence no optimum density, some explanation is required.

It must first be emphasized that the study was confined to the range of densities occurring naturally at Ravenglass. It is possible that only more extreme densities would be penalized and that a wide range of intermediate densities are all equally successful. This however cannot explain the marked peak of inter-nest distances at one meter—a much wider spread would be expected.

A more important point is the presence of a number of unusual features at Ravenglass. First, the habitat was rather atypical; Gribble (1962) lists only five colonies on coastal dunes and shingle in England and Wales, while 166 were in marshes, pools, reservoirs, etc. The accessibility of the dune habitat allowed an exceptionally high mortality by Foxes *Vulpes vulpes* (cf. Ytreberg 1956) so that breeding success was far below the level required to maintain the population (which is in fact declining—Appendix 6). As Kruuk (1964) suggests, the Black-headed Gull seems well adapted to protect its brood against aerial predators but may chiefly depend on an inaccessible habitat as a defence against mammal predators. It thus seems quite possible that the spacing of nests might be adaptive mainly against various bird predators, normally the most important, and that the very high Fox predation in the present study has obscured the results.

Considering these objections, it seems desirable to repeat the observations in a more typical marsh colony which has a lower mortality, and that due mainly to bird predators, before drawing conclusions on the function of nest spacing.

* * *

ACKNOWLEDGMENTS

I would like to thank Dr. N. Tinbergen F.R.S., who supervised this work for his help and advice throughout the study and Sir William Pennington Ramsden Bart and the Cumberland County Council for their permission to work at Ravenglass. Thanks are also due to Dr. J. M. Cullen and Dr. M. Bulmer for advice on statistics and to Dr. J. D. Delius, Mr. C. J. Henty and Dr. H. Kruuk for many helpful suggestions. The work was supported throughout by a Research Studentship from the Nature Conservancy, who also issued permits to catch gulls using stupefying baits. I.C.I. Ltd. kindly supplied dyestuffs, and Bayer Products division of the Winthrop Group supplied Avertin.

REFERENCES

Ashmole, N. P. (1961), The biology of certain terns: with special reference to the Black Noddy (*Anoüs tenuirostris*) and the Wideawake (*Sterna fuscata*) on Ascension Island, D. Phil. Thesis, Oxford University.

Coulson, J. C. and White, E. (1958), The effect of age on the breeding biology of the Kittiwake *Rissa tridactyla*, *Ibis*, **100**, 40–51.

Cullen, J. M. (1960), Some adaptations in the nesting behaviour of terns, *Proc. 7th Int. Orn. Congr.*, Helsinki, 1958: 153–157.

Darling, F. F. (1938), *Bird Flocks and the Breeding Cycle.* Cambridge: University Press.

Gribble, F. C. (1962), Census of Black-headed Gull colonies in England and Wales, 1958, *Bird Study*, **9**, 56–71.

Hensley, M. M. and Cope, J. B. (1951), Further data on removal and repopulation of the breeding birds in a spruce-fir forest community, *Auk*, **68**, 483–493.

Hinde, R. A. (1956), The biological significance of the territories of birds, *Ibis*, **98**, 340–369.

Huxley, J. S. (1934), A natural experiment on the territorial instinct, *Brit. Birds*, **27**, 270–277.

Kalleberg, H. (1958), Observations in a stream tank of territoriality and competition in juvenile Salmon and Trout (*Salmo salar* L. and *S. trutta* L.) *Rept. Inst. Freshw. Res. Drottningholm*, **35**, 113–117.

Kluyver, H. N. and Tinbergen, N. (1953), Territory and the regulation of density in titmice, *Arch. Neerl. Zool.*, **10**, 265–290.

Kruuk, H. (1964), Predators and anti-predator behaviour of the Black-headed Gull (*Larus ridibundus* L.), *Behaviour Suppl.*, **11**, 1–130.

Lack, D. (1943), The age of some more British birds, *Brit. Birds*, **36**, 214–221.

Lack, D. (1954), *The Natural Regulation of Animal Numbers.* Oxford: University Press.

Manley, G. H. (1960), The agonistic behaviour of the Black-headed Gull (*Larus ridibundus* L.), D. Phil. Thesis, Oxford University.

Marchant, S. (1952), The status of the Black-headed Gull colony at Ravenglass, *Brit. Birds*, **45**, 22–26.

Moynihan, M. (1955), Some aspects of reproductive behaviour in the Black-headed Gull (*Larus ridibundus* L.) and related species, *Behaviour Suppl.*, **4**, 1–201.

Paludan, K. (1951), Contributions to the breeding biology of *Larus argentatus* and *Larus fuscus*, *Uidensk. Medd. Dansk Naturh. Foren.*, **114**, 1–128.

Paynter, R. A. (1949), Clutch size and the egg and chick mortality of Kent Island Herring Gulls, *Ecol.*, **30**, 146–166.

Perrins, C. M. (1963), Some factors influencing brood size and population of tits, D. Phil. Thesis, Oxford University.

Phillips, G. C. (1962), Survival value of the white coloration of gulls and other sea-birds, D. Phil. Thesis, Oxford University.

Radford, M. C. (1961), British ringing recoveries of the Black-headed Gull, *Bird Study*, **8**, 42–55.

Ridpath, M. G., Thearle, R. J. P., McCowan, D. and Jones, F. J. S. (1961), Experiments on the value of stupefying and lethal substances in the control of harmful birds, *Ann., Appl. Biol.*, **49**, 77–101.

Stewart, R. E. and Aldrich, J. W. (1951), Removal and repopulation of breeding birds in a spruce-fir forest community, *Auk*, **68**, 471–482.

Tinbergen, N. (1956), On the functions of territory in gulls, *Ibis*, **98**, 401–411.

Tinbergen, N. (1957), The functions of territory, *Bird Study*, **4**, 14–27.

Tinbergen, N. (1959), Comparative studies of the behaviour of gulls (*Laridae*); a progress report, *Behaviour*, **15**, 1–70.

Weidmann, U. (1956), Observations and experiments on egg-laying in the Black-headed Gull, *Anim. Beh.*, **4**, 150–161.

Ytreberg, N. J. (1956), Contribution to the breeding biology of the Black-headed Gull (*Larus ridibundus*) in Norway, *Nytt. Mag. Zool.*, **4**, 5–106.

APPENDIX 1. MEASUREMENT OF BREEDING SUCCESS

Several sample areas of the Ravenglass colony were selected for detailed study (Fig. 2) and divided into strips, 10 ft wide, by lengths of green garden twine. Some of the areas were covered by uniform marram grass and some largely by nettles.

Nests were marked with numbered wooden pegs, which caused negligible disturbance. The effect of the markers on predators is not known, but can probably be assumed to be the same for all nests.

Egg mortality. The study areas were visited at 24-hour intervals, checking one 10 ft strip at a time, following a standard route. Each nest was marked when the first egg appeared, and the nest contents recorded on each subsequent visit. To minimize disturbance visits were made as short as possible by using a portable tape recorder in the field. In 1962 it was possible to estimate the effect of these visits on egg losses, using pairs which laid late in the season. At this time, in addition to the daily visit to check nest contents (lasting about one hour) a 2-hour visit was made every third day to estimate chick mortality. Total egg losses on those days showed a 6.4% increase over the others; but there was great variation in daily egg losses and the difference is not reliable. The daily egg-check seemed to disturb the birds very little, many continuing to incubate till I was less than 10 ft away. It thus seems probable that the recording method was not responsible for more than a small proportion of the egg losses.

Each nest was followed until the disappearance or hatching of all the eggs. An egg was recorded as hatched if the chick had fully emerged at the time of the visit. Eggs remaining unhatched beyond the normal maximum incubation period of 27 days (Ytreberg 1956) were examined and classed as rotten or damaged. In this way the laying date, clutch size, number of chicks hatched, and age of non-hatching eggs at disappearance were recorded. The method however fails to provide data on egg losses during the laying period since it was possible for an egg to be laid and lost during the 24-hour interval between visits and so fail to be recorded. Such losses, unless occurring at a particular time in the laying sequence, would not be made good (Ytreberg 1956, Weidmann 1956), and would lower the clutch size. In the present study, egg losses during laying were estimated by comparing the observed clutch size with that expected with no egg losses. This "true" clutch size is probably approached closely in Ytreberg's data, where egg losses were small. The true number of eggs laid by any group of birds has thus been estimated on the assumption that the mean clutch size was Ytreberg's figure of 2.89 eggs. This does not significantly alter any of the results but lowers the absolute level of breeding success.

Chick mortality. Within a few days of hatching, the chicks left the nest to hide whenever the colony was disturbed. They were then difficult to find and it was impossible to make a direct count of the number surviving at each nest. The method used by Paludan (1951) and Paynter (1949) was extended to overcome this difficulty. Chicks were given a numbered ring on hatching and from the first hatching until the end of the season regular searches for chicks were made in each study area every third day in 1962 and each day in 1963. I also searched a 10 ft wide strip round the boundary of each area. Each chick was thereby recorded on a proportion of the searches until it was found dead or ceased to be recorded after a certain date.

Some chicks were just able to fly at about 24 days, so that it was impossible to tell whether their disappearance after this was due to fledging or death. All chicks therefore recorded alive on or after their 21st day were classed as having been reared to fledging age.

Those chicks which disappeared from the record before reaching 21 days could have been either dead or overlooked. Very few were likely to be alive outside the study area, since in 1963 when 60% of the chick population of the colony was caught for ringing only two such birds were found. It was possible to calculate the likelihood of a chick evading observation. The number of chicks which were certainly alive on a given search day was found by counting the total number of chicks recorded alive on later searches. It was then possible to find, for that day, the proportion found out of

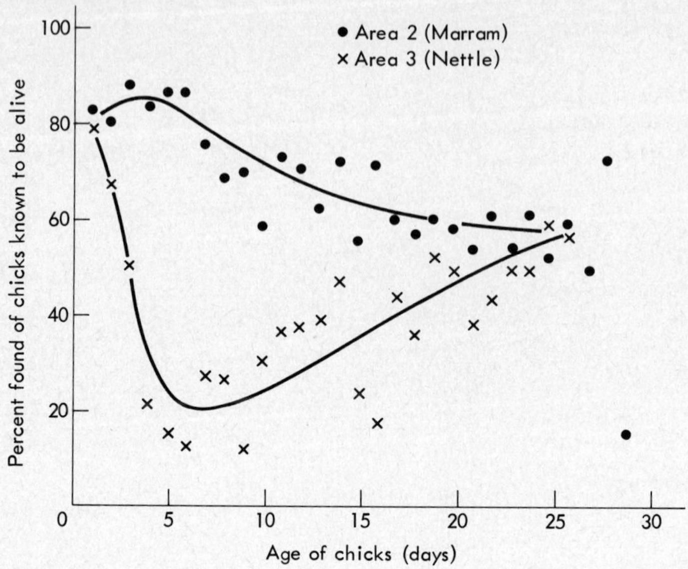

Fig. 8. Relationship between Black-headed Gull chick age and likelihood of being found by the observer.

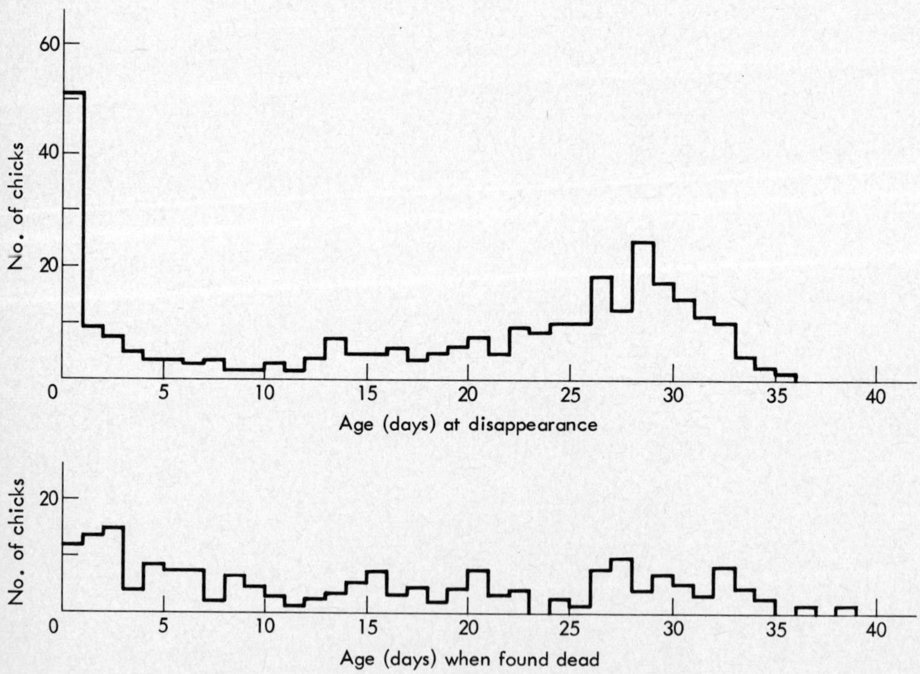

Fig. 9. Age distribution of Black-headed Gull chicks found dead or disappearing from the record.

the total number of chicks known to be present. In Fig. 8 this proportion is plotted against chick age. Search efficiency was high for a few days after hatching when the chicks were still in or near the nest, but then dropped as active hiding began. The drop was least in area 2 which had only sparse marram cover, and greatest in area 3 with thick nettle cover. Areas 1 and 4, with an intermediate vegetation, fell between these two curves. A rise in search efficiency followed, possibly due to the increase in the size of chicks. For the period just prior to fledging, about 50% of all live chicks were found on any one search. Thus the chance of a live chick not being found in four successive searches was less than 5%. Any chick missing for four or more subsequent days prior to 21 days of age was classed as dead (with 95% probability). A check was possible by finding the distribution of disappearances of different length in chicks known to be alive. Disappearances of over four days made up only 3.5% of the total. Only a small proportion of chicks, those which disappeared fewer than four search-days prior to fledging age, could not be classified.

* * *

Figure 9 shows the age distribution of chicks disappearing or found dead. There is a striking peak of losses in the first few days, the period when chicks were most easily found. The other peak at 28–29 days of age probably reflects the peak of fledging.

The data on chick survival (the number of chicks probably reared by each nest) could then be grouped by density, laying date, etc., and the breeding success of the various groups determined.

APPENDIX 2. MORTALITY AGENTS

Egg mortality factors. Very little quantitative evidence could be obtained on the causes of egg loss, eggs just disappearing between visits, leaving either no trace or only a few pieces of shell in the nest; almost all losses were however very probably due to predators. These could be identified from tracks left at a few nests by direct observation. The predators were Fox *Vulpes vulpes*, Hedgehog *Erinaceus europaceus*, Carrion Crow *Corvus corone*, Herring Gull *Larus argentatus*, Lesser Black-backed Gull *Larus fuscus* and other Black-headed Gulls. Many of the losses recorded as egg mortality would actually be losses of newly hatched chicks, since the gull predators were frequently seen to take chicks while still wet. Usually such chicks would not be recorded as having hatched. The data clearly showed a much higher rate of egg loss during the laying and hatching periods than during incubation.

Chick mortality factors. As with egg losses, most chicks classed as dead just disappeared. The remainder were found dead and, except for about one third not found until some time after death, were examined for external injuries, subcutaneous bruising, wounds, fractures and hemorrhage.

Some, mainly small chicks less than five days old, showed no injuries and the cause of death is unknown. Some had severe injuries, for example crushed skulls and broken necks often associated with tears in the pectoral muscles. Fox tracks were found beside many such dead chicks. Another group typically had the tail and the skin of the rump missing, with the oil glands and sometimes the pelvic muscles removed, and usually no other injuries. Hedgehog tracks were found beside a few such chicks (which often survived for some time) and a Hedgehog was found in the act of eating a chick in this way. Finally a small group of chicks had superficial wounds mainly on the head and back. These were probably caused by the pecks of adult Black-headed Gulls, since two chicks seen to be severely pecked had very similar injuries. Table 12 shows the proportion of chicks in each of these groups.

TABLE 12. Black-headed Gull chick mortality agents

	1962	1963	Total	%
Number of chicks disappearing	153	93	246	
Number of chicks found dead	143	110	253	
Number examined	77	78	155	
Number killed by Fox	27	72	72	46
Hedgehog	12	13	13	8
Black-headed Gull	15	26	26	17
Unknown agent	23	44	44	28

APPENDIX 3. CALCULATION OF THE RANDOM DISTRIBUTION OF INTER-NEST DISTANCES

Given the same number of nests n, distributed randomly with respect to each other in the same area A, then the number of nests having their nearest neighbour at a distance x is given by the expression $\exp\left[(-\pi n/A)(x-1/2a)^2\right] - \exp\left[(-\pi n/A)(x+1/2a)^2\right]$, where a is the unit of measurement used. Thus in the present study with 0.2 meters as the unit, the number of nests with the nearest nest at 1 meter is given by:

$$\exp\left[(-\pi n/A)(0.9)^2\right] - \exp\left[(-\pi n/A)(1.1)^2\right]$$

This formula was derived by Dr. J. M. Cullen and Dr. M. Bulmer.

* * *

Egg Shell Removal by the Black-headed Gull, Larus ridibundus *L.: A Behaviour Component of Camouflage*

N. TINBERGEN, G. J. BROEKHUYSEN, F. FEEKES,
J. C. W. HOUGHTON, H. KRUUK AND E. SZULC

The Black-headed Gull removes the empty egg shell shortly after the chick has hatched. The present paper describes some experiments on the function of this response and on the stimuli eliciting it. Carrion Crows and Herring Gulls find white eggs more readily than normal gulls' eggs; it is concluded that the natural colors of the eggs afford a certain degree of cryptic protection. When normal eggs are given an egg shell at 15 cm distance their vulnerability is greatly increased; this "betrayal effect" decreases rapidly with increased distance between egg and shell. We therefore conclude that egg shell removal helps to protect the brood from predators.

As reported by C. Beer (1960) the Black-headed Gull removes a surprisingly wide range of objects from the nest. Large scale tests with egg shell dummies in which color, shape, size and distance from the nest were varied showed that objects of all colors are carried but that "khaki" (the normal ground color of the egg) and white are particularly stimulating, while green elicits very few responses. Egg shells elicit more responses than cylindrical rings of the same color, and these are responded to better than "angles." Size can be varied within wide limits; very large rings elicit fear which interferes with removal. Various other indications are mentioned which show that the score as obtained in the mass tests does not accurately reflect the responsiveness of the reaction itself but rather the result of its interaction with other behaviour tendencies. The eliciting effect decreases rapidly with increasing distance.

On the whole, the gulls' response is very well adapted to its main function of selectively removing the empty shell, but the relatively high scores for objects which have very little resemblance to egg shells suggest that it is adapted to the removal of any object which might make the blood more conspicuous.

A pilot test showed that gulls which have incubated black eggs respond better to black egg shell dummies than normal gulls.

The lack of promptness of the response as compared with non-colonial waders (Ringed Plover and Oystercatcher) is adaptive, since it tends to reduce predation by other Black-headed Gulls which are shown to prey selectively on wet chicks. A hitherto unrecognized function of territory is suggested.

In a discussion of the entire anti-predator system of the Black-headed Gull its complexity and its compromise character are stressed: the safety demands of the individual clash with those of the brood; there are conflicts between the several safety devices which each benefit the brood; and there are clashes between the ideal safety measures required by each type of predator.

INTRODUCTION

Many birds dispose in one way or another of the empty egg shell after the chick has hatched. A shell may be built in or trampled down; it may be broken up and eaten; or, more usually, it is picked up, carried away and dropped at some distance from the nest.

Reprinted from *Behaviour*, volume 19, pages 74–118 (1962).

C. and D. Nethersole Thompson (1942), who have given a detailed summary of our knowledge of egg shell disposal in birds, emphasise the inter- and even intraspecific variability of the responses involved. Since, in addition, the actual response is often over in a few seconds, and happens only once or twice for each egg, it is not surprising that our knowledge is still fragmentary. On the whole, the presence or absence of the response and its particular form seems to be typical of species or groups of species; for instance, it seems to be absent or nearly so in Anseres and in Gallinaceous birds; Accipitres often break up and eat the shell; Snipe are said to be "particularly lax" (Nethersole Thompson); Avocets, *Recurvirostra avosetta* L., remove discarded egg shells anywhere in the colony (Makkink, 1936). In the many species which carry the egg shell away, the response, occurring as it does just after hatching, when the young birds need warmth and protection from predators, must be supposed to have considerable survival value.

The Black-headed Gull invariably removes the egg shell in a matter of hours after hatching; it is extremely rare to find an egg shell in the nest once the chicks have dried. We have only a few direct observations on the time lapse between hatching and carrying in undisturbed gulls, but the 10 records we have (1′, 1′, 15′, 55′, 60′, 105′, 109′, 192′, 206′, 225′) suggest that the response is usually not very prompt. The carrying is done by the parent actually engaged at the nest, never (as far as we know) by the non-brooding partner which may be standing on the territory, even when it stands next to the sitting bird. At nest relief either the leaving partner, or, more often, the reliever carries the shell. Often however it is the sitting bird who starts looking at the shell, stretches its neck towards it, takes it in its bill and nibbles it (sometimes breaking off fragments while doing so, which then are swallowed), and finally rises and then either walks or flies away with the shell in its bill. The shell is dropped anywhere between a few inches and a hundred yards from the nest. We have also observed birds which flew off with the shell, made a wide loop in the air, and descended again at the nest with the shell still in their bill which they then either dropped on the nest or carried effectively straight away. There is no special place to which the shell is carried, though there may be a slight tendency to fly against the wind, or over an updraught, or where the carrier is less likely to be harrassed by other gulls; almost always the shell lands well beyond the territory's boundary. On rare occasions a shell may land in a neighbour's nest—where the latter then treats it as one of its own shells, i.e. removes it.

STATEMENT OF THE PROBLEMS

During our studies of the biology and the behaviour of gulls this response gradually began to intrigue us for a variety of reasons. (1) The shell does not differ strikingly from an egg, since it is only the small "lid" at the obtuse end which comes off during hatching; yet it is treated very differently from an egg, and eggs are never carried away. This raised the question of the stimuli by which the gulls recognise the shell. Systematic tests with egg shell dummies could provide the answer. (2) What could be the survival value of the response? (a) Would the sharp edges of the shell be likely to injure the chicks? Nethersole Thompson raises this possibility, adding that poultry breeders know this danger well. (b) Would the shell tend to slip over an unhatched egg, thus trapping the chick in a double shell? (c) Would the shells interfere in some way with brooding? (d) Would the moist organic material left behind in the shell provide a breeding ground for bacteria or moulds? (e) Would egg shells, if left near the nest, perhaps attract the attention of predators and so endanger the brood?

The following facts, obtained earlier by our co-workers seemed to give some clues.

1. C. Beer (1960) found that Black-headed Gulls do not merely carry shells but a great variety of other objects as well if they happen to be found in the nest. It seemed that the best characterization of this class of objects would be: "Any object—perhaps below a certain size—which does not resemble an egg, or a chick, or nest material, or food"; in short: "any strange object." The very wide range of objects responded to suggests that the birds respond to very few sign stimuli; it might be that the response was adapted to deal with a much wider range of objects than just the egg shell.

2. C. Beer (1960), testing the gulls' readiness to show this response at different times of the season, offered standard egg shell dummies (halved ping-pong balls painted egg shell colour outside) to a large number of gulls once every day from the moment nest scrapes were formed (which is up to about three weeks before the laying of the first egg) till well beyond the hatching of the chicks. He found that under these conditions of standard (and near-optimal) stimulation the response could be elicited from at least 20 days before laying till 3 weeks after hatching. In this respect the response behaves rather like typical incubation responses such as sitting and egg retrieving which also develop gradually in the pre-egg period (Beer, 1960). In view of the heavy predation to which eggs are subjected (see below), and of the fact that the eggs are otherwise carefully guarded, this fact suggests that the response is important throughout the incubation period, and not merely during the few days when the chicks hatch.

3. Finally, E. Cullen (1957) found that the Kittiwake, *Rissa tridactyla* L. never carries the egg shell. The shells are just left in the nest until they are accidentally kicked off. It is true that this often happens in the first few days after hatching, but shells occasionally stay in the nest or on the rim for weeks, and at any rate they remain in or on the nest much longer than is the case with the Black-headed Gull. The Sandwich Tern, *Sterna sandvicensis* Lath. does not remove the egg shells either (J. M. Cullen, 1960).

These observations combined suggest that neither the avoidance of injury, nor of parasitic infection, nor of interference with brooding are the main functions of egg shell removal—if this were so, then the Kittiwake as the most nidicolous species of gull would not lack the response. The most likely function seemed to be the maintenance of the camouflage of the brood—neither Kittiwake nor Sandwich Tern can be said to go in for camouflage to the extent of the other gulls and terns.

Thus these observations naturally led to an investigation into the function of the response and to a study of the stimuli eliciting it. In the following we shall deal with the problem of survival value first.

THE SURVIVAL VALUE OF EGG SHELL REMOVAL

The assumption that egg shell removal would serve to maintain the camouflage of the brood presupposes that the brood is protected by camouflage. This basic assumption, usually taken for granted but—as far as we know—never really tested, was investigated in the following way.

First, we collected whatever observations we could about predation in the colony. While these observations are largely qualitative, they show convincingly that predation is severe throughout the season.

Very many eggs disappear in the course of spring. We did not make systematic counts but can give the following qualitative data. In the Ravenglass colony Carrion Crows, *Corvus corone* L., did not account for many egg losses, because, as we could observe time and again, they are easily chased away by the mass attack of the Black-headed Gulls. In fact we never saw a Crow alight in the colony. However, attacks by one, two or three gulls (which often occurred in our tests) did not deter Crows, and we must assume that nests on the fringe of the colony, and the dozens of nests we regularly find outside the colony and which do not survive, often fall victims to the Crows.

,Egg predation within the colony was due to the following predators. Three pairs of Herring Gulls, *Larus argentatus* Pont., which bred in the gullery levied a constant toll of eggs, and later of chicks. Although the Black-headed Gulls attacked them, they could not altogether stop them from snatching eggs and chicks. We observed hundreds of occasions on which non-breeding Herring Gulls and Lesser Black-backed Gulls, *Larus fuscus* L., (many of them immature) passed near or over the colony or over our tests. On only one occasion did we see any of these taking an egg; they usually were totally uninterested in the colony.

Black-headed Gulls prey on each other's eggs to a certain extent. Most of those who visited our experiments did not attack an undamaged egg (the few exceptions are mentioned in our tables) but finished an egg once it had been broken by other predators. Later in the season individual Black-headed Gulls specialized on a diet of newly hatched chicks (see the next section).

Foxes, *Vulpes vulpes* L., which regularly visited the colony, and killed large numbers of adults early in the season and many half or fully grown chicks towards the end, did visit the gullery in the egg season, but we have no direct evidence of the amount of damage done by them in this part of the year. The gulls were greatly disturbed whenever a Fox entered the colony, but they did not attack him as fiercely as they attacked Herring Gulls and Crows. They flew over the Fox in a dense flock, calling the alarm, and made occasional swoops at it.

Stoats, *Mustela erminea* L., and Hedgehogs, *Erinaceus europaeus* L., visited the gullery and probably accounted for some losses. The gulls hovered over them in a low, dense flock but did not quite succeed in deterring them. In spite of several thousands of man-hours spent in hides in the gullery by us and by our colleagues Dr. C. Beer and Dr. G. Manley we never saw either a Fox, a Stoat or a Hedgehog actually taking an egg; they stayed in the dense cover, and though we have been able to read a great deal from their tracks whenever they had moved over bare sand, this method is of no avail in the egg season since the gull's nests are situated in vegetation.

Surprisingly, the Peregrine Falcons, *Falco peregrinus* L., which were often seen near or over the gulleries left the Black-headed Gulls in peace, though they regularly took waders and carrier pigeons. The gulls panicked however when a Peregrine flew past.

In order to eliminate the effect of the gulls' social nest defence we put out eggs, singly and widely scattered (20 yards apart) in two wide valleys outside the gullery proper. These valleys had a close vegetation of grasses, sedges and other plants not exceeding 10 cm in height. Each egg was laid out in a small depression roughly the size of a Black-headed Gull's nest. Two categories of eggs were arranged in alternate pattern, and in successive tests exchanged position. While predators, particularly Carrion Crows, showed remarkable and quick conditioning to the general areas where we presented eggs, there was no indication that the exact spots where eggs had been found were remembered, and in any case such retention, if it would occur, would tend to reduce rather than enhance differential predation. We assumed (erroneously) that it might well be days before the first predator would discover the eggs, and in our first tests we therefore did not keep a continuous watch. We soon discovered however that the first eggs were taken within one or a few hours after laying out the test, and from then on we usually watched the test area from a hide put up in a commanding position, allowing a view of the entire valley. The tests were usually broken off as soon as approximately half the eggs had been taken.

Experiment 1

Not wishing to take eggs of the Black-headed Gulls themselves (the colony is protected) we did our first test with hens' eggs, half of which were painted a matte white, the other half painted roughly like gulls' eggs. To the human eye the latter, though not quite similar to gulls' eggs, were relatively well camouflaged. It soon became clear that we had underrated the eye-sight of the predators, for the "artificially camouflaged" eggs were readily found. In 9 sessions, lasting from 20 minutes to 7½ hours, we saw Carrion Crows and Herring Gulls take the numbers of eggs (out of a total presented of 104) given in Table 1.

We were astonished to see how easily particularly the Carrion Crows found even our "camouflaged" eggs. Each test area was usually visited every few hours by a pair of Crows. They would fly in at a height of about 20 feet, looking down. Their sudden stalling and subsequent alighting near an egg were unmistakable signs that they had seen it; they usually discovered even camouflaged eggs from well over 10 meters distance. Often the Crows would discover every single egg, whether white or camouflaged, in the area over which they happened to fly. They either carried an egg away in their bills without damaging it to eat it elsewhere, or opened it on the spot and ate part of the contents, or (usually after they had first eaten four or five eggs) they carried it away and buried it. In some cases we saw Crows uncover these buried eggs one or more days after they had been cached.

TABLE 1. Numbers of artificially camouflaged hens' eggs ("Artif. cam.") and white hens' eggs ("White") taken and not taken by predators

	Artif. cam.		White	
	taken	not taken	taken	not taken
Carrion Crows	16		18	
Herring Gulls	0		1 (+3)	
Total	16	36	19 (+3)	33 (−3)

N presentations: 2 × 52. Differences between "Artif. cam." and "White" not significant.*

TABLE 2. Numbers of normal Black-headed Gulls' eggs ("Natural") and Black-headed Gulls' eggs painted white ("White") taken and not taken by predators

	Natural		White	
	taken	not taken	taken	not taken
Carrion Crows	8		14	
Herring Gulls	1		19	
Black-headed Gulls	2		7	
Unknown	2		3	
Total	13	55	43	26

N presentations: 68 + 69: Difference between "Natural" and "White" significant at .1% level.

Of the numerous Herring Gulls and Lesser Black-backed Gulls living in the general area, many of which flew over our test area every day, interest in eggs was shown only by three resident pairs of Herring Gulls. Their eyesight was undoubtedly less keen than that of the Crows (as expressed in the scores in Table 1, and particularly Table 2); further they were remarkably timid. For instance in Experiment 1 we were certain on three separate occasions that a Herring Gull had discovered a white egg but did not dare approach it (the "3" in brackets in Table 1 refers to these occasions). Yet as soon as Crows began to search the area, Herring Gulls would appear and attack them. Often Crows and Gulls attacked each other mutually, swooping down on their opponents from the air, for minutes before either of them alighted near an egg. Usually the Gulls succeeded in claiming an egg first. There were also occasions on which the Crows had the area to themselves.

It seemed obvious that our "camouflage" was not effective at all. Now the artificially camouflaged eggs differed from real gulls' eggs in four respects: (1) the ground colour, though to the human eye matching the overall colour of the rather brownish background very well, was slightly different from the ground colour of most of the gulls' eggs; (2) the dark grey dots which we painted on the eggs were more

* The *P*-Values given in the tables were calculated by the X^2-method, except where stated otherwise.

uniform in size and distribution than those on the real gulls' eggs; (3) unlike the natural dots they were all of one hue; and (4) hens' eggs are considerably larger than Black-headed Gulls' eggs, and hence probably more conspicuous.

Experiment 2

Therefore, we next tested real, unchanged Black-headed Gulls' eggs against Black-headed Gulls' eggs painted a matte white. The tests were conducted in the same way as the previous ones. The results, obtained in 12 sessions lasting from 20 minutes to 4 hours, in which 137 eggs were presented, are summarized in Table 2.

Experiment 3

In order to get an impression of the parts played by size and by the nature of our "artificial camouflage," we painted Black-headed Gulls' eggs in the same way as the "camouflaged" hens' eggs, and compared their vulnerability with that of Black-headed Gulls' eggs painted white. The results of this test, to which we devoted only 5 sessions of from 1 to 3 hours, and in which 48 eggs were presented, are given in Table 3.

From these experiments, and particularly from Experiment 2, we conclude that the natural egg color of the Black-headed Gulls' eggs makes them less vulnerable to attack by predators hunting by sight than they would be if they were white; in other words that their colour acts as camouflage. The difference between the results of Experiments 1 and 3 on the one hand and Experiment 2 on the other indicates that we had underrated the eyesight of the predators, and also that their reactions to the different aspects of camouflage deserve a closer study: the parts played by over-all colour, by pattern and hue of the dotting, perhaps even by the texture of the eggs' surface we hope to investigate later—it was a pleasant surprise to discover that large scale experiments are possible. For our present purpose we consider it sufficient to know that painting the eggs white makes them more vulnerable.

TABLE 3. Numbers of artificially camouflaged Black-headed Gulls' eggs ("Artif. cam.") and Black-headed Gulls' eggs painted white ("White") taken and not taken by predators

	Artif. cam.		White	
	taken	not taken	taken	not taken
Carrion Crows	4		5	
Herring Gulls	4		8	
Black-headed Gulls	1		1	
Total	9	15	14	10

N presentations: 24 + 24. Difference between "Artif. cam." and "White" not significant (20% $< p <$ 30%).

Experiment 4

We can now turn to the question whether or not the presence of an egg shell endangers an egg or a chick it is near. For obvious reasons we chose to investigate this for eggs

TABLE 4. Numbers of Black-headed Gulls' "eggs-with-shell" and "eggs-without-shell" taken and not taken by predators; eggs not concealed

	Eggs-with-shell		Eggs-without-shell	
	taken	not taken	taken	not taken
Carrion Crows	6		7	
Herring Gulls	9		7	
Total	15	15	14	16

N presentations: 30 + 30. Difference between egg with shell and egg without shell not significant.

rather than for chicks. The principle of this experiment was the same as that of the previous ones. We laid out, again avoiding site-conditioning in the predators, equal numbers of single Black-headed Gulls' eggs with and without an empty egg shell beside them. The shells used were such from which a chick had actually hatched the year before and which we had dried in the shade and kept in closed tins. The shells were put at about 5 cm from the eggs which were again put in nest-shaped pits. Predators were watched during 5 sessions lasting from 45 minutes to $4\frac{1}{2}$ hours. Sixty eggs were used. The results are given in Table 4.

We did not consider this result conclusive because the circumstances of the experiment differed from the natural situation in two respects. (1) Although a nest in which a chick has recently hatched may contain unhatched eggs, there are chicks equally or rather more often; and chicks, apart from having a less conspicuous shape than eggs, do at a quite early stage show a tendency to crouch at least half concealed in the vegetation when the parent gulls call the alarm. The Crows and Gulls might have less difficulty finding eggs than seeing chicks. (2) Both predators were always vigorously attacked by many Black-headed Gulls whenever they came in or near the gullery. In avoiding these attacks their attention is taken up for the greater part of the time (as judged by their head movements and evading action) by keeping an eye on the attackers; in the natural situation they never have the opportunity to look and search at their leisure. And to predators searching for camouflaged prey leisure means time for random scanning and opportunity for undivided attention—both probably factors enhancing discovery, i.e., fixation of nonconspicuous objects. In other words our experiment had probably made things too easy for the predators, even though in the colony the nests themselves are often visible from a distance.

Experiment 5

We therefore decided to repeat this experiment with slightly concealed eggs. This was done by covering each egg (whether or not accompanied by an egg shell) with two or three straws of dead Marram Grass, a very slight change which nevertheless made the situation far more similar to that offered by crouching chicks. Most tests of this experiment were done without watching from a hide; we knew by now who the main

TABLE 5. Number of Black-headed Gulls' eggs-with-shell and eggs-without-shell taken and not taken by predators; eggs slightly concealed

Eggs-with-shell		Eggs-without-shell	
taken	not taken	taken	not taken
39	21	13	47

N presentations: 60 + 60. Difference between egg with shell and eggs without shell: $p < .1\%$.

predators were, and for our main problem it was not really relevant to know the agent. In 8 tests, lasting from 2 hours 40 minutes to 4 hours 40 minutes, 120 eggs were offered, 60 with a shell at 5 cm distance. The results are given in Table 5.

The conclusion must be that the near presence of an egg shell helps Carrion Crows and Herring Gulls in finding a more or less concealed, camouflaged prey, and that therefore egg shells would endanger the brood if they were not carried away.

Experiment 6

This was a series of pilot tests designed to examine whether the Carrion Crows would be readily conditioned to shells once they had found eggs near shells, and whether, if they would find shells without eggs, they would lose interest in the shells. These tests began on April 21st, 1960, after the Crows operating in the valley had already gained experience with white and camouflaged eggs, and, since they had broken many of those while eating them, could have learned that egg shells meant food. In the very first test, egg shells were laid out in the Western area of the valley, where in previous tests with whole eggs the Crows had never been seen to alight. At their first visit the Crows alighted near these shells, pecked at them, and searched in the neighbourhood. They left after a few minutes whereupon we took the shells away. A few hours later we again laid out egg shells without eggs, this time scattered over the whole valley. When the Crows next returned they flew round over the valley, looking down as usual, but they did not alight. Next morning semi-concealed eggs were laid out over the whole valley, each with an egg shell at the usual distance of 5 cm. This time the Crows did alight and took a number of the eggs. A few hours later we once more laid out shells only, scattered over the valley, and at the next visit the Crows came down near several of them and searched. The morning after this we laid out just shells, this time in the Eastern part of the valley. This time the Crows did not alight. When next, between this and their next visit we gave each shell an egg at 5 cm, the Crows alighted again when they returned and took several eggs. Later that same day just shells were given in the Western part, which could not induce the Crows to alight. Such tests were continued until April 30, and while they did not give sufficient information, they

TABLE 6. Record of "conditioning test" explained in text

Es + = eggs = shells offered, eggs taken; *es* − = eggs = shells offered, eggs not taken; *S* + = shells alone offered, crows alighted and searched; *S* − = shells alone offered, crows did not alight.

Eastern area					*S* −	*Es* +		*Es* +			
Entire valley		*S* −	*Es* +	*S* +							*S* −
Western area	*S* +						*S* −		*S* −	*S* −	
Date (April)	21	21	22	22	23	23	23	24	24	25	25
Eastern area	*Es* +	*Es* +			*Es* −	*Es* −	*Es* +		*Es* +		
Entire valley										*S* +	
Western area			*S* −	*S* −				*S* −			*S* +
Date (April)	25	26	26	27	27	28	28	28	29	29	30

strongly suggest: 1) that one experience with shells-without-eggs was sufficient to keep the Crows from alighting near shells the next time; and b) that renewed presentation of eggs with the shells attracted them again. Further, we had the impression (c) that the Crows later learned that egg shells in a part of the valley where shells only had been presented several times meant no food—in other words that they associated "shells only" with the locality. The full record of these experiments is given in Table 6.

Experiment 7

We next investigated the effect of distance between egg and shell. This was done in a mass test without direct observation from the hide, and in the same valley where we had regularly seen Crows and Herring Gulls take eggs. Half concealed gulls' eggs were laid out; one third of them had a shell at 15 cm distance, one third at 100 cm, and one third at 200 cm. A total of 450 eggs were presented in 15 tests, with the following results (Table 7).

Part of each group of eggs may of course have been found without the aid of the shell (and this may explain why so many eggs were found even of the 200-cm group), but the figures show that the "betrayal effect" is reduced with increased distance.

A second, similar test was taken with the shells at 15 cm and 200 cm; altogether 60 eggs were presented in 3 tests; the results are given in Table 8 and show a similar result.

The following test, to be mentioned only briefly because it did not contribute to our results, was in fact done before any of the tests mentioned so far in an attempt to short-circuit our procedure. We put out 50 hens' eggs painted in what we then hoped would be a good camouflaged pattern (see above), and gave each egg an egg shell dummy at 15 cm distance. These dummies were metal cylinders made by bending a strip of metal sheet measuring 2 × 10 cm as used in our later tests with Black-headed Gulls. Half of these were painted like the eggs, inside and outside (they were in fact

TABLE 7. Number of eggs taken out of the 450 offered with a shell 15, 100, and 200 cm away

15 cm		100 cm		200 cm	
taken	not taken	taken	not taken	taken	not taken
63	87	48	102	32	118

N presentations; 3 × 150. Difference between 15 cm and 200 cm significant at .1% level. Significance of the total result: $p < 1\%$.

TABLE 8. Numbers of eggs-plus-shell taken, not taken for different distances (in cm) between egg and shell; 60 eggs were offered.

15 cm		200 cm	
taken	not taken	taken	not taken
13	17	4	26

N presentations: 2 × 30. Difference between 15 cm and 200 cm significant at 2.5% level.

satisfactorily cryptic to the human eye); the other half were painted white. We assumed that predators would be slow to come, and that in the beginning one check per day would be sufficient. Upon our first check, about 24 hours after we laid out the eggs, we found that all 50 eggs had disappeared. Our first conclusion was that a whole horde of predators, such as a flock of Herring Gulls, had raided the valley, but we later saw that this valley was searched mainly by one pair of Carrion Crows and one, two, or sometimes three pairs of Herring Gulls. During the following weeks we discovered that the Crows kept digging up these eggs which they must have buried on the first day.

This failure forced us to check the effectiveness of our "camouflage" paint first (see Experiments 1–3) and to test the effect of the natural egg shell (Experiments 4–7); by the time these questions had been settled we had to start our tests on the stimuli eliciting egg shell removal in the Black-headed Gulls themselves, and so had to abandon for the moment any tests on the effect of the colour of the egg shell on the predators. However, the method having now been worked out, it is hoped to investigate this more fully.

The conclusion of this part of our study must therefore be that the eggs of the Black-headed Gulls are subject to predation; that in tests outside the colony the number of eggs found by Carrion Crows and Herring Gulls is lower than it would be if the eggs were white; that the proximity of an egg shell endangers the brood; and that this effect decreases with increasing distance. While it will now be worth investigating the predators' responsiveness to eggs and shells in more detail, the facts reported leave little room for doubt about the survival value of egg shell removal as an antipredator device. Whether or not the response has other functions is of course left undecided.

* * *

THE LACK OF PROMPTNESS OF THE RESPONSE

The predator tests reported in an earlier section demonstrated the intense pressure exerted at least by Carrion Crows and Herring Gulls against leaving the egg shell near the nest. Admittedly our observations refer to only one colony (which however contains a sizeable part of the British breeding population), but Carrion Crows are practically omnipresent and are notorious egg robbers. One should therefore expect that the Black-headed Gull would have developed a very prompt response and would remove the shell immediately after the chick has hatched. Yet, as we have seen, this is not usual. We cannot believe that the species has not been able to achieve promptness—egg shell removal is so widespread taxonomically that it must be an old response. The most likely reason is that there is a counteracting selection pressure—that too prompt removal is in some way penalized. At first we thought that the risk of carrying the chick with the shell before it had hatched completely might be responsible, but this risk is the same for gulls and other species. Yet we have observations (admittedly few in number) which show that Oystercatchers and Ringed Plovers carry the shell with far less delay—in spite of the fact that their chicks stay in the nest for a mere couple of hours, and shell removal therefore must be less urgent. In the course of 1959 and 1960 the reason for the delay became gradually clear: in both years there were a number of Black-headed Gulls in the colony which preyed selectively on nearly hatched eggs and on wet chicks. Although we are certain that not all gulls engage in this "cannibalism," this type of predation is very common, particularly towards the end of the season. In fact many of our efforts to observe the development of the behaviour during the first few hours of a chick's life (which were usually done late in the season) were time and again frustrated by the wet chicks being snatched away by such robber gulls immediately after hatching. We have not made systematic notes on this, but twenty is a conservative estimate of the number of occasions on which we actually observed such chicks being taken in front of the hide, while dry chicks only hours older (equally or even more available) were left alone. The number of occasions on which we lost wet chicks without actually seeing it happen is much higher still. On only three occasions did we observe a Black-headed Gull trying to swallow a dry chick. While a wet chick is usually swallowed in a few seconds (often even in flight) dry chicks of less than a day old took approximately ten minutes to swallow. There can be no doubt that chicks are practically safe from predation by neighbours (though not from Herring Gulls) as soon as their plumage becomes dry and fluffy.

It is interesting to see the behaviour of parent gulls sitting on hatching chicks while there are robber gulls about. The parents are aware of the latters' intentions;

they show signs of increased hostility whenever the robber comes near, and they are extremely loath to leave the nest. As soon as the gulls are disturbed by other predators or fly up for human visitors robbers snatch up the wet chicks in a fraction of a second. We had the impression that the robber gulls kept an eye on many nests and knew where chicks were hatching. On one occasion we saw a still wet chick being taken during the few seconds when the parent carried away the shell.

We feel justified therefore to ascribe the lack of promptness of the response to this tendency of some members of the colony to prey on wet chicks.

DISCUSSION

Removal of an egg shell lasts a few seconds. It is normally done three times in a year. Nothing would seem to be more trivial than this response, which at first glance might seem to be no more than fussy "tidying-up" by a "house proud" bird. Yet we have seen that it has considerable survival value, and that the behavioural organisation is complicated, and well adapted to the needs. In addition, our study has given us some insight in some more general problems of an ecological and evolutionary nature. The following discussion owes much to the stimulating studies of E. Cullen (1956) and M. Cullen (1960).

Territory

In 1956, N. Tinbergen listed the alleged functions of territory in gulls as follows. One component, site attachment, assists in pair formation, in homing to the nest site, and in providing known shelter for the chicks. Inter-pair hostility, the other aspect of territoriality, prevents interference with pair formation, and, by forcing breeding pairs and thus nests apart, renders mass slaughter by predators less likely (see also Tinbergen, 1952b). We believe that the facts mentioned in this paper show a hitherto unrecognized function of territorial hostility in the Black-headed Gull: reducing the likelihood of predation by neighbouring gulls. We distinguish this effect from the effect of spacing-out on interspecific predation for the following reason. L. Tinbergen (1960) has shown that some predatory birds can be guided for shorter or longer periods, by a "searching image": they can, through an unknown process, concentrate their attention on one particular type of prey while being less responsive to other types. The stability of this state of narrowed responsiveness seems to be controlled in part by the density of the prey species: a series of successes in a short time seems to strengthen or at least to maintain specialisation, but lack of success tends to widen their responsiveness again. This property of predators, which may well be wide-spread, puts a premium on spacing-out of potential prey animals, because it increases searching time between successes. In this connection it is distance which counts.

We suggest that predation by neighbouring Black-headed Gulls is reduced by another aspect of territoriality. Each gull learns, during the prolonged period of territorial fighting in the early part of the season, not to intrude into the territories of its neighbours. The factor which reduces predation in this case is the existence of barriers, irrespective of distances between the broods (density).

Of course the inhibition of trespassing as a consequence of acquired knowledge of the boundaries does not totally prevent predation, but the fact that gulls do not trespass except very briefly and on rare moments (such as during a general alarm caused by another predator, e.g., Man or Fox which require adult gulls to leave the ground) obviously reduces the amount of intra-species predation considerably.

The compromise character of colony density

As we have seen, the mass attack by Black-headed Gulls discourages at least one predator, the Carrion Crow, from penetrating into the colony. This demonstrates the advantage of colonial nesting. Crows were not deterred by attacks of one, two or even three Black-headed Gulls. On the other hand, spacing-out of breeding pairs within the colony and the establishment of knowledge of boundaries, both achieved by territorial hostility, have also distinct advantages. Thus the density of a Black-headed Gull colony has the character of a compromise between at least these two opposing demands. As E. Cullen (1957) and J. M. Cullen (1960) have shown, a study of the anti-predator devices of other species might help in elucidating the adaptedness of inter-specific differences in colony density.

Synchronisation of the breeding calendar

Some of our data strengthen the conclusion that predator pressure may be an ultimate factor in the synchronisation of breeding. Beer's data (1960) show that the scatter in time of the appearance of fledged young on the beach does not differ strikingly from the scatter of egg laying. Yet there are every season a large number of late broods, partly those of birds arriving later than the main body, partly repeat clutches of birds who have lost their first clutch. Our observations show that at least intra-specific predation of wet chicks is particularly severe towards the end of the season, and it is striking how many of the late broods disappear. It is clear that most of the successful broods come from pairs which arrive early and which have not been forced to relay; failure to synchronise is heavily penalized.

The anti-predator system as a whole

At this stage of our studies it seems worthwhile to review what is now known of the ways in which the Black-headed Gull protects itself against predators. Some of these devices protect the individual—for brevity's sake we will call these "egoistic" devices, even though they may at the same time protect others. Others protect the brood, often at the cost of danger to the individual, and such devices we will call "altruistic." Naturally these terms refer to the function, not to motivation.

The most obvious, and seemingly trivial response is escape. Even this, however, takes different forms, dependent on the nature of the predator and on the age of the bird. A gull can simply fly away, as it does at the approach of a human being. The response to a Peregrine Falcon, which we have observed in detail several times, is different: the gulls fly up, form dense flocks and fly at great speed low over the ground or the water, executing quick zigzag maneuvres. We believe that "erratic flights," in which individual gulls separate themselves from the flock and fly away, often downwards, at great speed and with very quick and sharp turns, are likely to be elicited by a Peregrine approaching gulls that are flying high. Chicks, in response to the alarmed behaviour of the adults in a colony, crouch, at first on the spot, but already after one day and occasionally even earlier they walk a little distance away from the nest and towards the surrounding vegetation. Each chick becomes soon conditioned to one or more individual hiding places; this conditioning allows them to reach safety more quickly than they would if they had to search for suitable cover.

Outside the breeding season Black-headed Gulls select wide open spaces: marshy meadows, open seashores, or water. This no doubt allows them to see an approaching predator in time.

In flocks of adult birds there operates at least one signaling system: The alarm call alerts other individuals.

The "altruistic" habitat selection in the breeding season shows signs of anti-predator adaptedness: In the open sand dunes gulls avoid nesting on the bare sand, even though males may start by taking up a pairing territory there. Once paired however they select a nest site in moderately dense vegetation. Black-headed Gulls breeding on inland lakes usually select islands; either large ones which accommodate large numbers, or smaller islands such as individual *Molinia* bushes which offer space for one nest only. Where Black-headed Gulls nest on tidal saltings, this inclination to select islands can be the undoing of their broods in high spring tides (Tinbergen, 1952a).

Several other behaviour patterns appear in the breeding season which, while endangering or at least not protecting the lives of the parents, do contribute to the safety of the brood. First, the scattering of the broods, which provides a certain degree of protection both from inter- and from intraspecific predation, is effected by the balanced attack-escape system, with its components of actual attack, withdrawal, and agonistic displays (Tinbergen, 1959; Manley, 1960). Further, unlike camouflaged species such as ducks, curlews and several other waders, and pheasants, the incubating gull leaves the camouflaged brood at the first sign of danger. The camouflage of the eggs depends on a specialized pigmentation system in the upper reaches of the oviduct.

Parent gulls attack predators; the fierceness of the attack, and the degree to which it is counteracted by escape tendencies depends on the type of predator, and the resultant seems highly adaptive. From the moment the first egg is laid, at least one parent stays on the territory and guards the brood. As we have seen, egg shell removal is also effective as an anti-predator device. There can finally be little doubt that the chick's plumage protects its bearer by being camouflaged.

Thus the picture that emerges is one of great complexity and beautiful adaptedness. It has further become clear that at least some of the different means of defence are not fully compatible with each other, and that the total system has the character of a compromise between various, in part directly conflicting demands. These conflicts are of different types. First, the safety requirements of the parents may differ from those of the brood. Thus the parent endangers itself by attacking predators. This is suggested by the fact that Foxes succeed in killing large numbers of adults in the colony. Though we have never seen a Fox killing an adult, their tracks in the sand cannot be misinterpreted. Often they kill many more birds than they eat. Some of these birds were "egg-bound" females (Manley, 1958), but in 1960, when we sexed 32 gulls killed by Foxes we found that 21 of these were males. Many of these gulls have their tails torn off and/or their legs broken. We believe that a Fox sometimes kills such birds by jumping at them when they "swoop." All this suggests that a certain balance between the tendency to attack a Fox and the tendency to flee from it is selected for.

The conflict between "egoistic" and "altruistic" behaviour is also very obvious in the time when the winter preference for wide, open spaces changes into the preference for the breeding habitat which, as we have just seen, is dangerous to the adults. The switch towards the breeding habitat selection is not sudden; there is a long period in which the birds show that they are afraid of it; even when, after long hesitation, they settle in the colony, there are frequent "dreads" when the birds suddenly fly off in panic; these dreads gradually subside (see also Tinbergen, 1953 and Cullen, 1956). Towards the end of the breeding season the adults begin to desert the colony in the evening to roost on the beach, leaving the chicks at the mercy of nocturnal predators.

Second, there are conflicts between two "altruistic" modes of defence, each of which has its advantages. Crowding, advantageous because it allows social attacks which are effective against Crows, has to compromise with spacing-out which also benefits the broods.

Finally there may be conflicts between the optimal ways of dealing with different predators. Herring Gulls and Crows might be prevented entirely from taking eggs and chicks if the gulls stayed on the nests, but this would expose them to the Foxes. While Herring Gulls and Crows exert pressure towards quick egg shell removal, neighbouring gulls exert an opposite pressure; the timing of the response is a compromise.

We cannot claim to have done more than demonstrate that egg shell disposal is a component of a larger system, nor are we forgetting that much in our functional interpretation requires further confirmation. It seems likely however that a more detailed study of all the elements of anti-predator systems of this and other species, and of the ways they are functionally interrelated, would throw light on the manifold ways in which natural selection has contributed towards interspecific diversity.

REFERENCES

Baerends, G. P. (1957), The ethological concept "Releasing Mechanism" illustrated by a study of the stimuli eliciting egg-retrieving in the Herring Gull, *Anatom. Rec.*, **128**, 518–519.

Baerends, G. P. (1959). The ethological analysis of incubation behaviour, *Ibis*, **101**, 357–368.

Beer, C. (1960), Incubation and nest-building by the Black-headed Gull, D. Phil. Thesis, Oxford University.

Cullen, E. (1957), Adaptations in the Kittiwake to cliff-nesting, *Ibis*, **99**, 275–302.

Cullen, J. M. (1960), Some adaptations in the nesting behaviour of terns, *Proc. 12th Internat. Ornithol. Congr.*, Helsinki, 1958.

Cullen, J. M. (1956), A study of the behaviour of the Arctic Tern, *Sterna paradisea*, D. Phil. Thesis, Oxford University.

Hinde, R. A. (1954), Factors governing the changes in strength of a partially inborn response, as shown by the mobbing behaviour of the Chaffinch (*Fringilla coelebs*). II. The Waning of the response, *Proc. Royal. Soc.*, B **142**, 331–358.

Magnus, D. (1958), Experimentelle Untersuchungen zur Bionomie und Ethologie des Kaisermantels *Argynnis paphia* L. (Lep. Nymph.). I., *Z. Tierpsychol.*, **15**, 397–426.

Makkink, G. F. (1936), An attempt at an ethogram of the European Avocet (*Recurvirostra avosetta* L.), *Ardea*, **25**, 1–62.

Manley, G. H. (1957), Unconscious Black-headed Gulls, *Bird study*, **4**, 171–172.

Manley, G. H. (1960), The agonistic behaviour of the Black-headed Gull, D. Phil. Thesis, Oxford University.

Nethersole Thompson, C. and D. (1942), Egg-shell disposal by birds, *British Birds*, **35**, 162–169, 190–200, 214–224, 241–250.

Schwartzkopf, J. (1960), Physiologie der höheren Sinne bei Säugern und Vögeln, *J. Ornithol.*, **101**, 61–92.

Siegel, S. (1956), *Nonparametric Statistics for the Behavioral Sciences*. New York.

Tinbergen, L. (1960), The natural control of insects in pine woods. I. Factors influencing the intensity of predation by songbirds, *Arch. Neerl. Zool.*, **13**, 265–336.

Tinbergen, N. (1951), *The Study of Instinct*. Oxford.

Tinbergen, N. (1952a), When instinct fails, *Country Life*, Feb. 15, 412–414.

Tinbergen, N. (1952b), On the significance of territory in the Herring Gull, *Ibis*, **94**, 158–159.

Tinbergen, N. (1953), *The Herring Gull's World*. London.

Tinbergen, N. (1956), On the functions of territory in gulls, *Ibis*, **98**, 408–411.

Tinbergen, N. (1959), Comparative studies of the behaviour of gulls (*Laridae*); a progress report, *Behaviour*, **15**, 1–70.

Villabolos, C. and J. (1947), *Colour Atlas*. Buenos Aires.

Weidmann, U. (1956), Observations and experiments on egg-laying in the Black-headed Gull (*Larus ridibundus* L.), *Brit. Journ. Anim. Behav.*, **4**, 150–162.

Ytreberg, N. J. (1956), Contribution to the breeding biology of the Black-headed Gull (*Larus ridibundus* L.) in Norway, *Nytt Magas. Zool.*, **4**, 5–106.

B. BEHAVIORAL ADAPTATIONS TO CLIFF-NESTING

This series of papers illustrates the "comparative" method of assessing function, with reference to one particular set of behavioral characteristics. The first paper introduces the problem: A species of gull that nests on cliffs differs in many ways from more usual ground-nesting gull species. The behavioral differences appear to be adaptive to the cliff-nesting habit. How can this suspicion be proved. The evidence lies in study of other cliff-nesting species, as illustrated by the subsequent papers, and particularly in intermediate species that would be predicted to show intermediate behavioral characteristics. The last paper reviews some methods of assessing the function of behavior patterns (see also Klopfer and Hailman, 1967: Chapter 9).

Adaptations in the Kittiwake to Cliff-nesting

ESTHER CULLEN

The Kittiwake is probably derived from a ground-nesting gull and the change to cliff-nesting was presumably an anti-predator device. Because predation is less, the species seems to have lost a number of behaviour patterns and morphological features possessed by other gulls. On the other hand it has to acquire a number of adaptations to suit its new life. Signal and non-signal movements have been altered: Some have been modified, like the beak-hiding or the nest-building movements, others have been lost, like several anti-predator devices or the aggressive upright posture. There also seem to be a few new acquisitions like the black neck-band of the young and the collecting of mud.

Adaptations to fit an animal for a new kind of life are of two kinds. Some are inherited, others are acquired as a result of the individual's experience and environment. The chief morphological peculiarities of the Kittiwake, such as the black neck-band in the young, are certainly due to inherited differences. The adult's claws, which help the Kittiwake to hold on to small ledges, are sharper than those of other gulls, and this may be partly because the tips are not abraded by walking as in the other gulls who spend the whole year close to land. On the other hand this difference is also apparent in the newly born young, so that here too it must have a genetic basis.

Reprinted from *Ibis*, volume 99, pages 275–302 (1957).

As I have shown, some of the differences in behaviour also are innate (the presence or absence of head-turning in the young, their readiness to run away when pecked). Others, for instance the use of mud as nest-material, are presumably also innate. Yet others, such as the social collecting of material, may be due to a complicated interaction of acquired and inherited factors. The existence of inherited factors leading to differences in behaviour should not blind one to the possibility that experience may be able to modify a pattern. This was shown in one of my experiments in which a newly-hatched Black-headed Gull learned after a day to take the food from the throat of its Kittiwake foster-parent in the way a young Kittiwake does from the first. Such modifiability is an objection to the existence of innate differences; it only shows that these differences may not be as rigid as has sometimes been supposed.

With adaptations described in this paper, I hope to have shown how this one change to nesting on tiny ledges on steep cliffs has had repercussions in many aspects of the life of the species and has led to morphological changes as well as a great many alterations in behaviour. In many animals adaptive differences between species have been described but I know of no other case where one relatively simple change can be shown to have been responsible for so many alterations.

INTRODUCTION

In 1952 I undertook a study of the breeding behaviour of the Kittiwake *Rissa tridactyla* in order to extend the scope of studies of various gulls, initiated by Dr. N. Tinbergen at Oxford. At that time accounts of the behaviour of this species were fragmentary (Selous 1905, Bent 1921, Perry 1940, Fisher and Lockley 1954), though recently the situation has been altered by a paper by Paludan (1955).

My observations were made while living on one of the Farne Islands off the Northumberland coast, where I was able to watch the birds through three seasons, 1953–1955, and more briefly in 1952 and 1956. Although my observations were all made in these colonies there seems no reason to doubt that the behaviour described is typical for the species.

It is well known that the Kittiwake, together with its close relative *Rissa brevirostris*, differs from most gulls in two ways: Outside the breeding season they live in the open sea rather than in the neighbourhood of land, and when they do visit the land to breed they nest on tiny cliff-ledges. Other gulls occasionally nest on ledges but these are far larger than the little shelves, sometimes only four inches wide, on which a Kittiwake can stick its nest. The present paper sets out some of the peculiarities of the Kittiwake's behaviour compared with the ground-nesting gulls and attempts to show that these peculiarities can be related to the cliff-nesting habit. A more detailed account of other aspects of the behaviour of the species is in preparation.

Since the great majority of gulls nest on the ground one may presume that this was also the ancestral breeding habitat of the Kittiwake. This view is supported by two facts: (1) Kittiwakes' eggs retain to some extent the cryptic pattern of blotching although this can be of little value, as every nest is marked conspicuously by a flag of white droppings; (2) the young Kittiwakes are able to run under suitable conditions, though not quite as well as young ground-nesting gulls. This is an unusual feature for a species nesting in such precarious sites.

The advantage of cliff-nesting is certainly that it reduces predation. The nests seem fairly safe not only from ground-predators but also from such aerial ones as large gulls. I never saw a single Kittiwake chick or egg taken from the nest by the Herring Gulls *Larus argentatus* or Lesser Blackbacks *Larus fuscus*, which frequently preyed on the eggs and young of the Eider Ducks *Somateria mollissima* and terns nesting on the ground nearby; Nicholson (1930) remarks on a similar immunity even in the Kittiwakes nesting within a foot or so of the nests of Iceland Gulls *Larus leucopterus*. Perry (1940) observed two pairs of Herring Gulls repeatedly robbing young Kittiwakes from their nests but such predation does not seem to be the rule. It is reported however that in some places in the Shetlands Great Skuas *Catharacta skua* are a greater danger to Kittiwakes' nests. Lockie (1952) on Hermaness found them feeding largely on Kittiwakes' eggs and young and Venables and Venables (1955) also report nest-robbing by them in Foula. But heavy predation of nests seems restricted to few places (L. S. V. Venables, personal communication) and in the colony in Noss for instance the Skuas seem to specialise more on catching the Kittiwakes in the air (Perry 1948).

The Kittiwakes' security from large gulls probably results from the difficulty which they would often have in landing on the small perches, and this is enhanced by the complicated wind-eddies in such places. In strong winds this turbulence is sometimes so awkward that even the agile Kittiwake has to circle again and again before it can alight where it wants to.

In addition of this protection from other species of gull the Kittiwake does not suffer from the depredations of its own species as do the ground-nesting gulls. Unlike the other gulls, the Kittiwake does not take eggs or young of other birds but eats mainly fish and plankton. This difference in diet is a subsidiary reason why the Kittiwake's eggs are safer than the ground-nesters'. But even if Kittiwakes did prey on their own species they would suffer less than the other gulls because their nest-sites are much less accessible to ground-predators who might cause the parents to leave their nests exposed to be robbed by neighbours.

Throughout this paper I have contrasted the behaviour of the Kittiwake with that of the ground-nesting gulls as a group. Only two species, the Black-headed Gull *Larus ridibundus* and Herring Gull have been at all fully studied, but quite a lot is known about the other gulls in a more fragmentary way. Although more studies are desirable, on the existing information there seems a good indication that in many respects the ground-nesters behave alike. For information about the behaviour of the ground-nesters I have used published information (recent references summarized in Moynihan 1955 and Tinbergen 1956) and drawn extensively on personal communications from N. Tinbergen, M. Moynihan, and R. and U. Weidmann. I have also had the opportunity of seeing something myself of the behaviour of Herring and Lesser Black-backed Gulls on the Farne Islands.

FIGHTING METHODS AND RELATED BEHAVIOUR

The ground-nesting gulls have various ways of attacking, the Kittiwake has specialized on one and this seems to have had a number of repercussions on the Kittiwake's behaviour.

According to Tinbergen (1953) and Moynihan (1955) an attacking Black-headed or Herring Gull tries to get above the enemy and peck down, or it may try to grasp any part of the other and then pull. If it happens to grasp the beak it sometimes tries to twist the other's head from side to side and thus may upset its balance. But this twisting seems rather rare: It is reported from the Common Gull *Larus canus* (Weidmann 1955), Tinbergen has seen it occasionally in the Herring Gull (personal communication) and I have observed it twice in the same species.

The Kittiwake on the other hand attacks by darting its head forward horizontally and always tries to grasp the opponent's beak. If it succeeds it twists the head from side to side. As most fights start on the ledges, which comprise the birds' whole territories, this method is very effective and often the opponent is twisted off the sill. It may however hang on to the other's beak and both tumble into the sea together, where they go on twisting with beaks still locked, ducking each other's heads under the water.

It would be difficult for the Kittiwake to use the other fighting methods of the ground-nesters; on their small perches they obviously cannot pull backwards and the structure of the cliffs often does not allow a bird to get above the other to peck downwards. On the water the ground-nesters' methods would probably be equally inefficient and it may be no mere coincidence that both times I saw the beak-twisting method used by Herring Gulls they were fighting on the water.

In connection with its specialised fighting method the Kittiwake seems to have lost a particular threat posture, which has been found in all the other 16 species of gull observed. In this posture, the "aggressive upright" (Moynihan 1955) or "upright threat" (Tinbergen 1953), the neck is stretched upwards and often a little forward and head and bill are pointed down. Tinbergen (1953) interprets the stretched neck and the attitude of head and beak as an intention-movement of pecking down at the opponent from above and later authors essentially agree with this interpretation in other species (Moynihan 1955, Weidmann 1955). As the Kittiwake does not fight in this way it is not surprising that it does not have the aggressive upright posture.

With the Kittiwake's specialisation on one way of fighting the beak seems to have become an important stimulus in aggressive encounters. The fact that an attacking bird always goes for the beak of its opponent shows that the beak is important in directing the attacks. But it also appears to be a strong releasing stimulus. This can be concluded from observations in three situations: (1) when a female joins an aggressive male and tries to stay beside him and to prevent him attacking her, (2) when two females compete for one male, each trying to drive the other away, and (3) when one bird is trying to conquer a nest from another bird or from another pair. In each of these situations the bird who is attacked, whom I shall speak of as the visitor or intruder, may fight back, but often it seems too frightened to attack and turns its beak away hiding it in the breast-feathers, at the same time erecting the feathers of the neck. The attacker tries to get hold of the opponent's beak from either side, twisting and turning its neck over the other's; but if the beak is sufficiently hidden it does not succeed and usually after a while the attempts to attack stop, though the aggressor could easily peck other parts of the intruder's body. Should the intruder's posture relax so that it shows its beak again the owner of the nest starts to attack once more and tries to grab the beak and throw the other off the ledge.

It should be added that the beak-hiding does not always succeed in pacifying an opponent and a very aggressive bird may peck the neck or body of an intruder whose beak is out of reach. There is no doubt however that by tucking its beak out of sight a visitor can stay on a ledge beside the owner for much longer than it otherwise could. On one occasion I saw an intruder crouching on a ledge in this way beside the owner pair for a whole hour. A pair may ignore the presence of a stranger on their ledge to such an extent that they perform various friendly displays together such as mutual head-tossing, sometimes even standing on the intruder as if unaware of it, whereas a stranger with its beak visible would never be tolerated.

To test the importance of the beak experimentally I used models of Kittiwake heads, which were cast of a plastic material ("Welvic" paste) from the same mould and then painted. I placed them on Kittiwakes' nests and counted the number of pecks delivered at them during a standard time after the bird had alighted on the nest. In some preliminary tests to determine the attack-releasing effect of the beak I presented two models in succession, one with a yellow beak like the adult's and one without a beak. Too few birds responded successively for a definite conclusion, but the results at least suggest that the head with the beak releases more attacks than the one without.

The experiments show more convincingly that the beak directs attacks. There were seven individuals who attacked the model with the beak, making 286 attacks in all. Of these attacks 98% (all but seven) were directed at the beak itself.

From general observations of fights it appears that the beak is not such an important stimulus in the ground-nesting gulls as it is in the Kittiwake. In particular the ground-nesters do not as a rule aim their attacks at the beak of the opponent, Huxley and Fisher (1940) and R. Weidmann (personal communication) made some model-experiments with Black-headed Gulls which give information on the attack-directing effect of the beak in this species. Huxley and Fisher used stuffed birds and a corpse which they decapitated after a few experiments, presenting head and body separately. They found that the whole specimens were attacked and so was the head alone, but only a few pecks were aimed at the beak of the latter and none at all at the beaks of the whole birds. Weidmann repeated the experiments and also found that a far smaller percentage of the attacks were aimed at the beak of the head-model in the Black-headed Gull than I found in the Kittiwake though exact figures are not available. From general observations it seems probable that the beak directs attacks equally little in the other ground-nesters.

In releasing attacks the importance of the beak in the ground-nesters is more uncertain. I have said earlier that in the Kittiwake the effect of the beak-hiding demonstrated the attack-releasing effect of the beak. The ground-nesting gulls have a head-turning movement rather similar to that of the Kittiwake, the so-called "head-flagging" (named by Noble and Wurm 1943), but its effect is less conspicuous than in the Kittiwake and no experiments have been made to investigate its function. It has however been previously inferred without actual experiments that the beak had a certain attack-releasing effect in the Black-headed Gull (Tinbergen and Moynihan 1952).

None of the ground-nesting species shows such an elaborate hiding of the beak as the Kittiwake, they merely turn or jerk the face away from the opponent in the head-flagging. There cannot be any doubt that the movements are homologous, especially as the Kittiwake at a lower intensity also merely turns the face away, but one can understand why the Kittiwake has the elaborate beak-hiding: firstly because the beak may have a stronger attack-releasing value, and secondly because the situation in which the Kittiwake does the elaborate beak-hiding does not occur in the ground-nesting gulls. There, a frightened bird which is attacked but wants to stay near the opponent or on the territory can always take a few steps away from the attacker to a place where it is not so frightened and does not provoke the other so much. But the Kittiwake has only the choice between staying very close to the other bird or leaving the territory altogether. Thus the intruding bird's greater fear and the attacker's increased aggressiveness might have led to the Kittiwake's exaggerated beak-hiding.

When young Kittiwakes are threatened they show the same head-turning movement as the adults and, as in the adults, it "appeases" the opponent, i.e., it tends to stop him from attacking. Young ground-nesters apparently do not head-flag at all. Adult gulls, Kittiwakes as well as ground-nesters, head-flag (or hide the beak) when they are pecked or threatened but want to stay. This situation hardly ever arises for young ground-nesters, for while they may be attacked by strangers, unlike young Kittiwakes, they can usually run away. This must be the reason that the Kittiwake is the only species known where the young shows the head-turning movement.

A young Kittiwake is not often pecked by a stranger. In the first weeks after hatching the parents guard the nest so well that a strange bird is seldom able to land there. But later the young are more often left alone and an adult who is searching for territory may land on their nest. When this happens the adult may attack the young. On the other hand fights between nest-mates over food are common and some kind of appeasement seems necessary then. There are usually two young in a nest, one a day or two older than the other. They may sit peacefully together for hours but hostility flares up when the parent comes to feed them. As soon as the younger starts to beg for food or tries to get the food, the older gives it a peck or prepares to do so and the smaller at once turns its head away. This stops the older from attacking and while the smaller looks away the older often can get the food for itself. Only when the older chick is satisfied is the younger one able to get food.

In the young ground-nesters fighting between brood-mates does not seem to occur in connection with food or for any other reason (personal communications, R. and U. Weidmann, N. Tinbergen). The need for appeasement does not therefore normally arise.

I created this need artificially for two Herring Gulls and a Black-headed Gull which were placed in Kittiwakes' nest. They were fiercely attacked by their Kittiwake foster-brothers and wire-netting fastened round the rim of the nest prevented them from escaping. Even in these circumstances the young ground-nesters showed no

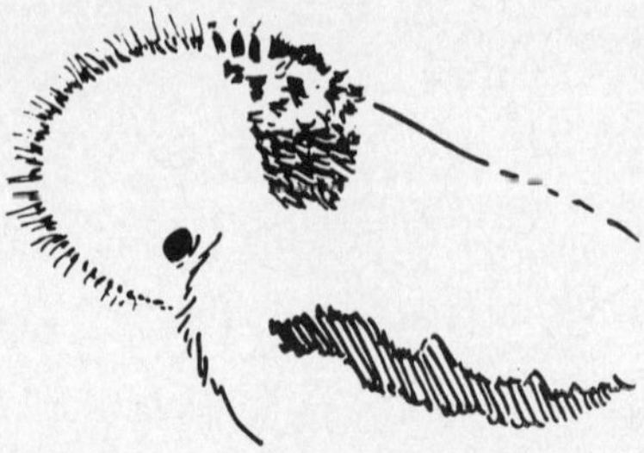

Fig. 1. Young Kittiwake beak-hiding.

trace of the head-turning. As all three chicks had been hatched in Kittiwakes' nest this difference in behaviour must be innate.

The young ground-nesters were attacked more fiercely by the Kittiwake foster-brothers than young Kittiwakes would be in the same situation. Even two Kittiwake chicks which from hatching were always in the company of a young Herring Gull and Black-headed Gull respectively remained extremely hostile towards their companions whereas two young Kittiwakes settle down more quickly. This continued hostility might well be due to the lack of head-turning in the ground-nester chicks.

It seems that the appeasement effect of the beak-hiding in the adult Kittiwake is mainly due to the removal of the beak from view. But when a young Kittiwake turns its head away it not only conceals its beak, but it also shows off the black band across the nape, which is present only in this plumage (Fig. 1). As the band is erected and displayed only in this situation one may infer that it is a special appeasement structure supporting the effect of hiding the beak. Of all the 44 species of gull which Dwight (1925) described, only the young Kittiwake and the young of *Rissa brevirostris* have such a band. This supports the idea of its function, since only the young of these two species must stay where they are however severely they may be attacked.

A slight difficulty remains. I have suggested that the head-turning of the young has been evolved in connection with fighting between the nest-mates, but this fighting is most fierce in the first days after hatching, before the neckband develops. In these early fights the tiny chicks will hide their beaks just like the adults, squatting low in the nest-cup and tucking the beak under the breast. This gesture tends to appease the opponent although the displaying bird presents nothing but the pink skin of the neck shining through the white down. These early fights establish a kind of peck-order between the chicks and later when the black band has developed actual fights between the young are rarer. But the head-turning remains extremely common, especially at feeding times. There is however reason to think that it is still important in preventing fights and an additional structure, the black band, may indeed be more necessary than at first, because the young are now much stronger.

We may conclude therefore that the young Kittiwake who has both the head-turning movement and the black neck-band has developed these characters as a consequence of cliff-nesting.

The head-turning in the Kittiwake seems to work because it conceals the weapon from the opponent and Tinbergen and Moynihan (1952) have suggested basically the same explanation for the head-flagging in the Black-headed Gull. But Lorenz has put forward another interpretation of similar movements, based mainly on his observations of dogs and wolves. He holds that the movements have appeasement effect because the inferior animal presents the most vulnerable part of its body to the enemy, who has a specific inhibition to attack the spot. This explanation cannot hold for the Kittiwake and probably not for the gulls in general, since in exceptional circumstances the neck may be pecked severely and the victim does not seem to take serious harm from it. It even appears uncertain at present whether Lorenz's explanation can hold for wolves and dogs, since, according to R. Schenkel (personal communication), Lorenz has misinterpreted the appeasement ceremony in these animals.

ADVERTISING DISPLAY OF THE MALE

The Kittiwake differs from the ground-nesters in the way in which the new pairs form at the beginning of the breeding season.

In the Kittiwake the males occupy their territories, i.e., their nesting ledges, as soon as they arrive at their nesting cliffs from the winter quarters. They advertise themselves there with a special display which attracts females to them and it is on the nesting ledges that the pairs form. Unmated ground-nesting gulls usually do not seem to go straight to their breeding territories after they have arrived in the nesting area, but may spend the first few days either on neutral ground, the so-called "clubs" (Tinbergen 1953) or "pre-" or "pairing-territories" (Tinbergen and Moynihan 1952, Moynihan 1955). Here as a rule the pairs form and only later do they seem to occupy the actual breeding territories. In some species pairs may form outside the breeding season (Drost 1952) or perhaps even away from the breeding grounds (Moynihan, personal communication).

There may be several reasons why the Kittiwakes go straight to their nesting places while other species apparently do not. Firstly the Kittiwakes appear to be frightened of the land: They avoid sitting on the top of the nesting cliffs in the first few days after the occupation of the colony and it seems that the nesting ledges are the only places on the cliffs where the birds dare to land at this time. Secondly the Kittiwakes may have to stake their claims early because the number of suitable ledges is restricted and the competition for nest-sites is severe. On the other hand for the ground-nesters there is probably little shortage of nest-sites and they can therefore afford to wander about in clubs and pairing-territories.

The Kittiwake males advertise themselves to the females by a particular display, the "choking" (names by Noble and Wurm (1943) in the Laughing Gull *Larus atricilla*). Figure 2 shows the posture of the body in a choking bird and in this position head and neck are rhythmically jerked up and down with a nodding movement of a frequency of about 3 nods per second. Unmated females are attracted by this display and land beside advertising males.

Fig. 2. Kittiwake "choking."

Choking occurs in other species of gulls, but the Kittiwake is the only species known regularly to use it when advertising for a mate. Unmated Black-headed Gulls use another display, the "oblique" posture with the "long call" (Moynihan 1955) and from recent observations it appears that unmated Herring Gulls also may use the long call as advertisement display (Tinbergen 1956).

On the other hand there are indications that both in the Herring Gull (Goethe 1937) and Black-headed Gull (R. and U. Weidmann, personal communication) an unmated male may occasionally choke in order to attract a mate. In these species and also in the Common Gull one can see much more often the attraction exerted by the choking display between a mated pair (Tinbergen 1953, Weidmann 1955). One bird, usually the male, walks away from its mate towards the nest or a prospective nest-site and chokes there; and this at once causes the female to approach and join in.

From these reports of the choking of the ground-nesters two points emerge. Firstly that choking attracts a mate and even though it may not be the usual display by which the unmated male first allures a female, it is not surprising that the Kittiwake has developed it for this purpose. Secondly it appears that the ground nesting gulls tend to go to a nesting place in order to choke and this may be the reason why the Kittiwake males, whose pairing territory is the nesting ledge, have specialized more than any other gull in advertising themselves by choking.

The link between choking and the nest-site can be partly understood when the origin of the display is explained. The origin of choking has been discussed by several authors, most recently Moynihan (1955). Two different explanations have been put forward: one is that most if not all elements of the display are derived from nest-building movements; the other is that the pattern is derived from feeding the chicks. For various reasons, which will be published in a later paper, I believe the first explanation to be true and that choking behaviour is mainly derived from a displacement nest-building activity. It has been suggested (Armstrong 1950, Tinbergen 1952, Lorenz 1953, and more recent authors) that in situations where displacement activities might be expected, external stimuli may play a part in determining which particular activity is chosen and it might well be that for this reason choking so often occurs at the nest-site.

It should however be stressed that this influence of the nest-site is not direct, because when a Kittiwake male advertises on the water or on the cliff-top, as happens occasionally, he still chokes as he would when advertising on the nest. The influence of the nest-site must be supposed rather to have acted phylogenetically in selecting for this particular type of advertisement display.

COPULATION

Almost all the Kittiwake copulations take place on the nesting ledge. Only exceptionally are pairs seen to copulate elsewhere, for instance on the cliff-tops, and these are birds who do not own a ledge.

During copulation the female sits down; and Paludan (1955) has pointed out that this is different from the behaviour of a female ground-nester, who remains standing during the whole performance, often moving a little to and fro, apparently to keep her balance. Paludan has also pointed out that this method of copulating would be most unsuitable in the Kittiwake, as the female has only the tiny ledge to move on.

That this adaptation is not immediately dependent on the external situation is shown by the fact that the female sits down even during copulations on flatter places, where she would have plenty of room to move about.

NEST-BUILDING

Nobody who has seen the loosely built nests of one of the ground-nesting gulls will expect that such a construction could be stuck onto the small shelves of the Kittiwake cliffs. And indeed the Kittiwake has a nest-building technique which is much more elaborate than the ground-nesters'.

To begin the nest the Kittiwakes mainly bring mud or soil, often mixed with roots or grass-tufts. Such trips for mud sometimes alternate with trips for grass, seaweed or other fibrous material. One bird usually deposits its material sideways over its shoulder with a rhythmic downward jerking movement of the head, performing about three to four jerks per second; at each downstroke the beak is opened and in this way the material, which often sticks to the beak, is jerked down piece by piece onto the nesting platform. Then the bird starts to trample on the ledge as if it were marking time. A bird trampling intensely does about six to seven steps a second and may go on trampling for up to half an hour with only short intervals. This behaviour is poorly directed as the bird often tramples a long time beside the nest-material on the bare rock, apparently without noticing the mistake. But as the pairs go on building in that way for several hours they succeed in the end in stamping the mud and the fibrous material to a firm platform, which sticks to the narrow ledges and enlarges them and produces horizontal shelves on amazingly steep slopes. In the later stages of building, the birds collect mainly material such as dry grasses, which they deal with in the same way as the mud: jerking it down and trampling on it, so as to form the nest itself on the reinforced mud-foundation.

In other gulls nest-building is simpler; they collect no mud but only fibrous material and they hardly trample on it at all. The nests produced are piles of material loosely interwoven and this is all that is necessary on the ground.

Published descriptions (e.g., Moynihan 1953) of the depositing movements in ground-nesting gulls do not mention the downward jerking movement which I have described. However the downward jerk sometimes occurs in the Black-headed Gull (R. and U. Weidmann, personal communication) and Herring Gull (own observation). These species however do not seem to jerk as vigorously as the Kittiwake, usually performing only one movement, which is therefore easily overlooked, and only occasionally a few jerks in succession. They apparently do not need to jerk as vigorously and as long as the Kittiwake, because the material they use does not stick to the beak while the Kittiwakes often have great difficulty in getting the sticky mud or humus onto the ledge. I counted the number of jerks per depositing bout in eight individuals, which all brought material of both kinds (sticky, containing mud or soil, and non-sticky, containing only grasses or seaweed), and found that 45 bouts with sticky material averaged 12.5 jerks, while 24 bouts with non-sticky material averaged only 2.6.

It is not known at present whether the differences found between species in the number of jerks are merely due to differences in the nest-material normally used or whether they are more intrinsic. But there are strong indications for other nest-building activities that the specific differences depend not only on the immediate external stimuli: for instance once the Kittiwake has covered the mud with grass its nest is similar to that of a ground-nester, yet it still tramples intensely on the dry stuff, while the ground-nesters show only traces of the movement. But this still does not decide whether the difference is innate or whether experience plays a part. On the other hand the fact that there is plenty of mud available to the ground-nesters and yet they do not collect it suggests that at least this difference between the Kittiwake and the other species is innate.

Another outcome of the different nest-building techniques is the different shape of the nest-cup; the Kittiwake's is relatively deeper than the ground-nester's, thus holding the eggs more safely. This is necessary because if the Kittiwake's are accidentally kicked out, they almost always fall and are broken, whereas the ground-nesting gulls retrieve theirs in such an event (Kirkman 1937, Tinbergen 1953).

The nest-building behaviour of the Kittiwake has another peculiar feature: The birds collect nest-material in groups. On the Inner Farne the collecting grounds were on the grassy top of the island and one would often see a single bird flying around hesitating to alight anywhere until it saw other birds alighting or already on the ground. In this way parties of twenty birds or more would quickly assemble at one place. A few birds after a while might detach themselves from the group and fly away with what they had gathered, while others joined the remaining body, so that there was a busy coming and going for a time, until suddenly the whole party would leave the place together. Late-comers who either had no time to collect material or were just going to land flew off with the others empty-handed. One could easily see that the birds were frightened on the flat ground even when they were in a flock. One could not approach them within forty yards or so, whereas on the nest one could almost touch them.

This social collecting of nest-material is known only in the Kittiwake. The ground-nesters set off alone; they find their material either in the breeding colony itself or in places similar to where they nest, whereas the Kittiwake has to land on exposed flat ground and one can therefore understand its greater reluctance to collect alone. As in many birds the flock seems to offer protection from predators and only the sense of security given by the flock seems to overcome the birds' reluctance to land on the ground.

There are other indications of the birds' fear of the land: On the Inner Farne they never flew over the island except when collecting nest-material and this avoidance of the land has struck other observers (Naumann 1820, Bent 1921). Further, early in the season they will not assemble on the preening places on top of the island and away from the colony as they do later. They restrict their visits to the ledges and rocks just above and below the nesting cliffs and only start to land on the preening places away from the colony when the gathering of nest-material has begun. It seems thus that only the birds' strong urge to collect nest-material can overcome their fear of the unfamiliar land. On the preening places, as on the collecting grounds, the birds always remain shy.

The birds do not gather nest-material all through the nest-building season. They have spells of building, during which collecting birds move to and fro between the colony and the collecting grounds and these active phases are interrupted by long spells when none or hardly any of the birds are building. This synchronisation is very remarkable and unusual for a gull-colony and might have the function of insuring that a bird finds companions on the collecting grounds. It might be brought about by a strong mutual stimulation to build or by a certain stimulus to which the birds respond simultaneously, or by a combination of both. While I have no evidence for or against the first possibility there is some evidence supporting the second. From my observations in the first two seasons I found that most of the big building outbreaks occurred during or after rain. In 1955 I noted weather and building-activities in a colony of about fifty nests during five days, each divided into four roughly equal periods, at the height of the nest-building phase. In ten of the thirteen periods without rain there was little or no building (i.e., building at not more than three nests) and in three there was a lot (i.e., building at four nests and mostly more), whereas in all seven periods with rain a lot of building was recorded.

Although these figures are small they show that rain is correlated with the building activity in a Kittiwake colony ($P < 1\%$). Apart from the function of the rain-stimulus in synchronising the birds' building, collecting during and after rain might be advantageous to the Kittiwake, since during that time it probably is easier to pick up mud and humus from the softened ground.

The Kittiwakes are very ready to steal nest-material from unguarded nests, to such an extent that half- or almost finished nests may be dismantled completely. Most pairs however do not leave their nests undefended once they have started to build seriously; the mates take turns in guarding the nest, in contrast to their behaviour in earlier days when they frequently were both absent from the colony, sometimes for several days together.

Ground-nesting gulls often leave their nest alone until the first egg is laid and they steal material much more rarely (personal communication, N. Tinbergen, R. Weidmann). This difference again seems to be connected with cliff-nesting; it is obviously more convenient for a Kittiwake to collect from a nearby source on the cliff than from collecting grounds of which it is more frightened.

We can thus summarize the adaptations connected with nest-building: The Kittiwakes have specialized in collecting their material in groups as a consequence of their fear of open ground, which is itself apparently the result of cliff-nesting. They collect mud and soil in addition to fibrous material and work the material further with jerking and trampling. In this way they build a fairly safe nest on their small perches, and they guard it well from robbery by their neighbours.

CONCEALMENT AND DEFENCE OF BROOD

I have mentioned earlier that the Kittiwakes' nests are much better protected from predators than the ground-nesters'. Because of this security the Kittiwake has been able to give up a number of behaviour patterns and morphological features which protect the eggs and young of the ground-nesting gulls.

When a predator approaches a colony of ground-nesting gulls, the birds which can see it will start to give the alarm call and fly up when it is still at a distance; other birds join them even when they cannot see the enemy. (As Tinbergen 1953, has pointed out, this wariness at the nest must be correlated to the fact that the adult birds are very conspicuous whereas the eggs and young, being cryptically colored and widely spaced, cannot be found easily.) Disturbed ground-nesters will moreover pursue an intruder and, particularly near hatching time and when they have young, they will swoop at it and may even strike it. This probably deters or at least distracts predators from searching diligently for the cryptic eggs or young (Tinbergen 1953).

In the Kittiwake it is more difficult to evoke these anti-predator reactions than in the ground-nesters. The birds allow an intruder to approach them far more closely before they react and very often they fly up without giving the alarm call at all. Some individuals can actually be lifted off the nest in broad daylight, even in the exceptional Kittiwake colony in Denmark where the birds nest on the ground (Paludan 1955). It is very striking how the alarm call is heard much less often in a Kittiwake colony than in a colony of ground-nesters, apparently because the Kittiwakes are safer and do not need to warn others so much.

The Kittiwakes also rarely attack predators at all. When I climbed among the nests I was hardly ever swooped at and never struck; and this applies also to the ground-nesting Kittiwakes studied by Paludan (1955). In one year on the Inner Farne a Herring Gull was seen several times to catch a young Kittiwake in the air. Even when it did so only a couple of feet from a nesting cliff the adult birds left their nests merely to hover in a completely silent cloud over the scene without interfering, while the chick was screaming and trying to defend itself against the powerful beak of its attacker. Such an assault in a ground-nesters' colony undoubtedly would have provoked violent attacks by the adult birds nesting nearby.

Although eggs and young ground-nesting gulls are cryptic, they would be easily detected if the nest itself was not cryptic also. During incubation the ground-nesting gulls leave the nest in order to defaecate and when the young hatch the parents carry away the egg-shells. Apart from having any hygienic function, these actions are probably important in concealing the nest, as a collection of the white droppings or the white inside of an empty shell might easily betray the nest to the predator hunting by eye. A comparison of this behaviour in some species of terns points to the same conclusion (J. M. Cullen, personal communication).

On the other hand a Kittiwake while standing on the nest simply lets its droppings fall over the rim of the nest and brooding bird merely gets up for defaecation. In this way soon a most conspicuous white flag forms below each nest. The egg-shells too are just left lying on the nest until they are accidentally knocked off. This usually happens soon after hatching, but may take several weeks.

Young ground-nesters are not only cryptically colored but all the species which have been watched also show cryptic behaviour. When a predator approaches they run to hide under cover, and crouch. By contrast the young Kittiwakes are most conspicuously coloured: In down they are white with a light grey back and in the juvenile plumage they are like the adults with the addition of some black bars. In typical nest-sites it is impossible for them to run when danger approaches and Salomonsen (1941) reports that even in a Kittiwake colony on the ground the half-grown young remained on their nests when he came near (while the young Black-headed Gulls from the nests around all ran away). Yet Kittiwake chicks will run when taken from the nest, put on the ground and frightened.

On the other hand two young Herring Gulls and one Black-headed Gull which were hatched and reared in Kittiwakes' cliff-nests would all have run over the edge of the nest had I not prevented them by fastening wire-netting round the nest-rim. From this and Salomonsen's observations we can conclude that the difference in running between the young of the different species must be innate.

CLUTCH-SIZE

Lack (1954 and previously) has suggested that in many birds the size of the clutch is adapted to correspond to the maximum number of young which the parents can, on the average, supply with food.

To test this hypothesis for the Kittiwake a number of nests were checked daily in a more or less random sample in the Kittiwake colony on the Inner Farne in 1953, 1954 and 1955 and the data were used to calculate clutch-size, nesting success, etc. In addition other nests were checked less often and might be expected to yield rather

TABLE 1. Survival of Kittiwake eggs and young in relation to (a) clutch-size and (b) and (c) number of young hatched per nest. Only first clutches included.

Clutch-size	No. of pairs	Young hatched No.	Young hatched Per cent hatching	Young surviving	
1	26	13	50	12	
2	130	182	70	164	(a)
3	18	40	74	32	
Size of family	**No. of pairs**	**Young surviving No.**	**Young surviving Per cent**	**Average no. young per pair**	
Undisturbed nests					
1 young	26	24	92	0.9	
2 young	91	164	90	1.8	(b)
3 young	9	20	74	2.2	
Both undisturbed and experimental nests					
1 young	31	28	90	0.9	
2 young	91	164	90	1.8	(c)
3 young	19	45	79	2.4	

different results since eggs or young might disappear between two checks. Clutch-size, hatching success and survival-rate were not significantly different for the two groups ($P > 5\%$) and the data have been therefore combined in Tables 1 and 2. A few young were not able to fly by the time I left the island, but in view of their age I have regarded them as surviving in compiling Tables 1 and 2.

The clutch-size distribution in 138 nests checked daily was 21 (15%) with one egg, 104 (75%) with two and 13 (9%) with three.

Table 1a shows that the hatching success increases with clutch-size. In order to discount this effect and to make comparison possible with the group of experimental nests mentioned below, the nesting success of the Kittiwake was calculated on the basis of the number of young hatched rather than on the number of eggs laid (Table 1b). Since the number of families of one and three were small, I adjusted the family size in some additional nests artificially, where necessary adding a young of a suitable age. Table 1c combines the results from these nests with those whose clutch-sizes were undisturbed and shows that families of three have a lower survival rate than families of one and two. (Survival of families of two is significantly higher than that of families of three ($P < 3\%$.) The results obtained so far from this single colony fit with one requirement of Lack's hypothesis, but it should be noted that the number of young produced per nest is nevertheless greatest for the families of three.

TABLE 2. Nesting success in different species of gull. Different methods of sampling may account for some of the differences in the results.

		No. of		No. of		Per cent hatched young surviving	Average young per pair surviving	Reference
		Pairs	Eggs	Hatching	Surviving			
Kittiwake	1953–55	179	348	240	211	88	1.18	this paper
Herring Gull	1936	59	126	108	46	43	0.78	Darling (1938)
	1937	68	198	189	77	41	1.13	ditto
	1943	90	270	243	less than 131	54	1.46	Paludan (1951)
	1944	87	371	206	41	20	0.47	ditto
	1947	100				c.51	c.0.51	Paynter (1949)
Lesser Black-backed Gull	1936	43	93	86	c.50	c.58	c.1.16	Darling (1938)
	1937	71	206	196	c.104	c.53	c.1.47	ditto
	1943	120	354	216	less than 32	15	0.27	Paludan (1951)
	1944	112	362	220	12	5	0.11	ditto

Lack has emphasized the importance of food-shortage in determining the number of young which can be raised and there is some circumstantial evidence that food-shortage may account for mortality of young Kittiwakes. As noted above, young nest-mates regularly establish a peck-order which is important mainly at feeding times and when the parents have only a little food the superior chick may be the only one to be fed. This food-fighting suggests that food must at times be short. Furthermore the Kittiwakes' eggs hatch asynchronously, a device in birds which is thought to lessen the bad consequences of food-shortage (Lack 1954).

In spite of this indirect evidence of food-shortage I was not able to see any direct signs of starvation except at a few nests and in some of these the parents seemed at fault, for nearby nests with the same number of young were being supplied adequately. Food-shortage might, however, be more evident in other colonies and under different conditions.

On the other hand food-shortage is probably not the only reason for a greater mortality in larger families, and of the young whose deaths are recorded in Table 1 I judged that no more than half can have been due to starvation. Chicks sometimes fall off their nests; or part of the nest, or even the whole of it, may collapse, precipitating the young into the abyss. One must suppose that the more young there are on a nest the more likely they are to fall, both because they have less room to stand and because their combined weight is greater. I recovered some of the young which had fallen and found them perfectly healthy and they were in a number of cases successfully reared by foster-parents. I seldom saw a chick actually fall from its nest, but when a chick was missing, one could get some idea of the likelihood that such an event had taken place from the condition of the nest. From observations in a group of about twenty nests which I studied most intensely over three years, it would seem that the disappearance of young is just as often due to falling as to starvation.

It appears therefore that both shortage of food and the collapse of nests might account for the greater mortality in larger families in the Kittiwake.

The Kittiwake usually lays two eggs and so does its close relative *R. brevirostris* (Bent 1921), but the usual clutch-size of the gulls is three, except the Ivory Gull *Pagophila eburnea* which usually lays two (Buturlin 1906, Stark and Sclater 1906, Bent 1921, Baker 1935, Witherby *et al.* 1944).

Many young ground-nesting gulls are killed by predators, including members of the same species (Goethe 1937, Kirkman 1937, Darling 1938, Paynter 1949, Paludan 1951). The Kittiwake young have a much lower mortality (Table 2) presumably thanks to their different nesting habit which protects them from predation and the disturbances which go with it. The mortality of the young Kittiwakes, at least on the Inner Farne, is so low that in spite of their clutch the birds produce on the average more young per nest than is in many cases reported of the other gulls with their larger clutches (Table 2).

Comparing my own data for the Kittiwake with Paynter's (1949) for the Herring Gull and Paludan's (1951) for the Herring and Lesser Black-backed Gull it appears that so many young ground-nesters die soon after hatching that within a week (Paludan) or from about three weeks (Paynter) there are fewer young per pair in these species than in the Kittiwake. Other observers also remark on this high early mortality in the Herring Gull, Lesser Black-backed Gull and Black-headed Gull without giving quantitative evidence (Goethe 1937, Kirkman 1937, Darling 1938, Lockley 1947, Tinbergen 1953, R. and U. Weidmann, personal communication).

The available data thus suggest that through much of the nestling period the Kittiwake parents have as many or more young to feed than the ground-nesters.

Lack has repeatedly stressed that clutch-size in a species cannot be adapted to adult mortality, but he points out (1954) that mortality at an early age might be a subsidiary factor affecting the clutch-size. Even if the number of young which could be raised by the Kittiwake and the ground-nesting gulls were the same, one would expect that with relatively little early mortality, a consequence of cliff-nesting, the Kittiwake would evolve a smaller clutch-size than the other species.

FEEDING YOUNG

The Kittiwake feeds its young in a different way from the ground-nesting gulls and this can be correlated with the different nesting habitats.

Young ground-nesters leave the nest a few days after hatching and may be fed at any place in the territory. Young Kittiwakes are confined to the nest until they can fly, that is, when they are about six weeks old, and have to be fed there. All gulls feed their young by regurgitation and in the ground-nesting species the parents very often drop the food to the ground and the young pick it up. But a young Kittiwake takes its food from the throat of the parent and food is rarely dropped. Gulls' food is often half-digested and therefore cannot be picked up completely and it is very likely that soon a little heap of rotting food would collect in a Kittiwake's nest if the birds fed in the ground-nesters' way. This would not only smear the young but it might also develop into a source of disease. For birds restricted to the nest it obviously is advantageous to have a more hygienic feeding method.

The Kittiwake keeps its nest clean in another way. It immediately picks up remains of food and other strange objects which may have fallen into the nest and either swallows them or, more frequently, flings them away with a vigorous head-shake. This habit has not been observed in the ground-nesters (N. Tinbergen, R. and U. Weidmann, personal communication), who do not seem to need it.

There are other behaviour patterns connected with the feeding of the young which seem to have changed in the Kittiwake. When a ground-nesting gull comes with food, its young are often hidden under cover, away from the place where the adult alights and if they are hungry they will run towards it with a characteristic food-begging movement: from a horizontal posture the so called "hunched posture" (Moynihan 1955) or "attitude of inferiority" (Tinbergen 1953) they quickly stretch their neck up vertically and withdraw it again, repeating these movements in succession so that head and neck perform a kind of pumping movement (Tinbergen 1953, Moynihan 1955, R. Weidmann, personal communication). In the stretched phase they open the beak widely and give the food-begging call as they withdraw the neck again. One might guess that the function of this alternation and particularly the stretching of the neck is to make the young conspicuous to the parent who may alight some yards away. This explanation is confirmed by the behaviour of a young Black-headed Gull which I reared, for the pumping movement was much more pronounced when some distance away than when close to its "parent." By contrast the young Kittiwake lacks both the pumping movement and the need to make itself conspicuous for it is always on the nest.

It may be noted that in the food-begging of adult ground-nesting gulls, before courtship-feeding and copulation, a time when a pair is standing close together on the ground, the pumping is absent.

A related difference is that the Kittiwake lacks a special feeding call such as that which attracts the young ground-nester to the parent.

FLYING MOVEMENTS IN THE YOUNG

Both the Kittiwakes and the ground-nesting gulls perform incipient flight movements when they are small, but this behaviour develops differently later in life.

From a very early age, before the wing-feathers have grown, the downy chicks of the ground-nesters flap their wings and at the same time jump up into the air. As the young grow these jumps get higher, the wing-flapping more vigorous and gradually it develops into proper flying.

The young Kittiwakes on the other hand, although they start wing-flapping from about the same age as the ground-nesting gulls (from the third day, Kittiwake, own observation; from the fifth day, Black-headed Gull, personal communication R. Weidmann; from the seventh to ninth day, Herring Gull, Goethe 1955), do not perform the movement as vigorously as the other species and they do not start to jump into the air as early as the young ground-nesters. When eventually they start to do so, a few days before they fly they never lift their feet for more than an inch or so from the nest, whereas much smaller ground-nesting chicks can be seen jumping a foot or two in the air.

This lack of vigorous flying movements in the young Kittiwake is certainly an adaptation connected with cliff-nesting for wing-flapping as vigorous as ground-nesters' would endanger their lives.

The orientation of the chicks when wing-flapping also differs with the species. The ground-nesters face into the wind, but the young Kittiwakes always face the wall. Kittiwake chicks spend much of their time orientated in this way, even when they are resting or sleeping and in the first days after hatching they hardly ever face outwards. This probably is a habit which prevents them from falling down. It is therefore not surprising that during a vigorous movement like the wing-flapping the young face the wall; moreover, they often would not have room to spread their wings if they were facing the other way. When the young Kittiwakes prepare for their first real flight, however, they have to face outward. Thus their wing-flapping and jumping does not grade imperceptibly into proper flying as in the ground-nesters but their first flight is much more of a new achievement.

RECOGNITION OF YOUNG

Several authors have claimed that ground-nesting gulls and terns are able to distinguish their own young from strangers (Watson and Lashley 1915, Culemann 1928, Dircksen 1932, Goethe 1937, Kirkman 1937, Palmer 1941, Tinbergen 1953, Pfeffer 1955). The experiments on which these claims are based are often unsatisfactory and in many cases do not distinguish between personal recognition by the parent of the young and other factors, such as the recognition by the young of the parent and possible differences in behaviour which this may lead to, which themselves may partly be the means by which the parent "recognises" its young. Most of the experiments however show that if young more than a few days old are exchanged from different nests the parents drive them away. Tinbergen's experiments with Herring Gulls show that the parents begin to distinguish between their own young and strangers when their own young are about five days old and Pfeffer-Hulsemann (1955) claims the same of the Common Gull. Tinbergen stresses that this ability to recognise the young is very different from the ability to recognise the eggs, which depends almost entirely on location, and that this difference corresponds to the behaviour of eggs and young, as the eggs remain in the nest while the young soon start to move about. Furthermore, it appears that the parents start to recognise their own young at the time when the young begin to move about actively. Thus one may infer that the recognition enables the parents to attend to their own chicks by the time they are old enough to wander about. According to this explanation one would expect that the Kittiwakes, whose young, like the eggs, do not move from the nest, would also lack the ability to distinguish between their own young and those of strangers.

To examine this possibility I exchanged a number of young from different nests for five minutes (Tables 3, 4, 5) or longer (Table 6). Chicks of one nest were sometimes presented to several strange parents, and chicks which had been exchanged before were sometimes used again in an older age-group. In the first series I exchanged broods of the same age and the same number of chicks (Table 3). In another series I substituted young of different ages either for eggs or for young without altering the number of young (Table 4) and in some further experiments the number of individuals in the nest was altered as well as the nature of the contents (Table 5). In all the tests except the one in Table 3 with a young bird of the fifth age-group the parents showed the same behaviour to the strange young as they usually do to their own which varies with the age of the chicks: Small young are brooded, larger ones just guarded. Sometimes a young one was preened a little or even fed. The young Kittiwakes behaved towards the strange parents very much as they did towards their own; they usually stood beside them and might preen or peck at the adult's beak, begging for food, etc.

TABLE 3. Exchange of young Kittiwakes from different nests. In each test, except the last, the whole brood was replaced by strangers, of the same number and age as the original young.

Age in days	Age-group	No. of tests	Results
0–6	1	2	accepted
	1–2	4	accepted
7–13	2	8	accepted
	2–3	2	accepted
14–20	3	12	accepted
21–27	4	5	accepted
28–34	5	1	pecked a little (see text)
Only one of two young exchanged	4	1	accepted

TABLE 4. Young Kittiwakes of different age-groups exchanged. The whole brood was replaced by the same number of strangers. All accepted.

Age-group of own brood	Age-group of replacing brood	Number of tests
2nd clutch ca. 1 week before hatch	3	1
hatch near hatch	2	2
1	2	2
1–2	4	2
2	1	1
4	1–2	3
4	5	1
5	4	1
6	4	1

TABLE 5. Number of individuals and nature of contents of nest changed in Kittiwakes' nests. All accepted.

Owner's brood	Change made
1 young (age-group 1)	added 1 Kittiwake (age-group 1).
2 young (age-group 1)	added 1 Kittiwake (age-group 1).
1 young (age-group 4)	added 1 Kittiwake (age-group 4).
1 egg (pipped)	added 1 Shag (newly hatched).
1 young (2–4 days)	added 1 Shag (newly hatched).
2 young (1–3 days)	older replaced by a Herring Gull (newly hatched).
1 young (6 days)	added a Herring Gull (1 day).

TABLE 6. Strange young left in experimental nests to be reared after tests listed in Tables 3 and 4. (The whole brood was replaced without altering the number of young.)

Number of exchanged young per nest	Age-group of own young	Age-group of strange young	Number of tests	Result
1	egg (ca. 1 week before hatch)	3	1	reared
1	1	1	2	ditto
2	1–2	1–2	2	ditto
2	3	3	2	3 reared, 1 died prematurely
1	6	4	1	reared

It appears therefore that under the experimental conditions the parents do not recognize their offspring up to an age of at least four weeks either by number or the state of development, let alone by more subtle individual characteristics. On the other hand Paludan (personal communication) found that the parents may attack their own young outside the nest, which confirms that they "recognise" their brood by location.

It is difficult to exchange young more than about four weeks old because from this age they are liable to jump off their ledges when approached. Of the three tests I made with young of this class the stranger was accepted in two. Once however the foreign chick was slightly pecked although its behaviour appeared not to differ from that of the parents' own young. These experiments are supported by other observations in which young who could fly (and were therefore at least $5\frac{1}{2}$ weeks old) landed on strange ledges. Of twelve cases where a young bird landed on a nest with young, the stranger was pecked by the parent in six in a clearly hostile, although mostly inhibited way. In six cases the young was not molested and once it was even fed. In four of these intrusions the parent saw the chick landing and in three of the four it attacked; in four cases the young was on the nest first and in three of the four the parent accepted it. This and further observations suggest that a juvenile approaching the nest provokes attack more strongly than a juvenile already on the nest and this may apply even to the birds' own offspring. There are however signs that the parents can really distinguish their young from strangers at this age: An adult may threaten one of its own young who lands but does not actually attack it and I have seen a parent responding to the calls of its own young which landed on a strange nest. I suspect therefore that also in the cases when the strange fledglings were treated like their own young the birds might be aware of the change.

Summarising these facts, it can be said that there is no evidence of personal recognition between parent and young in the Kittiwake up to four or five weeks after the young have hatched, but that the parents seem to discriminate later. Thus the behaviour of the Kittiwake agrees with what was expected and confirms the idea about the function of recognition in the ground-nesters.

So far I have discussed mainly the recognition between adults and young, but the reaction of the chicks themselves to strange nest-companions is also interesting. Nest-mates know each other and fight strange young fiercely. Because of this hostility, it is difficult to foster a strange chick into a family with young, if at least one of the young is more than a few days old. Small strange young usually get used to each other after some fighting, but in larger ones the squabbling goes on and even after the loser has adopted the appeasement posture the other sometimes goes on pecking him fiercely for a long time.

This hostility of the young towards strangers is particularly surprising in view of the fact that the parents apparently do not distinguish their own young from strangers before the young fly, but the most probable explanation is that the young birds' hostility towards a stranger has another function from that of the parents'. In the ground-nesting gulls the parents' aggression is presumably a means of spacing out the families and ensuring that a stranger is not fed. Of this the Kittiwakes have no need. On the other hand the fighting between the nest-mates is probably concerned with establishing the peck-order, already mentioned, whose function it is to determine the order in which the young are fed. In the normal course of events the peck-order is settled at an early age, when both the young are small and weak and easily tired, but if two strange young encounter one another on a nest (as happened during experiments) this may lead to a longer battle before one is subdued. It may be remarked that the outcome of these battles between strangers was a foregone conclusion as it always happens, in my experience, that the larger chick wins, whether or not it is the rightful owner of the nest.

To summarise this section we may say that the apparent inability of parents and young to recognise one another in the Kittiwake is another effect of cliff-nesting and the consequent isolation of the nests of different pairs from one another.

Ground-nesting gulls	**Kittiwake**
High predation-rate in nesting colonies.	Predation pressure relaxed on cliffs.
Alarm-call frequent.	Alarm-call rarer.
Adults leave nest when predator some way distant.	Remain on nest until predator very close.
Vigorously attacks predator intruding in colony.	Very weak attacks at most at intruding predator.
Brooding birds disperse droppings and carry egg-shells away from nest.	Neither droppings nor egg-shells dispersed.
Young cryptic in appearance and behaviour.	Young not cryptic either in appearance or behaviour.
Clutch-size normally three eggs.	Clutch size normally two eggs.
Suited to life in colony on ground.	Adapted to life on cliffs.

A.	Several fighting methods.	More specialized to fighting in one way (grabbing beak and twisting).
	Upright threat posture occurs, derived from preparation to peck down at opponent.	No upright threat.
	Beak does not specially direct attacks. Not known if it is such a strong releasing stimulus as in the Kittiwake.	Beak releases and directs attacks.
	Beak turned away in appeasement but not elaborately hidden.	Beak turned away in appeasement and elaborately hidden.
B.	Young run away when attacked.	Young do not run when attacked.
	No head-flagging in young.	Head-turning and hiding of beak in young when pecked and appropriate behaviour in attacker.
	No neck-band.	Possess black neck-band.
C.	Number of nest-sites, probably less restricted and therefore probably less competition for nest-sites.	Number of nest-sites restricted, probably more competition.
	Often first occupy pairing territories before nesting territories and pairs form away from nest.	Occupy nesting ledges at arrival in breeding area and pairs form on the nest.
	Choking not normally used by unmated males as advertisement display.	Choking normal advertisement display of unmated males.
D.	Copulation on the ground, female stands.	Copulation on the tiny ledge or nest, female sits on tarsi.
E.	Nest-material collected near nest, building not synchronized, individual collecting.	Nest-material collected in unfamiliar places, synchronisation of building and social collecting.
	Little stealing of nest-material.	Birds very ready to steal nest-material.
	Nests often unguarded before laying of first egg.	Nests guarded.
	Nest-building technique relatively simple.	Nest-building technique more elaborate.
	Mud not used.	Mud as nest-material.
	Only one or, at most, very short series of depositing jerks.	Prolonged jerking of head when depositing nest-material.

	Only traces of trampling on nest-material.	Prolonged trampling on nest-material.
	Nest has relatively shallow cup.	Nest has deeper cup.
F.	Young leave nest a few days after hatching.	Young have to stay on nest for long period.
	Young fed by regurgitation on the ground.	Young fed from throat.
	Nest-cleaning absent or less conspicuous.	Young and adults pick up and throw away strange objects falling into nest.
	Parents have feeding call, probably to attract young.	Parents have no feeding call.
	Hungry young make themselves conspicuous to parents by head-pumping.	Head-pumping absent in young.
	Parents learn to recognise own young in a few days.	Parents do not recognise own chicks at least up to the age of four weeks.
G.	Young face any direction. Vigorous wing-flapping in young.	Young face wall much of the time. Flight movements much weaker.
H.	Weaker claws, cannot hold on so well.	Strongly developed claws and toe-musculature.

ACKNOWLEDGMENTS

I am grateful to the Janggen-Pohn Stiftung, St. Gallen, Switzerland, for a research grant which permitted this work and to the Nuffield Foundation for providing part of the equipment. Acknowledgments are also due to the Natural History Society of Northumberland, Durham and Newcastle-on-Tyne and to the National Trust for amenities for working on the islands. I would like to thank Dr. N. Tinbergen for his encouragement and most helpful advice and also Prof. A. C. Hardy for his hospitality in the Zoology Department in Oxford. Among others to whom I am grateful for helpful criticism are R. E. Moreau and my husband, who also helped in the field and revised the English text.

REFERENCES

Armstrong, E. A. (1950), The nature and function of displacement activities, *Sympos. Soc. Exp. Biol.*, **4**, 361–384.

Baker, E. C. S. (1935), *The Nidification of Birds of the Indian Empire* 4. London.

Bent, A. C. (1921), Life histories of North American gulls and terns. *U.S. Nat. Mus. Bull.*, 113.

Buturlin, S. A. (1906), The breeding-grounds of the Rosy Gull, *Ibis* (8), **6**, 131–139.

Culemann, H. W. (1928), Ornithologische Beobachtungen um und auf Mellum vom 13. Mai bis 5. September 1926, *J. Ornithol.*, **76**, 609–653.

Darling, F. F. (1938), *Bird Flocks and the Breeding Cycle.* Cambridge.

Dircksen, R. (1932), Die Biologie des Austernfischers, der Brandseechwalbe und der Kustenseechwalbe nach Beobachtungen und Untersuchungen auf Norderoog, *J. Ornithol.*, **80**, 427–521.

Drost, R. (1952), Das Verhalten der mannlichen und weiblichen Silbermöwen (*Larus a. argentatus* Pont.) ausserhalb der Brutzeit, *Vogelwarte*, **16**, 108–116.

Dwight, J. (1925), The gulls (Laridae) of the world; their plumages, moults, variations, relationships and distribution, *Bull. Amer. Mus. Nat. Hist.*, **52**, 63–408.

Fisher, J. and Lockley, R. M. (1954), *Sea-birds.* London.

Goethe, F. (1937), Beobachtungen und Untersuchungen zur Biologie der Silbermöwe (*Larus a. argentatus* Pont.) auf der Vogelinsel Memmertsand, *J. Ornithol.*, **85**, 1–119.

Goethe, F. (1955), Beobachtungen bei der Aufzucht junger Silbermöwen, *Z. Tierpsychol.*, **12**, 402–433.

Huxley, J. S. and Fisher, J. (1940), Hostility reactions in Black-headed Gulls, *Proc. Zool. Soc. London* (A), **110**, 1–10.

Kirkman, F. B. (1937), *Bird Behaviour.* London and Edinburgh.

Lack, D. (1954), *The Natural Regulation of Animal Numbers.* Oxford.

Lockie, J. D. (1952), The food of Great Skuas on Hermaness, Unst. Shetland, *Scot. Nat.*, **64**, 158–162.

Lockley, R. M. (1947), *Letters from Skokholm.* London.

Lorenz, K. (1953), Die Entwicklung der vergleichenden Verhaltenforschung in den letzten 12 Jahren, *Zool. Anz. Suppl.*, **17**, 36–58.

Moynihan, M. (1953), Some displacement activities of the Black-headed Gull, *Behaviour*, **5**, 58–80.

Moynihan, M. (1955), Some aspects of reproductive behavior in the Black-headed Gull (*Larus ridibundus ridibundus* L.) and related species, *Behaviour Suppl.*, **4**, 1–201.

Naumann, J. A. (1820), *Naturgeschichte der Vogel Deutschlands.* Leipzig.

Nicholson, E. M. (1930), Field-notes on Greenland birds, *Ibis* (12), **6**, 280–313, 395–428.

Noble, G. K. and Wurm, M. (1943), The social behavior of the Laughing Gull, *Ann. N.Y. Acad. Sci.*, **45**, 179–220.

Palmer, R. S. (1941), A behavior study of the Common Tern (*Sterna hirundo hirundo* L.), *Proc. Bost. Soc. Nat. Hist.*, **42**, 1–119.

Paludan, K. (1951), Contributions to the breeding biology of *Larus argentatus* and *Larus fuscus*, *Vidensk. Medd. Dansk. Nuturh. Foren.*, **14**, 1–128.

Paludan, K. (1955), Some behavior patterns of *Rissa tridactyla*, *Vidensk. Medd. Dansk. Naturh. Foren.*, **117**, 1–21.

Paynter, R. A. (1949), Clutch-size and the egg and chick mortality of Kent Island Herring Gulls, *Ecol.*, **30**, 146–166.

Perry, R. (1940), *Lundy, Isle of Puffins.* London.

Perry, R. (1948), *Shetland Sanctuary*. London.

Pfeffer-Hulsemann, K. von (1955), Die angeborenen Verhaltsensweisen der Sturmmowe (*Larus c. canus* L.), *Z. Tierpsychol.*, **12**, 443–451.

Salomonsen, F. (1941), Tretaaet Maage (*Rissa tridactyla* L.) som Ynglfugl i Danmark, *Dansk. Orn. Foren. Tidsskr.*, **35**, 159–179.

Selous, E. (1950), *The Bird Watcher in the Shetlands*. London.

Stark, A. and Sclater, W. L. (1906), *The Birds of South Africa*, **4**, London.

Tinbergen, N. (1952), "Derived" activities; their causation, biological significance, origin and emancipation during evolution, *Q. Rev. Biol.*, **27**, 1–32.

Tinbergen, N. (1953), *The Herring Gull's World*. London.

Tinbergen, N. (1956), On the functions of territory in gulls, *Ibis*, **98**, 401–411.

Tinbergen, N. and Moynihan, M. (1952), Head-flagging in the Black-headed Gull; its function and origin, *Brit. Birds*, **45**, 19–22.

Venables, L. S. V. and Venables, U. M. (1955), *Birds and Mammals of Shetland*. Edinburgh and London.

Watson, J. B. and Lashley, K. S. (1915), Homing and related activities of birds, *Pap. Dept. Mar. Biol.*, **7**, Publ. Carnegie Inst. Washington, **21**, 1–104.

Weidmann, U. (1955), Some reproductive activities of the Common Gull *Larus canus* L., *Ardea*, **43**, 85–132.

Witherby, H. F. *et al.* (1944), *The Handbook of British Birds*, **5**, London.

Determinants of Cliff Edge and Escape Responses in Herring Gull Chicks in Nature

JOHN T. EMLEN, JR.

Herring gull chicks from nests on (a) cliff ledges and (b) vegetated plateaus showed different types of escape and edge response when tested on an elevated platform. The birds from cliff sites, though bolder at the platform edge, were more reluctant to jump. Those from the plateau nests leapt to the ground with relatively little hesitation. Chicks from cliff eggs, hatched and reared in the plateau by foster parents, resembled native plateau chicks in their escape responses while chicks of plateau origin reciprocally transplanted to and reared on cliff sites showed the reluctance typical of cliff chicks. The possibility of two genetic types reproductively isolated on the two breeding situations is thus essentially eliminated. The likelihood of a stable genetic polymorphism is considered and tentatively discarded as less plausible than an explanation based on learning. Although conditions favorable for simple instrumental conditioning along divergent lines are present in the two rearing situations, a theory is favored which attributes the divergence to major difference in the amount and variety of experience

Reprinted from *Behaviour*, volume 22, pages 1–15 (1963). Original numbering of figures is retained.

afforded by the two environments. Modest support for this theory is provided by experiments in which plateau chicks confined by fencing to the immediate nest vicinity tended to resemble the naturally deprived cliff chicks in their reluctance to jump. In conclusion the concept of species-characteristic responses is considered inapplicable to the cliff edge and escape patterns of herring gull chicks except as it incorporates recognition of a determining role in factors of the rearing environment.

INTRODUCTION

Cullen (1957) and others have noted that chicks of the cliff-nesting kittiwake gull, *Rissa tridactyla,* characteristically "freeze" in position when approached by a human observer while those of plateau-nesting gull species run for cover. Observations on chicks of the herring gull, *Larus argentatus* at Kent Island, New Brunswick, suggest that both of these distinct types of escape behavior occur within this single species according to the nature of the substrate on which the birds are hatched and reared. Accordingly, a series of tests and experiments was made during the summers of 1960 and 1961 designed to analyze the natural determinants of escape and cliff edge responses in chicks from the two situations.

A majority of the herring gulls on Kent Island nest on the gently rolling, turf-covered interior of the island where their nests are hidden in dense low stands of forbs irregularly broken by small rock outcropping, barren trails, muskrat foraging areas and hillocks trampled by the gulls. Peripherally, the colony extends over the rim and down onto the rocky cliffs which border the island's south and south-east shores. Here the birds nest on narrow rocky ledges essentially devoid of vegetative cover. Chicks resting in or near their nests in these two contrasting situations provided the subjects for the study.

The plan of procedure was to test all birds on site using an elevated translucent platform on which behavior could be observed from concealment below. The drop from the platform edge constituted a psychological barrier restricting the subject's freedom to escape. Separate tests were devised for evaluating edge withdrawal reactions and escape reactions.

I am indebted to Dr. C. E. Huntington and Mr. T. Skaling for assistance in the field, and to Drs. Huntington, J. Neess and W. Welker for reading and criticizing this report.

METHODS AND PROCEDURES

The apparatus used in all the tests described in this report consisted of a 16″ × 17″ platform fixed to the top of a tripod at a height of 78 inches above the ground. The apparatus was carried into the field and was set up for each bird at a convenient point on level ground as close as possible to the nest or capture site. Subjects were tested only once.

The platform was made of one-eighth inch plexiglass retaining the protective gummed-paper covering on its upper surface. This provided a translucent "one-way"

Fig. 3. Position of chick on test platform during the visual cliff edge test. The shaded area indicates the opaque, paper-covered part of the platform.

film through which the activities of the test bird could be watched against the sky while the observer remained invisible below. Portions of the gummed paper covering were removed to provide "visual cliff" conditions for certain of the tests.

Birds to be tested were approached quietly as they rested in their nests or in the vegetation or rocks. All birds of appropriate size were used as they were encountered in order to minimize bias in the selection of test subjects. Very few potential subjects escaped. A hood, generally a tin can, was quietly slipped over the head and the bird quickly and gently lifted to the platform above the observer's head. After placing the bird in the first test position on the translucent platform the hood was carefully removed and observations started. In the 1960 observations responses were recorded for three "edge tests" followed by a graded series of six "jump tests" as described below.

Edge Test A (visual cliff)

The bird was placed with its head directly over and parallel with the line dividing the paper-covered portion and the transparent (visual cliff) portion of the platform (Fig. 3). The tendency for the bird to turn its head or its body toward the opaque side was evaluated at four levels: no response (0),—slight (1),—moderate (2),—strong (3) in three repeated tests, the bird being replaced in the test position each time by the observer's hand reaching up from below as soon as it had made a recordable response.

Edge Test B (push resistance)

Immediately after the completion of Test A the bird was slowly pushed by a vertical shield of dark opaque pressboard to the edge of the platform and its resistance to being pushed evaluated at four levels as above. Again the test was repeated three times for each bird.

Edge Test C (edge withdrawal)

The tendency of the bird to withdraw from the platform edge to the vertical wall of the pusher as it was moved back or after it had been moved back about six inches was evaluated at four levels (as above) following each of the three pushing tests.

These edge tests were effective only in birds less than about 14 to 16 days of age. In older birds the edge responses were masked by developing escape responses to the experimenter's hand and the pusher. No objective procedure was devised for avoiding bias in an observer's evaluation of edge responses other than the postponement of data analysis. It should be noted, however, that no hypothesis on the nature of the results had been formulated at the time of testing.

Jump Tests

Following the edge response tests, the birds, still on the platform and unable to see the observer below, were watched in their tendencies to escape by jumping to the ground during six stages of increasing provocation as described below:

Stage 1. The pusher was quietly removed and the bird permitted to adjustment period of 10 seconds.

Stage 2. The bird was allowed to explore the platform and its edges undisturbed for 30 seconds.

Stage 3. The observer quietly reached up with his hand and repeatedly placed the bird with its toes at the edge of the platform during a period of 30 seconds.

Stage 4. The observer handled the bird roughly for 10 seconds, then stepped back to make himself visible for an additional 30 seconds.

Stage 5. The observer walked away to a distance of 30 to 50 feet where he sat quietly for 30 seconds in full view of the subject.

Stage 6. The observer remained seated for an additional two minutes, then ran noisily back to the platform with arms waving.

If the bird jumped from the platform during the first stage (or during the edge tests) it was graded with a score of 0. If it remained during the sequence of tests it was rated with progressively higher scores, the highest score of 5 being given to birds which were still on the platform at the end of the series. The ratings used in 1961 were based on a slightly modified and supposedly more sensitive test sequence involving eight stages.

These testing and scoring procedures were, of course, arbitrarily designed, and the stages must not be interpreted as equal increments along a single continuum. The values derived are useful only for comparing birds of different origin by nonparametric techniques.

Data recorded with each test included: (1) subjective categorization of the topography of the nesting substrate into 5 classes from level to precipitous; (2) estimated distances to the nearest significant drop or rise in topography, the nearest beach or

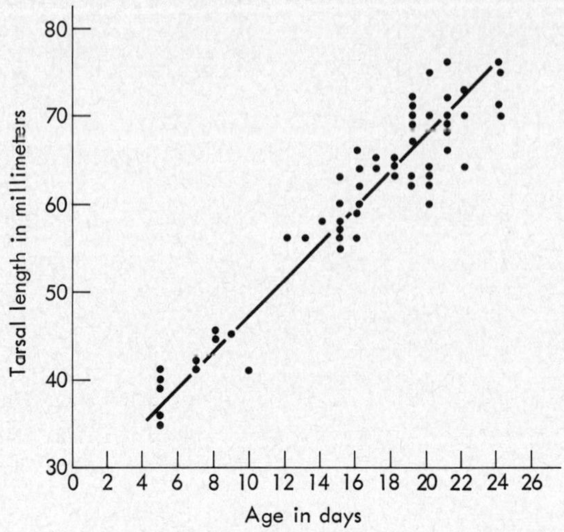

Fig. 6. Correlation of tarsal length and absolute age in 56 free-living gull chicks marked during the first few days of life for subsequent capture and measurement. The line for the mean, drawn by eye, indicates a fairly even growth rate of approximately two millimeters per day.

coastline and the nearest vegetated area; (3) an indication of substrate texture, whether loam, peat, sand, gravel or rocks; (4) a formula description of the vegetation at the nest site giving height, density and foliage type; (5) distance of the test chick from its nest at time of capture; (6) distance from nest or capture site to the test site (where pertinent, the texture and topography of the substrate at the site of testing was noted and the direction selected by the jumping bird with respect to these features); (7) a subjective rating of the "temperament" of the bird based on responses to being captured and handled; (8) records of the occurrence of regurgitation or defecation on the platform; (9) a crude evaluation into three categories of the clamor and general excitement of local adult gulls during the test; (10) a record of the time of day, the sky cover, wind and temperature conditions at the time of testing, and (11) an age index—the tarsal length measured at the conclusion of each test and subsequently translated to absolute age (Fig. 6).

RESULTS

Part I. Edge response

A negative edge response apparently develops during the first day post-hatching in all gull chicks. In six incubator-hatched chicks in which the age was known to within an hour the withdrawal response to the platform edge was detected shortly after the birds could stand and walk at ages varying from 3 to 18 hours. By the second day the response was universal and readily detectable in the edge tests.

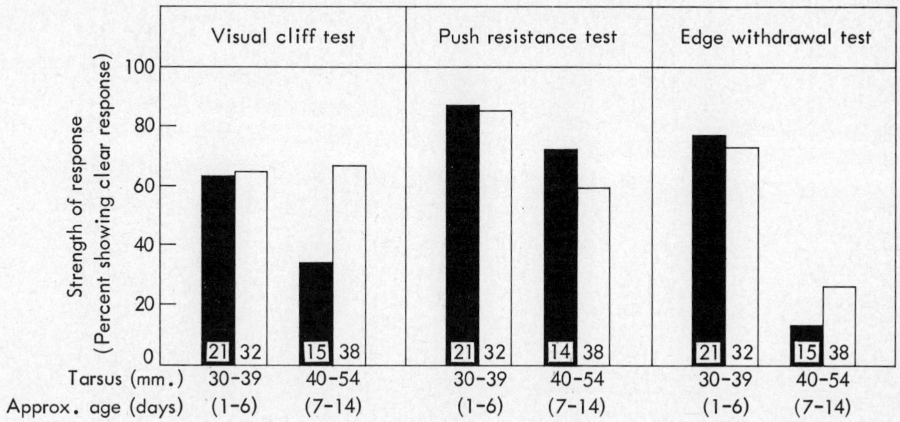

Fig. 7. Visual cliff, push resistance and edge withdrawal responses of chicks from cliff nests (black bars) and plateau nests (white bars) at two stages of development. The height of the bars indicates the percent of birds in each size (age) category which showed moderate or strong responses. The numbers in the bars indicate sample size.

In edge tests with chicks hatched in nature (Fig. 7) the level of responsiveness declined with advancing age, a decline which coincided with and was presumably related to the development of locomotor skills and of the escape response. In the visual cliff and edge-withdrawal tests, the two considered most valid as indicators of edge response, the declining trend was more pronounced in the birds from cliff nests than in those from plateau sites. The difference was significant (chi square test) in the visual cliff tests but not in the others.

Part II. Escape response

In the jump tests all gull chicks old enough to stand and walk but with tarsi less than 35 mm (ca 1–3 days of age) remained on the platform through at least five of the six successive stages (Fig. 8). A marked drop in this hesitancy to jump occurred about the fourth or fifth day in chicks from plateau nests with less than a third of them remaining on the platform through stage five of the testing. Chicks from nests in cliff situations showed no comparable change of response, and, except for a few individuals, continued in their refusal to jump until about 17 days of age. The difference between the two groups during the intermediate stages (4th to 17th day) was highly significant by the chi square test.

Examination of the recorded data on the condition of the birds and the nature of the habitat and weather at the time of testing failed to reveal any correlations of interest other than those described above for topography at the nest or capture site. Test birds were deliberately selected from the extreme types of topography. No correlations of edge or escape behavior with topography and substrate texture at the

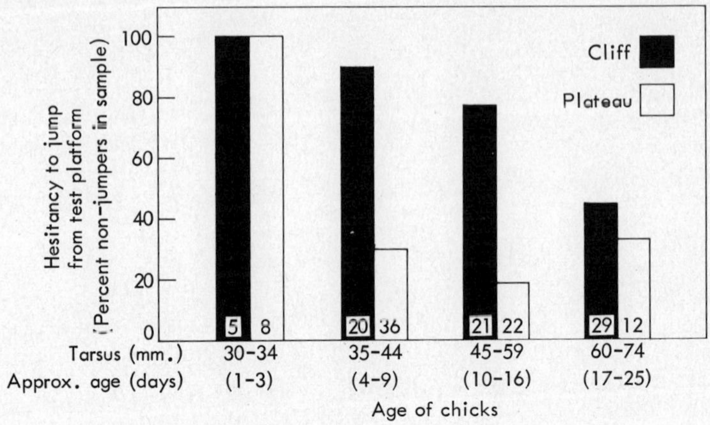

Fig. 8. Decline in reluctance to jump from the test platform with advancing age in chicks from cliff nests (black bars) and from plateau nests (white bars). The height of the bars indicates the percent of birds in each size (age) category which were non-jumpers, i.e. remained on the test platform through stage 5 of the test series. The numbers in the bars indicate sample size.

testing site, as opposed to nesting site, could be detected, and no consistent tendencies to select a jumping direction towards or away from a cliff or plateau situation were found.

Part III. Egg transplant experiments

An experiment was conducted in the summer of 1961 in which eggs from cliff nests and plateau nests were inter-changed during the early incubation stage of nesting, and the chicks allowed to develop in the environment of their foster parents. All eggs from each of the designated cliff nests were removed shortly after the clutch had been completed and immediately replaced by an equal number of eggs of plateau origin. Experimental nests were visited at frequent intervals during the period of hatching and the chicks marked for subsequent identification when one or two days of age by punching a diagnostic pattern of holes in the toe webs. The birds were then left undisturbed with their foster parents until they had reached an appropriate age for testing. At that time 53 toe-clipped chicks were recovered and tested on the jump platform, 28 of them from the plateau and 25 of them from the cliff area. Without exception these chicks were found in and had presumably been reared entirely within the habitat type and topographic situation in which they had hatched.

Chicks hatched on the plateau from eggs laid in cliff nests showed the free jumping characteristics of native plateau birds, while birds hatched and reared on cliff sites from eggs of plateau origin behaved like their cliff-dwelling neighbors (Table 1).

TABLE 1. Comparison of platform jump scores of chicks from eggs transplanted between cliff and plateau nest sites and their controls. (Because of different scoring techniques used in the 1961 tests, these figures are not directly comparable to those shown in Figs. 3 and 4.)

Site of origin	Site of rearing	Non-jumpers (6–8)	Inter-mediates (4–5)	Jumpers (1–3)
Cliff	Plateau	4(14%)	4(14%)	20(72%)
Plateau	Plateau (control)	3(13%)	6(25%)	15(62%)
Plateau	Cliff	8(32%)	10(40%)	7(28%)
Cliff	Cliff (control)	14(31%)	20(44%)	11(24%)

Part IV. Experiment with experience deprivation

In the summer of 1961 an experiment was performed to test the hypothesis that the hesitancy to jump in cliff-reared chicks is related to restrictions on movement imposed by narrow ledge situations during early stages of behavioral development. Circles of poultry netting, 4 feet in diameter were placed around 30 plateau nests during the early stages of incubation.

In 15 of these all vegetation within the fence was cleared away so that the chicks were hatched into a spatially restricted environment without sheltering foliage. In the other 15 the vegetation was left intact and the chicks, on hatching, had the natural cover of the plateau but were restricted in their freedom to move more than two feet from the nest. Standard jump tests were made after the chicks had reached an appropriate age. Chicks from unfenced nests in the same area served as local controls.

The number of penned chicks available for testing at the appropriate age was small, and the results are quite inconclusive (Table 2). If we adopt the questionable procedure of adding the data from the plateau controls of the transplant experiment to those of the local controls, there is a slightly higher refractoriness in the penned birds comparable to that of cliff birds. No difference is detectable in the scores of penned birds from cleared as opposed to vegetated pens.

TABLE 2. Comparison of platform jump scores of plateau chicks restricted to the close vicinity of their nests with controls not so restricted

a—Restricted in cleared pens	5(33%)	3(20%)	7(47%)
b—Restricted in vegetated pens	3(30%)	5(50%)	2(20%)
c—Unrestricted on vegetated plateau (local control)	3(23%)	7(54%)	3(23%)
d—Unrestricted on vegetated plateau (total controlled)	6(16%)	13(35%)	18(49%)

DISCUSSION

Habitat-response correlations

Avoidance of cliff edges and reluctance to jump are behavior traits with obvious survival value for chicks of any species living in precipitous nesting situations. Herring gulls examined in this study all showed such negative responses in elevated platform tests during the first few days of life regardless of their site of origin. After that, however, responses changed variously according to the topography of the substrate on which the bird had its home. These changes may be summarized as follows:

1. Negative responses declined with advancing age in all platform tests.
2. Cliff chicks dropped their withdrawal responses at the platform edge more rapidly than plateau chicks.
3. Cliff chicks retained their reluctance to escape by jumping longer than plateau chicks.

The decline in the withdrawal response in the platform edge tests indicates a growing refractoriness or insensibility to edge situations with advancing age. It suggests an adaptive adjustment of behavior to conditions to which a cliff-resident bird is continuously exposed, an habituation to a stimulus situation consistantly presented without negative reinforcement. The visual cliff and edge withdrawal tests (A and C in Fig. 7) would seem to be the best indicators of this changing response; the failure of the push resistance test (B in Fig. 7) to show the same differential decline may be due to complications introduced with the direct disturbance of the subject during testing.

The results of the jump tests graphically shown in Fig. 8 indicate that after the first few days of life chicks from cliff nests were more reluctant to jump than chicks from plateau nests. Thus, the birds which showed the greatest boldness in the platform edge tests were the most hesitant in the jump tests. This apparent contradiction, if real, can only be resolved by attributing invalidity to the weaker set of data—the edge test data. A more satisfactory explanation is that there is no contradiction, and that the two tests in fact measure two different response categories. Support for this latter interpretation is found when we consider that the suggested habituation to edge situations in chicks from cliff nests could be highly destructive if it increased the probability of hasty escape responses under provocation. Under these circumstances natural selection would strongly favor a dissociation of the escape response from the edge response.

Available information on mortality factors operating in the two nesting situations under consideration suggest survival values for each of the escape responses according to the situation in which it occurs. Predation by adult gulls is the principal factor among plateau nesting herring gulls where from 50 to 88% of all chicks may succumb before fledging (Paynter, 1949; Paludan, 1951; Darling, 1938). The causes of mortality in herring gull chicks on cliff sites is not known, but studies by Cullen (1957) on ledge nesting kittiwake gulls suggest that accidents associated with the hazard of the physical situation are more important. Insofar as these observations apply to the nesting situations of Kent Island, a chick situated on an exposed ledge overlooking a precipice would do well to stay put, while another perched in the center of a clearing on the plateau would be exposed to danger until it had moved to cover.

The differences between plateau and cliff-reared chicks as observed could conceivably be based upon genetically determined behavioral characteristics fostered by an intense natural selection of adapted phenotypes. On the other hand it is possible that the differences in escape response are acquired through learning processes operating at different levels or in different ways in the two nesting situations. Each of these possibilities is considered below.

Genetic considerations

A study of the visual cliff responses of a variety of neonate animals led Walk and Gibson (1961) to conclude that "an animal's response to the lack of visual support is unlearned, a reflex that is characteristic of the species." If this is true for species it could also be true for genetic varieties, and the response difference between gull chicks from the two nesting sites examined in this study could have its basis in a behavioral dimorphism of origin. Such a possibility gains plausibility when one considers the divergent selective pressures operating in the two nesting situations. Under the conditions described above natural selection would favor cliff clinging variants in cliff situations and more mobile wanderers and shelter seekers in the vegetated plateau areas of the colony.

The maintenance of two genetic types in a gull colony would depend on one of two mechanisms: (1) reproductive isolation or polytypy—a segregation of breeding adults of the two genotypes in the appropriate rearing situations; or (2) stable polymorphism—a broad range of genetic variation in each successive generation spanning the phenotypes required for survival in the two rearing situations.

Reproductive isolation of the genotypes responsible for a phenotypic dimorphism in juvenile behavior would require that the birds on maturing return to nest on the same type of site on which they were reared several years before. A tendency to return to the rearing locality for nesting has been demonstrated for a number of bird species and a predilection for a similar nesting situation has been indicated for several (Miller, 1942; Thorpe, 1945; Hochbaum, 1958). There is no evidence, however, for a tendency to return to a site sufficiently developed to provide the level of inbreeding necessary to maintain effective genetic segregation as here postulated. The results of the egg transplant experiments (Table 1), furthermore, provide rather convincing evidence against the existence of distinct breeding genotypes in the two nesting situations.

Stable polymorphism requires that enough individuals be produced at the extremes of variation to insure survival of an adequate breeding stock in each of the environmental situations to which these extreme variants are adapted. No segregation of breeding types is required, only the repetitive production of appropriately adapted phenotypes in each generation of chicks. Such stable polymorphism has been demonstrated in nature in the snail (*Cepea nemoralis*) by Cain and Sheppard (1950) and in various moths by Kettlewell (1956). Its occurrence in the Kent Island herring gull

TABLE 3. Percent representation of birds in the low score, median score and high score categories of the 1960 jump tests at four age (tarsal length) levels

	Tarsal length (age)			
	30–39	40–49	50–59	60–74
High score birds	70	52	45	39
Median score birds	15	25	32	41
Low score birds	15	23	23	20
Sample size	40	39	34	39

colony is problematical, but the high levels of overproduction and juvenile mortality which characterize gull colonies could conceivably be a reflection of the expensive selection process inherent in this mechanism. Evidence against the occurrence of stable polymorphism is seen in the recorded increased number of birds in the median score categories with advancing age (Table 3). In a stabilized polymorphism based on selective juvenile mortality one would expect a decrease rather than an increase in the numbers of intermediate performers.

Early learning

The alternative to a genetic explanation of the observed behavioral differences between cliff and plateau chicks is that the respective characteristics of the two groups are acquired as a result of divergent individual experience in the two contrasting habitats during the early days of life. We have seen that the hesitance to jump declined during the first week in plateau birds while it remained high in cliff birds (Fig. 8). Insofar as learning is involved, this difference must have been associated with provocative factors in the plateau situation, inhibitory factors in the cliff situation, or a combination of the two. Since there is no way to determine which of the two courses, if either, is "normal", we can do no more than compare and evaluate the relative significance of distinctive environment factors in the two situations as possible determinants of the respective behavioral characteristics. Modern learning theories suggest three processes which have been operating on the birds under study:

1. A negative conditioning process in birds from cliff sites in which injuries received in falls reinforce the reluctance to jump.
2. A negative conditioning process in birds from vegetated plateau sites in which attacks from adult gulls in the nesting area reinforce tendencies to flee from exposed situations.
3. An increase in general adaptability and boldness in chicks from vegetated plateau situations as a result of wider experience in a varied and relatively unrestricted environment.

The first process, a simple learning to refrain from jumping in edge situations as a result of negatively reinforcing experience loses its initial logical appeal when one considers that in the precipitous situations of this study most trial jumps or falls would be lethal either directly through injuries or indirectly through difficulties in returning to the home ledge to which the parent birds bring food. The possibility that negative reinforcement is generated by simple visual exposure to cliff edges seems to be refuted by the results of the edge and visual cliff tests discussed above. It is also interesting and perhaps pertinent that of six birds which jumped when the test platform was set up near the brink of a precipice, three selected the outward direction.

The proposition that chicks from vegetated plateau sites learn through experience the value of protecting shelter and hence tend to flee from exposed situations such as the test platform, was examined directly in the cleared pen experiment in which chicks were reared on small plots at the nest from which all vegetation had been removed (Table 2, groups a and b). The evidence provided by this experiment is weakened by the fact that the encircling fences incidentally excluded marauding gulls and thus may have lowered the incentive value of vegetative cover. The results, depreciated by this consideration and by the small sample size, fail to provide evidence for the proposition.

The possibility that jumps from the platform were motivated by attraction to sheltering cover rather than escape from exposure was tested unsystematically by deliberately placing the platform in contrasting situations during many of the tests and by noting the directions selected for jumping by the test birds. No useful correlations were detected. In the twenty instances where the platform was placed over the boundary line between a vegetated and a barren, rocky surface, the birds jumped without regard to the nature of the substrate even in those cases where the bird spent a minute or more circling the platform edges and peering down before jumping.

The third proposition links the dropping of the initial reluctance to jump in plateau birds with a general adaptability or boldness acquired through rearing in a relatively complex environment. Chicks from plateau nests characteristically venture many yards from their nests' sites within a few days of hatching and in so doing inevitably encounter a wide variety of objects and situations including hostile attacks from neighboring birds. Chicks reared on a narrow cliff ledge are, by comparison, closely restricted. Schaller and I (1962) have presented experimental data showing that visual exposure to a relatively rich environment during the first few days of life reduces the level of escape responses to strange objects and situations in chicks of the domestic fowl. The varied background of experience acquired by gull chicks from plateau situations may, in a similar way, increase the boldness with which they respond to unfamiliar situations such as that presented by the elevated jump platform; the lack of it may underlie the hesitancy of cliff birds on the platform.

Data obtained by experimentally restricting the environmental experience of plateau chicks by confining them within a small area near the nest (Table 2, lines b, c and d) are disappointingly meagre. The results suggest, however, that such restrictions may influence the birds to respond conservatively as proposed in this hypothesis.

GENERAL CONCLUSIONS

Where all or essentially all members of a species living in nature respond to a given stimulus situation in a more or less stereotyped way, the response is referred to as a natural or species-characteristic response. The role of environmental factors in the ontogenetic development of such responses has been difficult to assess, and two more or less opposing schools of interpretation are generally recognized. One group follows the nativist approach and holds that the uniformity and uniquity of species-characteristic pattern are due primarily to genetic homogeneity, and that the form of the pattern is the inevitable result of an innately determined maturational process. Environmental factors are regarded as secondary modifiers rather than determinants of the natural pattern. The other group emphasizes the importance of environmental factors throughout development and attributes the uniformity and uniquity of the natural response to the uniformity of the natural environment of development as much as to genetic homogeneity. The natural pattern is thought to be natural mainly because it was developed in, and hence molded by factors of the natural environment.

Experimental approaches to the resolution of this disagreement have been thwarted by semantic problems. The simple procedure of rearing birds in the presence of experimental (unnatural) environmental factors fails to provide an answer so long as one group interprets these factors as secondary modifiers of a genetically predetermined pattern and the other regards them as essential determinants of the pattern.

The occurrence of two behavioral phenotypes in a single population provides a situation in which the matter of unnatural, secondary factors is bypassed and in which environmental influences during development can be analysed and compared on equal terms. On this basis the two distinct response patterns found in the present study of herring gull chicks are interpreted as ontogenetic equivalents, and the concept of natural or species-characteristic response patterns is considered inapplicable except as it incorporates recognition of the role of environmental determinants.

REFERENCES

Cain, A. J. and Sheppard, P. M. (1950), Selection in the polymorphic land snail, *Cepaea nemoralis*, *Heredity*, **4**, 275–294.

Cullen, E. (1957), Adaptations in the kittiwake to cliff nesting, *Ibis*, **99**, 275–302.

Darling, F. F. (1938), *Bird Flocks and the Breeding Cycle*. Cambridge University Press.

Hochbaum, H. A., *Travels and Traditions of Waterfowl*. Minneapolis: Univ. of Minn. Press.

Miller, A. H. (1942), Habitat selection among higher vertebrates and its relation to intraspecific variation, *Amer. Nat.*, **76**, 25–35.

Paludan, K. (1951), Contributions to the breeding biology of *Larus argentatus* and *Larus fuscus*, *Vidensk. Medd. Dansk. Naturh. Foren.*, **114**, 1–128.

Paynter, R. A. (1949), Clutch size and the egg and chick mortality of Kent Island herring gulls, *Ecology*, **30**, 146–166.

Schaller, G. B. and Emlen, J. T. (1962), Ontogeny of avoidance responses in various precocial birds, *Animal Behaviour*, **10**, 370–381.

Thorpe, W. H. (1945), The evolutionary significance of habitat selection, *J. Animal Ecology*, **14**, 67–70.

Walk, R. D. and Gibson, E. J. (1961), A comparative and analytical study of visual depth perception, *Psychological Monographs*, **75** (15), 1–44.

The Black Noddy Anous tenuirostris *on Ascension Island Part 2, Behaviour*

J. M. CULLEN AND N. P. ASHMOLE

At a breeding colony on Ascension, many Black Noddies remained throughout the year. The main part of the paper describes their behaviour, with special reference to daily rhythm, sunning, absence of "panics," nesting (with possible value of the birds' own guano for cement), fighting, and courtship (with details of ground and aerial displays), copulation (apparently rare) and the behaviour of parents and young.

On Ascension the birds nest on small cliff-ledges, elsewhere they also use trees, both kinds of sites being relatively inaccessible to ground predators, unlike the majority of terns, which nest on the ground. As with the Kittiwake, another cliff-nesting species which has diverged from ground-nesting stock, a number of the Black Noddy's peculiarities can be related directly or indirectly to its nesting site. Indeed the occurrence in just these two species of two features absent in their ground-nesting relatives indicates that there are some cliff-nesting adaptations which had previously been overlooked. The Black Noddy's adaptation in this respect can be summarized as follows:

Cryptic adaptations lost. Cryptic plumage of nestling and juvenile. Dispersal of egg-shells and droppings.

Adaptations acquired, and their consequences. Shortage of nest-sites, focussing territory on nest, with consequent increasing, though not exclusive, importance of Bridling as "song." Allo-preening between mates. More elaborate nest-building, with gathering of material (if available), which is cemented together with guano; side-stepping and depressing tail when shaping nest. Young intolerant of others; crouch if threatened instead of running; peck at feet of intruders, or beak-hide to appease attacks, precocious development of adult (threat?) plumage.

The Brown Noddy is the closest relative of the Black, and what little is known of its breeding biology suggests it is less specialized for nesting above ground, and that its ability to nest on the ground as well as in bushes and trees allows it to breed in localities where the Black Noddy could not.

Reprinted in part from *Ibis*, volume 103b, pages 423–446 (1963).

INTRODUCTION

The general biology of the Black Noddy *Anous tenuirostris* on Ascension Island has been treated by Ashmole (1962); the present paper embodies the results of the work on its behaviour, about which little appears to have been published. The main references are Henshaw (1901), Campbell and White (1910), Macgillivray (1926), Murphy (1936), Warham (1956) and Moynihan (1962).

It may be helpful at the outset to point out that among terns the species is unusual, as it is a cliff- or tree-nester. Most of the gulls and terns nest on the ground, and E. Cullen (1957) has shown that many of the peculiarities of the Kittiwake *Rissa tridactyla*, a cliff-nesting gull, are the consequence of a change of nesting habitat. This example was continually at the back of our minds when comparing the Black Noddy with the other known species of terns, and even during this short study a number of convergences with the Kittiwake were found. All references to the Kittiwake are taken from the above paper or unpublished work by E. Cullen.

On Ascension most of the nests of the Black Noddy are difficult to reach, but in the early stages of the expedition J. M. C. (and D. F. Dorward) watched at an inaccessible, but easily observed, colony on the south coast of Ascension at Cocoanut Bay. The later observations were all made on Boatswain Bird Island. Here many thousands of noddies nest on the seaward-facing cliffs, but others breed on smaller rock-faces which can easily be watched at a distance of a few yards. Most of the behaviour observations to be described were made on a section of cliff accommodating about 200 pairs, and on which only a few nests were accessible but all were easily viewed.

* * *

The breeding periods of the Ascension Black Noddies during the expedition have been described elsewhere (Ashmole 1962). The observations on behaviour were started in December 1957, near the end of a breeding period when only a few young were still unable to fly. J. M. C. left Ascension in February, well before the start of the next laying period, at the end of April. However, many birds frequented the cliff during the non-breeding period, and much display was seen, though little courtship feeding and only one copulation. N. P. A. started observing Black Noddies only in April 1958, but he was able to continue through one laying period (April to August 1958) and intermittently until the start of the next in April 1959.

OCCUPATION OF LEDGES

The Black Noddies on Ascension occupy the nesting ledges all the year round, at least for roosting. As Ashmole (1962) has shown, birds which have nested previously often resort to their former breeding ledges. Some individuals perch and roost on ledges which would not be suitable for nesting, but these are almost always birds which do not "own" (i.e., regularly resort to and defend vigorously) better ledges elsewhere.

There seems to be competition for good nest-sites, at least for sites sufficiently near to other members of the species to satisfy this strongly colonial bird. Fighting for ledges is not confined to the breeding period. Since an established bird has a better chance of winning a fight over a ledge when the breeding period starts and the ledge assumes its roles as a nesting place, there may well be survival value in a bird securing a ledge as early in its life as possible. As the species is evidently able to remain around the island throughout the year the territorial behaviour occurs at all seasons.

Panics

An interesting point was the absence of "panic flights," which are such a well-known feature of the colonies of other species of terns: In a "panic flight" the birds over the whole (or part) of the colony suddenly fly away from their nesting area towards the sea, remaining away from their nests for a matter of seconds or minutes. These panics are particularly common at the beginning of the season (Cullen 1956 for *Sterna macrura*; Ashmole, pers. obs., for *S. fuscata*), which fits with Tinbergen's suggestion long ago (1931) that they are alarm flights due to the terns resorting to nesting habitats which are very different in appearance from the open mud flats and seascapes where they spend most of the year. Their unfamiliarity with the nesting area generates a nervousness which very readily sparks off a panic towards the more familiar sea. This view is supported by the absence of panics in the Black Noddy, where the birds remain at their breeding sites throughout the year.

* * *

Clubs and sunning

The resting places or "clubs," close to the colony, which are used by off-duty gulls and terns throughout the breeding season, were less obvious in the Black Noddies. There were certain parts of the cliff within the main study area which were sloping and had no proper ledges on them. These were not persistently defended by pairs like nesting ledges, and often half a dozen birds or more would stand together there. On these neutral areas the birds showed relatively little aggressive or sexual behaviour, but usually rested or preened or sunned themselves, while occasionally a bird would fly from a nesting ledge to one of them to rest. However, during the 1958 breeding period there was a shortage of nesting ledges on the cliff, and some of the small perches in the clubs were appropriated by pairs which attempted to build nests and breed there, usually without success. In some colonies certain areas are probably used as clubs throughout the year, but in our study area clubs were not a conspicuous feature of the life of the colony during the breeding period. However, there were places around the island, quite unsuitable for breeding, where the Black Noddies would assemble to sun, and these seemed to be neutral areas like clubs.

* * *

NESTING

Nest-sites

The Black Noddy nests on cliff ledges at all its breeding stations in the Atlantic, except in British Honduras, where the nests have been reported in mangroves (Salvin 1864). In the Pacific and Indian Oceans the species usually nests on the branches of trees. Unlike its relative, the Brown Noddy *Anous stolidus*, it seems never to nest on flat ground, though sometimes resorting to low bushes in the absence of anything better (Munro 1960). (Wilson and Evans (1890–1899) were informed by Palmer, Rothschild's collector, that in Laysan, Lisiansky and Midway the species laid its eggs on the sand like the Brown Noddy, but a photo in Rothschild (1893–1900) shows the birds nesting in bushes. And several recent papers dealing with this area refer only to cliff-, tree- or bush-nesting (Fisher and Baldwin 1946, Richardson and Fisher 1950, Munro 1944, Richardson 1957).) On Ascension no trees or bushes are available except a mile or two inland, (and they have most been introduced within the last 150 years—Stonehouse 1960), and the Black Noddy nests exclusively on small ledges and protuberances on steep cliffs.

Nest-material

Apart from guano, the importance of which will be described below, feathers were the only nest-material used in significant quantities by the noddies on Boatswain Bird Island. No other suitable material is readily available, as both land-plants and sea-weed are virtually absent. It is evident from the literature that more substantial nests are sometimes made, the most convenient local material being used: for instance, for Hawaii, "Rubbish gathered from the sea" (Henshaw 1901), at St. Paul's Rocks green seaweed torn from the rocks by the surf (Moseley 1879), at Clipperton Island "algae from the lagoon" (Gifford 1913), on the Capricorn Islands seaweed and the leaves and sticky seeding-heads of the tree Pisonia (Macgillivray 1926), and on Christmas Island the leaves and inflorescences of *Messerschmidia argentea*, sometimes with seaweed, sticks or grass, but not feathers (Gallagher 1960).

* * *

Importance of guano

The guano of the birds themselves was an important constituent of the nests of the Black Noddy on Ascension, and every regular perch soon acquired a covering of guano. It seems certain that in the course of time some ledges which were originally only perches, had become large enough to nest on. The guano often projected well beyond the original ledges, and hung down in curious stalactite forms beneath them. It accumulated because the birds defaecate while standing on the ledges, and the coastal rainfall is only a few inches in the year. One group of ledges on the cliff, some feet below a projecting rock, suffered serious damage from drips during the occasional periods of rain. On the other hand many of the nest-sites, especially those on the leeward side of Boatswain Bird Island, are entirely sheltered from rain, and it was here that the most impressive guano stalactites, sometimes more than two feet long, were found. The guano masses which formed the foundations for some of the nests occasionally broke away from the cliffs, leaving rock ledges on which nesting was impossible or hazardous. This happened to one ledge shortly before the start of laying. The owners were left with a ledge quite unsuitable for nesting, but after steadily gathering feathers in the next few weeks they somehow achieved a platform on which they laid and hatched an egg successfully.

Guano appears to be a common constituent of the nests of the Black Noddy in all parts of its range. Of 32 apparently independent accounts which we have found in the literature, 13 particularly mention droppings "fouling" the nest, while three others show it in a photo without comment. (Of the remainder none say that droppings were absent, and all close-up photos we have seen of the species' nest show droppings.) By contrast, and this is a point we shall return to later, of 38 accounts of the nests of Brown Noddies only two mention droppings. There must clearly be a species difference with regard to the quantity of guano found in the nest, and it may well be that it is an invariable component of the Black Noddy's nest, even though not always thought worthy of mention.

The role of guano was extremely important on Ascension in enlarging the ledges and binding together the other nest-material, such as it was—just as mud, specially gathered for the purpose, is used by the Kittiwake—and it would seem very probable that it has a similar significance in other parts of the Black Noddy's range. Since this seems to be an adaptation of the species to its nesting habitat, it raises the question whether the defaecation habits of the Black Noddy are in any way modified to secure this advantage. Our scanty observations were inconclusive. Sometimes the noddies left their ledges and defaecated while circling near the colony, but this was difficult to observe. Birds defaecating on the ledges usually did so with their tails over the side of the ledge, but the action was often casual, so that the droppings sometimes fell clear and sometimes caught on the ledge and helped to widen the platform. A more "deliberate" action was frequently seen: the bird backed towards the edge of the platform, flexed its legs, lowered the tail and deposited the faeces on the rim of the nest. However, we have seen this method also employed by other gulls and terns on the ground, so that it is not a method restricted to the species, though careful observations might show that it was commoner in the Black Noddy.

* * *

Nest-building actions

Like other larids, the Black Noddies have characteristic nest-building actions which one could recognise on Ascension even though they could probably be studied more satisfactorily in an area where more nest-material was available. These were often performed by the female as she waited at the nest while her mate was gathering material, and some of the actions were seen also after the egg was laid (again like other larids).

Bracing. This was clearly equivalent to the "scraping" well known among Charadriiformes. The bird tilts the body forward, so that the tail is cocked up, with the legs braced backwards so that the breast is pressed against the inside of the nest-saucer. None of the Black Noddies was ever seen to kick backwards with its legs as in the scraping of other species; instead the position was merely held for a few seconds and the bird then stood up. We have therefore preferred the term "bracing" to describe the action, since scraping implies a vigorous movement with the legs. During the bracing the bird frequently moves round in the nest, standing up, turning round and then again bracing.

While in the bracing position the tail is sometimes bent down for a moment towards the nest rim; this is a rather slow and deliberate independent action of the tail—not a tilting of the whole body. This movement of the tail during bracing is puzzling. It also occurs commonly in scraping Kittiwakes, where its function is not obvious. In the Black Noddy the tail sometimes, but by no means always, touches the nest rim. There was no indication that defaecation on to the nest rim ever accompanied the movement. It seems possible that in habitats in which nest-material is more plentiful the action could help to consolidate the edge of the nest as it may do in passerines (Deckert 1955), but on Ascension it had no evident function.

Both sexes brace, but the female perhaps more frequently, especially while her mate is collecting feathers. Sometimes both members of the pair were seen to brace in the nest within a short period. Bracing is commonest in the building period, right up to the day of laying, and was not noticed afterwards except on one occasion within a few hours of the loss of the egg (which had been laid less than two days before). Bracing was also seen in members of prospecting pairs, visiting ledges where they were not established as the owners.

Side-stepping. A movement we called "side-stepping" was sometimes associated with a bout of bracing and occurred occasionally at other times. While standing in the nest cup the body was slowly turned to point in a different direction by a rather rapid picking-up and putting-down of the feet. Although in its most conspicuous form the frequency of the foot movements was much greater than was necessary merely to achieve the turning of the body, it was often less striking, and must often have passed unnoticed Macgillivray (1926) was probably referring to this movement when he described how the Black Noddies on the Capricorn Islands make the foundations of their nests by folding limp *Pisonia* leaves on a branch "where they are arranged with the beak and trampled down with the feet." Where leaves or seaweed form the bulk of the nest, the behaviour may well serve to consolidate its structure, and even on Ascension to trample the feathers into the guano of the nests before they are blown away. It is interesting to note that, though the action is virtually absent in the ground-nesting gulls and terns, the Kittiwake also possesses it—trampling: E. Cullen (1957). It must therefore be a convergent adaptation to help fasten the nest-material onto the tiny ledges selected by these aberrant representatives of a typically ground-nesting group.

Sideways-building. All the larids, in common with waders, have a characteristic "sideways-building movement" by which, in its most complete form, nest-material is picked up in front and deposited close to the bird's flank. We saw this behaviour by the Black Noddies on Ascension only six times in all, never during the nest-building before laying, nor indeed by established pairs. At least twice the bird doing it was an unmated male temporarily separated from the partner with whom he had been circling or making display flights. A general fiddling with material, picking up and dropping it, without the sideways component, was much commoner, for instance by females at the nest while the male was away collecting, and this might be regarded, by analogy with other larids, as merely an incomplete, low-intensity form of the sideways-building, but the completed movement is much rarer than in other species. This cannot have been only due to the poverty of nest-material, since, in the absence of anything more suitable. Arctic Terns, for instance, will contrive to do the sideways movement with a tiny particle. We would therefore infer that the complete sideways-building movement is much rarer in the Black Noddy than in most gulls and terns, even when there is plenty of nest-material. We do not know why this should be so, nor whether it is true also of the Brown Noddy.

FIGHTING

Both sexes of the Black Noddy defend their ledges by displays and actual fighting. Usually an intruder flees as soon as the occupying bird threatens it, and is rarely pursued after leaving the ledge. If the occupier has to resort to force, the attack usually takes the form of a peck at the other bird with its beak, and this is sometimes directed at the feet of the intruder. This latter mode of attack is the one used by young noddies in defending their ledges against strange adults. As in other terns, the wings are not used to deliver blows.

Four or five times we saw a serious fight for possession of a nesting ledge. Only two birds were engaged at any one time, though twice a pair was trying (unsuccessfully) to displace a bird who was at the time alone on a ledge, though not necessarily unmated. Fighting consisted of pecking at any part of the body, with occasional interlocking of bills. Frequently, birds fell off the ledge during fighting. On one such occasion a bird had a firm grip of the other's leg, and held on in spite of savage pecks by the other at his head, neck and feet. They fluttered almost to the ground before they separated and continued fighting in the air and back on the ledge. No adult was seen to be injured during fighting.

"GROUND" DISPLAYS

A noddy occupying a ledge does not tolerate another adult on the ledge apart from its mate or a potential mate. Display is usually seen whenever one bird joins another on a ledge. The displays and calls used differ to some extent according to whether the new arrival is the mate of the occupier, or is an intruder, but so many of the displays occur in several contexts that it seems best to treat the main displays and their associated calls in turn, describing them and discussing the situations in which they appear . . .

Croaking. When one approaches a breeding cliff the noddies may remain on their ledges until one is a few yards away, but they usually start to repeat a rolling frog-like croak: "kik-krrrrr."

* * *

Nodding. If an intruder arrives on a ledge occupied by another, one commonly sees the "Nodding" display, which may or may not be the action which Watson (1908) refers to by this name in the Brown Noddy. Moynihan (1962) has also used the word to describe *Anous* displays, though in a very broad sense, as explained later.

* * *

Bridling. Another conspicuous display of the Black Noddy is "Bridling," homologous with the Choking of gulls (see discussion later). Unlike Nodding, this display is usually performed by only one bird; it is used when another bird is present on the ledge, or under certain circumstances when a partner or antagonist is in the neighbourhood. During Bridling the body is tipped forward so that the tail is high and the shoulders are low, and the bill is pointed downward at an angle of rather more than 45°. The action consists of a regular rhythmical backwards and forwards movement of the head, once or twice a second, often repeated for many seconds; as it is pulled back the bill becomes slightly more horizontal, and it is opened rather widely at the end of the backstroke. The bill is then closed sometimes with an audible click as the head is thrust forwards again to the original position. At the end of a bout the birds may remain in the Bridling position for some time before again performing the action. Sometimes the movements are continued for several minutes, with only short interruptions. The display is usually silent (but see later).

Bridling occurs in several different situations, and its causation is not easy to understand. Most commonly it is seen between the members of a pair—Friendly Bridling; but it is also seen in hostile encounters, when it is more exaggerated, with the neck rather elongated and sometimes with short sharp calls at the top of the upstroke which we did not hear in the friendly form.

Both sexes perform the action; occasionally they do it together for a short time, but usually one does it while the other watches. When one of the mates (normally the male) leaves the ledge to collect nest-material, remaining within sight of the ledge, the partner commonly stands in the preparatory Bridling posture, and may or may not make the typical movements as well. Unlike most of the displays, Bridling often starts without any obvious external stimulus: A pair are standing quietly on a ledge resting or preening when one starts Bridling, and the other usually stops what it is doing to watch. Bridling then usually dies down to nothing and the pair relax again. At other times the Bridling is set off by a particular event: for instance the return of a bird's mate (when it usually follows a bout of Nodding), or the arrival of another bird on a nearby ledge. When a pair are together the Bridling seems to have little effect beyond causing the mate to pay attention, though we have seen it make the mate approach closer to the displaying bird.

Bridling seems to play a part in pair formation, as shown by several observations on color-ringed "flirting" birds, i.e., birds which were displaying together but were not members of established pairs. On one such occasion two ringed birds which did not subsequently pair, but which had been displaying together, landed a few feet apart on a sloping area unsuitable for nesting. They "Foot-looked" (see below), and then the male started an exaggerated form of Bridling. The female then moved gingerly up to the Bridling male, and they Nodded together for a long time before leaving on a display flight.

We have also seen, though less frequently, that Bridling may have a frightening effect on the mate. Our most convincing observation concerned an established pair. On one occasion, shortly after arrival on their ledge the male was preening the head of the female, who was in the Bridling position. Then he stopped preening her and started the hostile Bridling, often uttering a hard, sharp call at the finish of the backward movement. The female responded by immediately abandoning the Bridling position for a more upright one, and then leaving for the next ledge. Later she returned and they Foot-looked and Bridled together.

When an intruder is driven off a ledge by the owner, the actual attack is sometimes preceded by the owner. More often the attack is preceded by Nodding, but sometimes Bridling may occur interspersed with it. A very striking situation in which Bridling occurs, is in intervals of a serious fight for a nesting ledge. Just like the "aggressive" Choking of gulls (Tinbergen 1960), it is then used by the bird left in possession of the ledge when its attackers have temporarily withdrawn. Bridling soon stops if the attackers go right away, but otherwise may be continued until the next assault is launched. Attackers were not seen to Bridle while waiting on nearby ledges in the gaps in the fighting.

When the members of a pair are together on a ledge and one is Bridling, the other may stand quietly by its mate, but it sometimes utters a short metallic "kek" every second or two in an irregular way, from no particular posture. This may continue as long as the Bridling continues, and at the same time the kekking bird may show an interest in the mate's beak as if it would take any food regurgitated. However this call was not noticed during Courtship Feeding.

Chin-up. Another common display of the noddies we term "Chin-up with Rattle." In the typical form the bird stretches the neck upward and points the bill at an almost equally steep angle. It then often, but by no means always, utters the Rattle which consists of a regular metallic "kuk-kuk-kuk-kuk—, . . ." whose frequency is about six syllables per second, continued for up to three or four seconds. At lower intensities the neck is less stretched, and the bill is raised only slightly above the horizontal.

* * *

Gaping. Another display which intruders evoke is the "Gape." With neck stretched upwards, the bill pointing down and widely open to show the bright orange-yellow inside, and the base of the tongue raised, the bird turns towards or advances at a newcomer.

* * *

Foot-look. Perhaps the commonest of all the social responses of the Black Noddy is the "Foot-look." A bird suddenly glances down at the ground in front of or between its feet, holding the position for two or three seconds.

* * *

Head-shake. After noticing the regularity of an upward toss of the head when a pair of Brown Noddies come together (the Head-flick of Moynihan), we re-examined the Black Noddies and noticed for the first time that they often made a Head-shake, just like the action of a bird with a wet head, but without the upward throw of the Brown Noddy's Head-flick.

* * *

Occurrence of different agonistic displays. To give a better idea of the frequency of the different displays in the different situations, we noted the actions seen in a sample of encounters when two birds came together on a ledge. These encounters were classified as Hostile, when one bird drove the other away, and Meeting, when the two birds relaxed or went off together in a display flight (see later). The hostile encounters were further divided into the performance of the attacking and fleeing birds. In the Meeting the performances of the two birds were usually alike and are not therefore distinguished. Table 1 shows the percentage of performances in each class in which a display occurred at least once. (Landing Gapes are omitted because they are not part of the encounter situation. Headshakes are omitted because they were overlooked until too late.)

All the actions described so far were seen at least occasionally both between mates and between territorial rivals. But it will be clear from what has already been said that some of the actions are much commoner in hostile situations than others. A detailed account of the motivation of the displays was not possible from our brief study, but much qualitative observation together with the assumption, which seems valid for the gulls (Tinbergen 1960), that the early displays in the Meeting Ceremony are more aggressive than later ones suggests the following general conclusions for the commoner displays. The Chin-up and Rattle seemed the most aggressive, and was

TABLE 1. Relative frequency of various displays (percent) when two birds came together on a ledge in Hostile and Meeting situations.

	Hostile		Meeting
	Attacking	Fleeing	
Chin-up and Rattle	15	2	1
Croak	1	0	0
Gape (not Landing Gape)	1	19	3
Nodding	32	36	28
Foot-look	20	20	27
Bridling	5	9	19
"parp"	5	11	12
"kahhh"	0	0	4
"kek"	0	2	6
No. of performances on which percentages are based	75	44	221

rather seldom seen between mates except when elicited by a third bird. At the other extreme, Bridling was very common between well established mates, but it also occurred, less frequently and in a slightly different form, in clearly hostile situations. Nodding and Foot-looking were common both between mates and between rivals, and in the latter case were used both by the subsequent winner and the subsequent loser of the encounter. The Gape was rarer between mates, and though both winner and loser did it frequently, it was differently oriented, the winner doing it so that the brilliant coloured mouth was displayed to its opponent, the loser tending to display it to one side, or even away from the other. The Croak and the associated posture was common neither in territorial fights nor as a reaction by the mates to each other, and the motivation was not clear. The other calls were not recorded sufficiently often to say anything about their motivation beyond what is mentioned in their descriptions.

The remaining postures, to be described now, were not noticed in obviously hostile situations between rivals, but only between actual or potential mates.

Courtship feeding and elongate posture. Courtship Feeding was a conspicuous feature of the behaviour of the Black Noddies in the period shortly before egg-laying. It occurred almost exclusively between the members of established pairs, was most common during the two weeks before laying and was hardly ever seen more than three weeks before. Macgillivray (1926) reported that Courtship Feeding continued during incubation: We did not see this, but on one occasion a bird (presumed female) begged for an instant just after her mate had arrived to relieve her on the egg, which had been laid a few days previously.

* * *

Copulation. We rarely observed Copulation, which surprised us as it is common in the other terns we are familiar with. Moynihan (1962) also notes its rarity in the Brown Noddy and the Fairy Tern *Gygis alba*. We did observe that in the two weeks before they laid many of the females had white marks round the cloaca, which were not seen at other times and presumably resulted from attempted Copulations.

* * *

Allopreening. The preening of one member of a pair by the other is common among the Black Noddies, and is not restricted to the breeding season. It seems to occur most frequently in the early morning, when ordinary preening is also commonest. In Allopreening the active bird usually confines its attention to the head of its partner; the latter remaining passive, even when the preening bird seems very rough. Either partner may preen the other, but there is a suggestion from our notes that the male Allopreens more than the female. Allopreening is not a normal feature of the behaviour of "flirting" birds, but on one occasion it was seen between two birds which had formed a very temporary liason. For its occurrence in pre-copulatory behaviour, see above.

Other calls. Black Noddies have a considerable vocabulary, and apart from the various calls already mentioned, several others could be identified. No doubt more could be separated, but the detection of calls was sometimes difficult among the sounds of wind and sea.

Two more or less distinct calls which were heard both in hostile situations and when a pair came together (Table 1) are a short strident "parp" and a more drawn-out "kahhhh." Neither was common, but the former at least seemed to be associated with the nodding display; it was uttered with the bill pointing obliquely upwards.

As a noddy flies in to alight on its ledge after an absence from the colony, it may utter a rapid "kyer-kyer-. . ." which we termed the "Arrival Call." This call seems to be used only by a bird returning to a ledge which it owns. It is frequently heard as a bird arrives to relieve its mate on the egg, or to feed its chick, but it is also used on returning to an empty ledge. A more prolonged rattle-like variant of it is heard if a bird arrives to find its ledge occupied by an intruder. It is then clearly aggressive, and is usually followed immediately by the stranger's fleeing.

Another call which was fairly frequently heard we named the "Hear Me" call because of its apparent function. It was harsh, easily distinguished from other calls, and was always uttered by a bird (usually, but not always, a male) making one of the short circling flights in front of the cliff which were a characteristic feature of life in the colony. The use of this call very often resulted in another bird flying out to join the one which called. Sometimes the attracted bird seemed to be a stranger, but often the two birds had been flirting together beforehand; members of established pairs did not often use this call. If a bird, after giving the "Hear Me" call, was joined by another, the two birds often circled together, and then might land on the cliff together or separately, or might go for a display flight. It seems likely that this call is important in attracting or maintaining contact between prospective mates.

The white cap. We could not be sure of the function of the white cap. It is displayed conspicuously at the opponent in some threat displays, but also in the less hostile Foot-look, though then often with a downward orientation. There is good evidence in the typical *Sterna*-type terns, where the arrangement of black and white on the head and body is exactly the reverse of the noddies, that the black cap enhances threat behaviour and is turned away in non-aggressive approaches to other birds (Cullen 1956, Tinbergen 1959). It may well be that in both *Sterna* and *Anous* the demonstration of the contrasting cap intimidates other birds.

If this is correct, the downward pointing of the white cap of the Black Noddy in the non-aggressive Foot-look might represent one of the four possible methods of turning away the threat-signal structure which Tinbergen (1959) has listed for appeasement displays, additional to the "upwards and away" and "sideways and away" of other terns (references in Cullen 1956, also Moynihan 1962, Lind 1963).

* * *

AERIAL DISPLAY

High flight

Like other terns (Cullen 1960b) the Black Noddy has a common aerial display, the "High Flight" usually involving two birds, which consists of an ascent followed by a fast glide to sea level. During the ascent, which may last only a few seconds or until they are specks in the sky, the birds keep close together and often look at each other. Much of the time they fly with peculiar wing-beats, more regular, rather faster, and of greater amplitude than in normal flight, which carry them forwards more slowly than wing-beats of this speed usually do; at times the birds almost seem to rise head to wind without progressing forwards at all. As the birds climb they usually sway a little from side to side about the flight path, the swaying being more or less synchronized between the two partners. The ascent is not in large circles as is usual in other terns. During many ascents when one was close enough, one could hear a "tik" call uttered about five or six times a second. The call was noted as more regular than the Rattle, but perhaps it is related.

The ascent changes gradually to the descent, with the birds flying at first horizontally or slightly downwards. After a few seconds the ascent may be resumed for a while, before the final descent, at first shallow and then more steeply, often in a fast glide, with the pair all the time keeping very close together.

During the ascent and the descent it can often be seen that one of the pair is continually adjusting its position to keep near the other and to follow the side-to-side swaying movements which the other initiates. Sometimes the same individual shadows the other both in the ascent and in the glide, but on other occasions the roles change after a while. A rather similar swaying occurs during the High Flight glide of other terns and Cullen (1960b) has suggested that it is due there to an overcompensation in the efforts of the shadowing bird to follow its partner's movements quickly enough. This is certainly not the explanation in the Black Noddy, for the swaying is initiated by the steering bird, as we called it, and not by the shadower.

* * *

Other aerial display

Most terns that have been studied appear to have two more or less distinct aerial displays, of which the High Flight, as just discussed, seems to be homologous throughout, the other—the "Low Flight"—being less distinctive (Cullen 1960b). The Black Noddy has no conspicuous second aerial performance, but the flights involving the "Hear Me" call show certain resemblances to the Low Flights of the other species: an unmated bird would leave its territory (ledge) and use a characteristic call during a short flight. This performance sometimes attracted another bird (often one who had previously been flirting with the performer), and the two birds would fly together, and would sometimes land and display, or go for a High Flight together.

* * *

PAIR FORMATION

The details of pair formation in the Black Noddy were not worked out, but its general course could be established by piecing together fragments of the process. Recognisable pairing ceremonies take place only on the ledges (except the aerial display) and not on the clubs, where Herring Gulls *Larus argentatus* sometimes pair (Tinbergen 1960). The Elongate Posture and Purring, together with Bridling, seemed important in attracting females. Perhaps they are not the earliest displays to arouse the attention of unmated females but we have not seen that females commonly landed beside males doing other displays. When a female joins a male, the displays of the pair together include some (e.g., Nodding) which are seen during hostile encounters. As in many other territorial birds, the male sometimes shows signs of aggression mixed with his display towards the newly arrived female, and she may be easily alarmed by him. The display flight by both partners presumably cements the pair bond, and after a time the mutual distrust subsides and the pair are more relaxed together. Even after the pair have been flirting together regularly, they often visit different sites together (cf. Macgillivray 1926). The final choice of nest site may evidently be deferred until close to egg-laying. After the pair is formed, perhaps not for a long time, Courtship Feeding (and presumably Copulation) become commoner.

This account is too incomplete to draw many comparisons with other species, but one thing is clear. Unlike the cliff-nesting Kittiwake, the Black Noddy does not always select the definitive nest-site before pairing. A Black Noddy sometimes acquires a mate when its territory is a spur of rock scarcely suitable for nesting, and the pair then tries to get a more adequate site later.

The Brown Noddy differs from the Black in its pairing behaviour, for according to Moynihan (1962) unmated males alternate between a potential nest-site (in a bush or small tree at the colony studied) and a small clear area on the ground, both of which were treated as territories and defended (cf. the nesting and pairing territories of gulls). This suggests, as will be discussed later, that the Brown Noddy may be less specialized than the Black as an "above-the-ground" nester. (It should be mentioned that where we watched the Black Noddies there were plenty of relatively level areas within a few yards of the nest sites which they could have used for ground territories if they had wanted to.)

E. Cullen (1957) argued that because the Kittiwake advertised for a mate on its nest site, this had contributed towards the selection of Choking as the "song" of the species (the display which attracted females and warned other males that a site was owned), rather than the Long Call which other gulls use. It is a confirmation of this argument that whereas the Brown Noddy's song is the Long Call according to Moynihan, in the Black Noddy the Bridling, which is homologous with the Choking of gulls, has partly taken over this function.

PARENTAL CARE AND THE BEHAVIOUR OF THE YOUNG

The incubation behaviour of the Black Noddy has already been described (Ashmole 1962). The disposal of egg-shells will be mentioned later. The chicks, which are not cryptic but even in their downy plumage are patterned just like parents, are fed by regurgitation, like those of the Brown Noddy (Watson 1908). When begging for food the young peck at the bill of their parents. The feathered young withdraw the neck, tilt the body forward a little and utter a plaintive monosyllabic piping call every second or two. Smaller young pipe and peck, but the hunched posture is less distinct. The way the chick takes food from the parent's beak is the same as that used by the female in Courtship Feeding, with the beak inserted crosswise.

Young noddies continue to depend on their parents for food, at least to some extent, after they can fly. On a few occasions when we saw marked juveniles begging after they could fly, it was always on their own ledges, and from the behaviour of unmarked individuals too it seems probable that while the young are at the colony they are fed only on their nests. The young sometimes beg from adults who are not their parents, but we doubt if they are ever fed, as the adults drive away such young if they approach.

In general, the young Black Noddies do not leave their own ledges until they can fly and during this time are intolerant of any juvenile or adult other than their parents on their ledges—in contrast with juvenile *Sterna fuscata* which, while keeping away from strange adults, associate freely with each other (N.P.A., pers. observation). If approached by a human the young noddies either crouch, usually facing away from the front of the ledge, or, if older, usually stand in an upright posture, raise the wings to some extent, and peck at an intruding object. The young are only brooded for a few days after hatching, and after this they are left unattended for most of the day, and wandering adults, househunting pairs, or single birds which have lost their eggs or chicks, may land by them.

If a bird other than its parent lands on its ledge, a small chick will usually turn towards the back of the ledge and crouch. In this position the intruder will often ignore the chick, or he may poke or peck at the chick's head. If pecked, the chick remains passive, but keeps its bill pointed right down and away from the adult, the nape fully exposed. The bill may be held to one side of the body, or be tucked down the front of the body and pressed into the breast feathers. Although the adult may continue to peck a chick which had adopted this posture, the attack is less vigorous than when the chick does not hide its beak, and the posture may therefore be said to function as appeasement. This turning away of the head is almost identical with what the Kittiwake does in the same circumstances (Facing Away—Tinbergen 1960) but only young noddies were seen to use the posture, while in the Kittiwake it is employed by adults as well.

The "beak-hiding" posture may be used by chicks up to at least four weeks old, but older chicks, and sometimes much younger ones will defend their ledges against intruding adults. They may threaten the adult, but if the adult does not flee they will peck at its legs and feet, and although the adult may peck back, vigorous young are generally successful in driving intruders away. Small or weak young however, frequently have to endure continual though often rather desultory pecking by adults, and some are certainly killed in this way. Many more chicks, however, are found alive but weak below the nesting colonies, often with plucked and bloodied crowns. Watching the rough jostling and pecking to which one submissive chick was subjected by two displaying strangers on its ledge, it was easy to see how chicks could be pushed off their ledges by intruders. On the other hand some chicks probably left voluntarily: Some of the nests are connected by narrow ledges with other parts of the cliff, and before the young can fly they begin to explore around the nest. At times they get themselves to places from which they cannot return, and may be severely pecked by birds onto whose ledges they find their way. Strong chicks which fall or are pushed off their ledges try to climb up again. A typical case is that of a juvenile which appeared one day at the foot of the cliff having fallen from higher up. It could flutter a little but could not fly well enough to regain its own nest. During the next two days the unfortunate bird scrambled about at the bottom of the cliff, sometimes managing to get quite high up but always being pecked or falling off. The third morning it was dead, its scalp torn open in two places.

Without watching the young for several hours a day it is difficult to discover the exact age at which they are first capable of flight. However we have no record of a chick flying at less than six weeks old, and the five cases for which we have good information all indicate that most young start to fly in the early part of their seventh week.

The young noddies are very unsteady in their early flights, particularly in returning to their own ledges, which sooner or later they must do, since they are not fed by their parents anywhere else (see above). They commonly make the mistake of trying to land when still some way above the ledge. Perhaps it is because these juveniles so often cause trouble by landing on the wrong ledge, that when one starts to fly around the cliff, it often sets the adults croaking.

DISCUSSION

Adaptions to cliff- or tree-nesting

The habit of nesting on cliffs and in trees has apparently led to some of the peculiarities of the Black Noddy, such as the absence of a cryptic plumage in the young. When the species breeds on sheer cliffs, as it does on Ascension, the nests are inaccessible to nearly all predators, and there is presumably no need for the young to escape observation. In birds nesting in bulky nests in trees, a cryptic plumage in the young is probably of little value, since most predators will approach from below, and will see the nest before the young.

It has been argued elsewhere (Cullen 1960a) that the dispersal of droppings and the hatched empty egg-shells is a cryptic adaptation in some larids to conceal the position of the nest from predators hunting by sight, and this has recently been demonstrated experimentally for the egg-shells of the Black-headed Gull *Larus ridibundus* by Tinbergen *et al.* (1962). The Black Noddy defaecates often without leaving its ledge, as already mentioned. Macgillivray (1926) writes that egg-shells of the Black Noddy were found beneath their nests, which suggests they are not carried away, but to confirm this we set up a standard type of situation which has been used in other species (Cullen 1956), in which an egg-shell is presented concave upwards in the nest of the individual to be tested, whose own egg(s) have been removed. Arctic or Common Terns will fly off with such a fragment and drop it yards away, and it is also convenient for experimentation that they do this throughout the incubation period (see also Beer 1962). Other species, on the other hand, like the Sandwich Tern, will not carry away the shells and brood them instead. We tested the responses of four individual Black Noddies (six tests in all). In none of them was the shell carried away though in one test it was picked up and dropped beside the nest so that it fell down the cliff. In the others the bird brooded the shell. Thus with regard to droppings and the disposal of shells the Black Noddy behaved like those species of terns which nest non-cryptically and like the Kittiwake among the gulls.

Although Black Noddies' sites may be safer from predators and so permit various cryptic adaptations to be given up, their use involves certain disadvantages: a shortage of suitable ledges and the need for a better constructed nest. It has been suggested earlier that there is a certain amount of competition for good ledges between the Black Noddies on Ascension, and this has presumably led to the greater importance of the nest-sites as the focus of the pairing territory compared with the Brown Noddy (see above) or with *Sterna* spp. (Cullen 1956), though not as exclusively as in the Kittiwake. This in turn has promoted the development of Bridling as the "song" display by which the male attracts females, for like its homologue, Choking, it occurs characteristically at potential nest-sites in all gulls and terns (Tinbergen 1960). This trend parallels the evolution of Choking as "song" in the Kittiwake, through in the Black Noddy, Bridling has not become the *exclusive* song, as has Choking in the Kittiwake.

The precarious sites of the Black Noddy require more elaborate nests than the shallow depressions of the ground-nesting species fashioned in a few minutes and lined, if at all, only with fragments of straw and shell. The special features of the Black Noddy's nesting have already been described and include the gathering of nest material (probably more efficacious at some colonies than it was at Ascension) and side-stepping (apparently to compact the material collected) similar to, though less marked than the trampling of the Kittiwake. The birds' droppings are also important, at least on Ascension, in enlarging the foundations of the nest, a remarkable parallel with the mud which is specially gathered by the Kittiwakes for the same purpose.

Another disadvantage of the cliff sites is their small size, which has presumably led to the intolerance of the young noddies for strange adults (potentially hostile) or juveniles (potentially food competitors) which might endanger the survival of the chick. When pecked by another bird, a young ground-nesting gull or tern either fights or runs away. Young Black Noddies, like Kittiwakes, crouch instead of running and adopt the beak-hiding posture which reduces the chance of their being attacked by the intruder. Furthermore when they are larger they have an efficient method of keeping adults off their ledges by pecking at their feet. The striking similarity of their plumage, especially the white cap, to that of the adults is perhaps related to this aggressiveness, for it will be remembered that the cap is displayed prominently at an opponent in certain threat displays. It is perhaps worth stressing that the downy and first juvenile plumage of the young are more than mere noncryptic patterns; in the nestling, there is white on the crown, and even the tiny white eye-mark is reproduced arguing that these features have survival value at an early age, whether or not in connection with hostility.

The Black Noddy and Kittiwake show two other peculiarities distinguishing them from their ground-nesting relatives, and whose adaptive significance for cliff-nesting is less obvious. E. Cullen (1957) did not include these in her list of cliff-nesting adaptations in the Kittiwake, but the very fact that we now see they are also found in the Black Noddy is strong evidence that some link, direct or indirect, exists. These peculiarities are (1) the depression of the tail when bracing or scraping, and (2) allopreening:

(1) The way the Black Noddy and Kittiwake depress their tails when forming the nest-cup has already been mentioned, together with the suggestion that it could help in shaping the nest-rim in localities where, unlike Ascension, plenty of nest-material is available. As the shaping of the nest is obviously of greater importance in species with precarious sites, this may account for its occurrence in these two species.

(2) Comparing different species of birds, allopreening between the mates appears to be correlated with their having to stand close together. For instance herons, cormorants and pigeons do it—where the female in pair formation must come right up to the male on his small defended nest-site—but waders, gamebirds and grebes do not—where the mates can stand a little way apart. We are indebted to Dr. F. McKinney for

a particularly fine example in the waterfowl, none of which commonly allopreens except for the tree-ducks *Dendrocygni,* in which the mates perch side by side on restricted perches. The correlation seems fairly clear, but not the reason for it. Allopreening might act as appeasement to help to reduce the disruptive, aggressive and fleeing behaviour between the mates which is aroused by their proximity to each other and which cannot be "equilibrated" spatially, i.e., by an individual moving a few inches or feet away from its partner until its approach-avoidance tendencies with respect to the partner are in balance. Alternatively the behavioural equilibration may be achieved in some other way, perhaps by some other displays, and the allopreening might happen merely because the birds regularly find themselves in such close proximity. In some species some of the allopreening is an extension of autopreening—"toilet" allopreening, the bird preening itself and then running over, as it were, on to its neighbour. But even in such cases there has evidently been further specialisation as toilet behaviour since particular attention always seems to be given to the head and neck of the other bird, just those parts which a bird cannot reach itself. Besides toilet allopreening, there is in many species "display allopreening," which is not mixed with autopreening, and is confined to special social situations where it seems to have been specialized as a signal, presumably tactile. The Shag *Phalacrocorax aristotelis* illustrates both kinds of preening (J.M.C., pers. observation). The Black Noddy and Kittiwake certainly have toilet allopreening, and further study may show them to have display allopreening as well. Thus although the reason for the allopreening of the Kittiwake and Black Noddy may not be clear, it appears very probable that it is somehow linked with cliff-nesting.

Our knowledge of the behaviour of the Black Noddy is still far from complete, but the above discussion shows that already a number of its differences from the ground-nesting species can be related to cliff-nesting and the indirect consequences of this. As repeatedly mentioned, the parallel with the Kittiwake compared with ground-nesting gulls is very striking, and has confirmed the view, which is becoming more and more clear, that many apparently "arbitrary" species differences in behaviour may be correlated with changes in some key feature, such as nesting habitat, itself a new, or at least different solution to the problem of survival.

Pending further studies, the Black Noddy seems a less perfected cliff-nester than the Kittiwake. We observed losses of eggs and young due to the poor construction of a number of nests (Ashmole 1962). How much of this was due to the lack of materials on Ascension and how much to the inefficient behaviour of the species remains to be seen. On the other hand it may be that some of the apparent "imperfections" on the cliffs are the result of adaptations to the tree-nesting habits of the species in other parts of its range.

Nesting adaptations in the Brown Noddy

The Black Noddy's closest relative is the Brown Noddy; indeed these two species (or three if *A. tenuirostris* and *minutus* are distinguished) comprise what most taxonomists have regarded as the genus *Anous*. Moynihan (1959) has enlarged the genus to include *Procelsterna* and *Gygis*, but we have followed the more usual grouping.

For a comparison of nest habits, information on the Brown Noddy is rather meagre and mainly in Watson (1908), Murphy (1936), van Bemmel and Hoogerwerf (1940), Gibson-Hill (1947, 1951), Warham (1956), and Dorward and Ashmole (1963), while the displays have been treated in detail by Moynihan (1962).

Whereas the Black Noddy nests only on cliff-ledges or in bushes or trees, the Brown will use a wider variety of sites. Where bushes or trees are present, it usually nests in them, and it also uses cliff-ledges (e.g., Gibson-Hill 1951). A number of observers mention crevices on cliffs or among rocky outcrops on the ground—the latter being the sites used on Ascension—but more open sites on flatter ground seem also to be acceptable (e.g., Ridley 1957). When nesting in trees or bushes the Brown Noddy collects a substantial nest of twigs, seaweed or whatever vegetation is available, and builds a bulky nest. In more rocky sites where vegetation is scanty a few pebbles or shells may be all there is of a nest. As has already been mentioned, guano is not a normal constituent of the nest.

The Brown Noddy is thus not an obligatory above-the-ground nester like the Black, and apparently never nests on the tiny brackets which the latter species is able to use. Less specialized nesting in precarious sites and its ability to nest on flat ground in the absence of anything raised permits it to inhabit more of the low, often arid islets, lacking ground predators, which dot the tropical and subtropical seas. On the other hand Gilbert (quoted by Whittell 1942) mentions the severity of lizard predation for ground-nesting birds, including the Brown Noddy, on Houtman Abrolhos, from which the Black Noddies, nesting in the upper branches of trees, were protected.

If the Brown Noddy is less specialized as an above-the-ground nester, is this reflected in other differences in its biology? The juvenile Brown Noddy, like the Black, looks very like the adult without any trace of the barred or speckled cryptic plumage of most juvenile terns and gulls. The nestlings are polymorphic (Dorward and Ashmole 1963) and at least on Ascension the pale chicks seemed better camouflaged than one might expect against rocks where the species nested. Any generalisation would be premature, but it seems likely that the chicks are on average less conspicuous than the young Black Noddies, and perhaps more than the young of typical *Sterna* terns. The fact that the white cap appears to be virtually absent in the downy plumage fits with the view that its earlier appearance in the Black Noddy is linked with the cliff site, perhaps via the enhanced aggressiveness of the young (see above).

Warham (1956) and Dorward and Ashmole (1963) have drawn attention to an interesting difference between the behaviour of young Black and Brown Noddies. When approached by a person, the former tend to crouch, as already mentioned, but the latter are much more ready to run. Watson (1908) and Gibson-Hill (1951) have earlier remarked how Brown Noddy chicks leave their nests when alarmed, even before they can fly. Such a practice must be a danger for a species which nests high above the ground, and Gibson-Hill saw young precipitating themselves into the sea from cliff-nests on Christmas Island when he approached; on the other hand it must be advantageous when the nests are near the ground where they would be accessible to predators, and under these conditions Watson found in the Dry Tortugas that the young can leave the nest and reach the ground safely.

Dorward and Ashmole also found an indication that the Brown Noddy first flies when its wings are less well developed than the Black, a difference which, if confirmed, could be correlated with the danger for the Black of leaving its precarious site before fully competent to fly. Presumably connected with this danger is the fact that young Black Noddies—but not the Brown (Watson 1908)—are fed only at their nests throughout their period of dependence. The same is true of the Kittiwake, while the ground-nesting gulls and terns continue to feed their young right away from the colony.

Three other features of the Brown Noddy's behaviour which have already been mentioned suggest that it is less specialized as an above-the-ground nester than the Black. (1) The absence of guano in the nest, (2) the existence of a ground "pairing territory" where a male advertises for a mate, in addition to the nesting territory and (3) a consequence of the last, the absence of Bridling from the "song" display.

As regards the nesting habits of the other genera included by Moynihan (1959) within *Anous*, virtually nothing is known of *Procelsterna*, while for *Gygis*, though Moynihan writes that they are "only an exaggeration of the arboreal nesting habits of such species as the Black Noddy," Dorward's (1963) account shows that the species has become specialized to tree (and cliff) sites in a quite different way. It may well be that *Gygis* is more closely related to *Anous* than to the *Sterna* terns, but we have felt that its difference from *Anous* in colouration, displays and nesting adaptations justifies the retention of generic rank. Verheyen's (1959) suggestion, with a massive accretion of morphological data, that *Anous* (but not *Gygis* or the other terns) is closely related to the skuas is startling but, as an analysis of his own data shows, without serious foundation.

ACKNOWLEDGMENTS

It is a pleasure to thank the Expeditions Committee of the British Ornithologists' Union for the opportunity to visit Ascension, and the many people, including the other members of the party, who made the expedition a success and with whom we have since profited in discussion. During this period we were receiving grants from the D.S.I.R. (N.P.A.) and Nuffield Foundation (J.M.C.) and during the preparation of this paper also from the Nature Conservancy (J.M.C.). A number of people, especially Esther Cullen, have helped to form the views which have been developed in this paper, and we should like to thank her also for permission to use unpublished observations.

REFERENCES

Ashmole, N. P. (1962), The Black Noddy *Anous tenuirostris* on Ascension Island, *Ibis*, **103b**, 235–273.

Beer, C. G. (1962), The egg-rolling of Black-headed Gulls *Larus ridibundus*, *Ibis*, **104**, 389–398.

Bemmel, A. C. V. van and Hoogerwerf, A. (1940), The birds of Goenoeng Apri, *Treubia Buitenzorg*, **17**, 421–472.

Bryan, W. A. (1903), A monograph of Marcus Island, *Occ. Pap. Bishop Mus. Honolulu*, **2**, (1), 77–139.

Campbell, A. J. and White, S. A. (1910), Birds identified on the Capricorn group during the expedition of the R.A.O.U. 8th to 17th October 1910, *Emu*, **10**, 195–204.

Cullen, E. (1957), Adaptations in the Kittiwake to cliff-nesting, *Ibis*, **99**, 275–302.

Cullen, J. M. (1956), A study of the behaviour of the Arctic Tern *Sterna macrura*, D. Phil. Thesis, Oxford University.

Cullen, J. M. (1960a), Some adaptations in the nesting behavior of terns, *Proc. 12th Int. Orn. Congr.*, 1958, 153–157.

Cullen, J. M. (1960b), The aerial display of the Arctic Tern and other species, *Ardea*, **48**, 1–37.

Cullen, J. M. (1962), Allo-, auto- and hetero-preening, *Ibis*, **105**, 121.

Deckert, G. (1955), Beitrage zur Kentnis der Nestbautechnik deutscher Sylviiden, *J. Ornithol.*, **96**, 186–206.

Dorward, D. F. (1963), The Fairy Tern *Gygis alba* on Ascension Island, *Ibis*, **103b**.

Dorward, D. F. and Ashmole, N. P. (1963), Notes on the biology of the Brown Noddy *Anous stolidus* on Ascension Island, *Ibis*, **103b**.

Fisher, H. I. and Baldwin, P. H. (1956), War and the birds of Midway Atoll, *Condor*, **48**, 3–15.

Gallagher, M. D. (1960), Bird notes from Christmas Island, Pacific Ocean, *Ibis*, **102**, 489–502.

Gibson-Hill, C. A. (1947), Notes on the birds of Christmas Island, *Bull. Raffles Mus.*, **18**, 87–165.

Gibson-Hill, C. A. (1951), Notes on the nesting habits of seven representative tropical sea birds, *J. Bombay Nat. Hist. Soc.*, **48**, 214–235.

Gifford, E. W. (1913), Expedition of the California Academy of Sciences to the Galapagos Islands, 1905–1906. VIII. The birds of the Galapagos Islands, with observations on the birds of Cocos and Clipperton Islands, *Proc. Calif. Acad. Sci.*, **2**, (4), 1–132.

Goethe, F. (1957), Das Herabstarren, eine (Ubersprungbewegung bei den Lariden, *Behaviour*, **11**, 310–317.

Henshaw, H. W. (1901), On the habits and haunts of the Noio or Hawaiian Noddy Tern, *Ibis*, **1**, (8), 196–200.

Lind, H. (1963), The reproductive behavior of the Gull-billed Tern, *Sterna nilotica, Gmelin. Vidensk. Med. Dansk Naturh. Foren.*, **125**, 407–448.

Macgillivray, W. (1926), Birds of the Capricorn Islands, *Emu*, **25**, 229–238.

Moseley, H. N. (1879), *Notes by a naturalist on the Challenger*, London: Macmillan.

Moynihan, M. (1959), A revision of the Laridae (Aves), *Amer. Mus. Novit.*, 1928, 1–42.

Moynihan, M. (1962), Hostile and sexual behaviour patterns of some South American and Pacific Laridae, *Behaviour, Suppl.*, **8**.

Munro, G. C. (1944), *Birds of Hawaii.* Honolulu: Tongg.

Murphy, R. C. (1936), *Oceanic Birds of South America.* New York: American Museum of Natural History.

Richardson, F. (1957), The breeding cycles of Hawaiian sea birds, *Bishop Mus. Bull.*, **218**, 1–41.

Richardson, F. and Fisher, H. I. (1950), Birds of Moku Manu and Manana Islands off Oahu, Hawaii, *Auk*, **67**, 285–306.

Ridley, M. W. (1957), Terns of the Seychelles Islands, *J. Bombay Nat. Hist. Soc.*, **54**, 920–924.

Rothschild, Lord (1893–1900), *The Avifauna of Laysan and the Neighbouring Islands.* London: Porter.

Salvin, O. (1864). A fortnight amongst the seabirds of British Honduras, *Ibis*, **6**, (1), 372–387.

Stonehouse, B. (1960), *Wideawake Island.* London: Hutchinson.

Stonehouse, B. (1963), Egg dimensions of some Ascension sea-birds, *Ibis*, **103b**.

Tinbergen, N. (1931), Zur Paarungsbiologie der Flusseeschwalbe (*Sterna hirundo hirundo* L.), *Ardea*, **20**, 1–18.

Tinbergen, N. (1959), Einige Gedanken an Beschwichtigungsgebarden, *Z. Tierpsychol.*, **16**, 651–665.

Tinbergen, N. (1960), Comparative studies of the behaviour of gulls: a progress report, *Behaviour*, **15**, 1–70.

Tinbergen, N. *et al.* (1962), Egg-shell removal by the Black-headed Gull, *Larus ridibundus* L.; a behaviour component of camouflage, *Behaviour*, **19**, 74–117.

Verheyen, R. (1959), Notes sur la systematique de base des Lariformes, *Bull. Inst. Sci. Nat. Belg.*, **35**, 1–16.

Warham, J. (1956), Observations on the birds of Pelsart Island, *Emu*, **56**, 83–93.

Watson, J. B. (1908), The behavior of Noddy and Sooty Terns, *Pap. Tortugas Lab.* Vol. 2., Publ. Carnegie Inst., **103**, 185–225.

Whittell, H. M. (1942), A review of the work of John Gilbert in Western Australia, *Emu*, **41**, 289–305.

Wilson, S. B. and Evans, A. H. (1890–1899), *The Birds of the Sandwich Islands.* London: Porter.

Cliff-nesting Adaptations of the Galapagos Swallow-tailed Gull

JACK P. HAILMAN

The Galapagos Swallow-tailed Gull (*Larus* (*Creagrus*) *furcatus*) nests on shallow to steep cliffs. In some respects (i.e., reduced nesting space, danger of falling over cliff) its environmental conditions resemble those of the cliff-nesting Kittiwake (*L.* (*Rissa*) *tridactylus*). Unlike *tridactylus*, *furcatus* has abundant nesting materials available, as do ground-nesting gulls such as *L. argentatus* and *atricilla*. In some aspects of its ecology (availability of nest sites, amount of nest predation) *furcatus* is intermediate between the *tridactylus* and ground-nesting gulls.

Many behavioral and morphological characteristics of *furcatus* were noted in field study and experiments. Thirty of these are unambiguous enough for comparison with the other species. Of those characters presumably adaptive to the environmental conditions shared with ground-nesting gulls, all five resembled the characters of the ground-nesting species. Of seven characters presumably related to the "intermediate" ecological conditions, five resembled characters of ground-nesters and two were intermediate. Finally, of 17 characters presumably adaptive to conditions shared with the Kittiwake, 11 resembled those of the Kittiwake, one was intermediate, and 6 resembled those of ground-nesting species.

Thus, Cullen's (1957) hypothesis that the Kittiwake's unusual characters are adaptive to special ecological conditions accompanying cliff-nesting is, in general, confirmed.

INTRODUCTION

E. Cullen (1957) showed that the cliff-nesting Black-legged Kittiwake (*Larus* (*Rissa*) *tridactylus*) differs from "typical" (i.e., ground-nesting) gulls in many respects. The species' unique morphological and behavioral characters, Cullen cogently argued, have resulted from adaptation (either directly or indirectly) to cliff-breeding. Epistemologically, the correlation between cliff-nesting and unusual characters constitutes a hypothesis that must be "tested" independently on a relatively unrelated cliff-nesting gull. Therefore, while I was studying the chick-feeding behavior of the cliff-nesting Galapagos Swallow-tailed Gull (*Larus* (*Creagrus*) *furcatus*) I noted the general habits of this species for comparison with the Kittiwake.

METHODS

The result of observations of the colony on southern Plazas Island off Santa Cruz (Indefatigable) Island and of several colonies on Tower Island, made during November, 1962, are presented in tabular form with explanatory comments in the text. The observations are compared with characteristics of the Kittiwake and "typical" ground-nesting gulls.

Reprinted from *Wilson Bulletin*, volume 77, pages 346–362 (1965).

The horizontal distance from the outer edge of the nest to the edge of the nesting ledge was measured with a tape measure in the beginning, and later estimated by eye; the vertical height of the nest above the sea was estimated by eye. Behavioral observations were made with binoculars and in some cases recorded photographically with still and motion pictures. Notes on the nocturnal habits, breeding cycle, and displays of *furcatus* are presented elsewhere (Hailman 1964c, 1964a, and in prep., respectively).

In this and other publications on gulls I have followed the latest family revision (Moynihan, 1959), which assigns all species of gulls to the genus *Larus*. Except where noted, all information on the Kittiwake's adaptations has been taken from Cullen (1957). Information for comparisons with "typical, ground-nesting gulls" has come primarily from Cullen (1957), my unpublished notes on *Larus atricilla*, Tinbergen (1953), and accounts in Bent (1921).

In the "visual cliff" experiment reported below, a standard, albeit makeshift, visual cliff apparatus was made from a wooden box 16.5 inches long, 11 inches wide, and 9 inches deep. Across the glass top ran a center strip of black tape (3.5 inches wide) upon which the chick stood. To one side of the strip was the plain glass ("deep" side), under which the inside of the box lined with square-ruled paper (0.9 mm squares) could be seen. On the other ("shallow") side, ruled paper lined the underside of the glass. Each chick was placed in the center of the strip under a small translucent box for a 30-second habituation period, after which the box was lifted and timing with a stop-watch begun. Ten newly hatched *furcatus* chicks raised from the egg in a dark incubator were tested. The chick was scored as having chosen a side (i.e., deep or shallow) if it placed one foot on that side so that the foot did not touch the center strip. If no choice was made within minutes, the chick was scored as "no choice" and was gently pushed toward the deep side or pinched in order to force a choice.

CLIFF-NESTING OF THE SWALLOW-TAILED GULL

The actual cliff habitat

There are certain important differences between the "cliff" habitat of the Swallow-tailed Gull and the Kittiwakes. (1) The Swallow-tail nests on lava ledges or barancas whose angle varies from vertical to nearly horizontal, while the Kittiwake nests almost exclusively on vertical cliffs. (2) The two gulls nest at different heights, the Kittiwake sometimes very high (130 meters), the Swallow-tail at variable heights (1 to 25 meters), rarely higher than 8 meters (Fig. 1). (3) The Kittiwake's cliff almost always overlooks the sea, while that of the Swallow-tailed Gull may overlook land near the water (e.g., the colony in NW corner of Darwin Bay on Tower Island). (4) Similarly, flat land at the top or foot of the nesting cliff, or at least near it is available to Swallow-tails for display activities; this is usually not so true for Kittiwakes. (5) Finally, the distance from the nest to the edge of the cliff gives some idea of the restriction of living space and of the likelihood of eggs or chicks falling off the cliff. Minimum distances from the center of the nest to the edge are shown in Fig. 1 for a sample of 41 nests of the

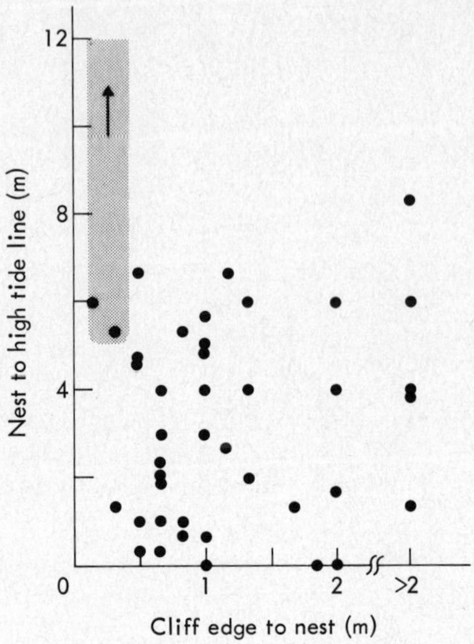

Fig. 1. Sites of 41 active nests of *Larus furcatus* on South Plazas Island, November 1962. The median height is 4 meters above the sea. The shaded portion indicates nest sites of *Larus tridactylus*, which go to above 130 meters, with median values of 15–35 meters above the sea (Coulson, 1963).

colony at Plazas. Apparently all Kittiwakes nest on ledges which just barely hold a nest and two standing adults, so each nest is placed at about the shortest distance found for the Swallow-tailed Gull (25–50 cm). (Recently, however, Kittiwakes have begun nesting on flat ground; see Paludan, 1955; Coulson, 1963).

Possible selective pressures producing cliff-nesting habits

Cullen (1957) believes that the Kittiwake's cliff-nesting is an adaptation to avoid predation on the eggs and chicks, and even upon the adults. Predation is probably unimportant in the Swallow-tailed Gull (see below), since its nest predators would be primarily aerial (and thus would have access to the nest) no matter where the gulls nested in the Galapagos (Hailman, 1964c). However, even aerial predators, such as Frigatebirds (*Fregata* spp.) may have difficulty landing on narrow ledges; aerial predators also have difficulty landing on the Kittiwake's ledge. The Swallow-tailed Gull's cliff-nesting might be linked in some way with pelagic habits; such an idea, if correct, would probably also apply to the Kittiwake. Also, the *furcatus* population may merely have exploited an unoccupied niche. The number of "typical" nesting sites for gulls (high grass in sand dunes or marshy area) is restricted in the Galapagos, which may account for the small population of the endemic Lava Gull (*L. fuliginosus*), a "typical species." (Food may also limit *fuliginosus* due to competition with other species for refuse; Hailman, 1963.)

ENVIRONMENTAL CONDITIONS AND CORRELATED CHARACTERISTICS

"Selective pressures" accompanying cliff-nesting

The multiple differences from typical gulls shown by the Kittiwake are presumably the result of several separate selective pressures. Probably only some of these selective pressures act on *furcatus*, partly because of the habitat differences (above). Thus, if Cullen's hypothesis be correct, the relatively unrelated Swallow-tailed Gull should prove convergent with the Kittiwake in those characters presumably related to situations that both species experience; in other respects the Swallow-tail should resemble typical gulls.

Specifically, the special environmental conditions experienced by the Kittiwake as a result of its cliff-nesting habits are: (1) reduced room for nesting; (2) scarcity of nesting sites; (3) scarcity of nest materials; (4) relaxation of predation on nest and eggs; and (5) danger of eggs rolling off cliff and chicks falling off cliff. (I have altered this classification somewhat from Cullen's presentation.)

Reduced room for nesting

The nesting space of individual pairs of cliff-nesting birds is limited to small ledges, particularly if all possible nest-sites are utilized by the species. To hold as large a territory around its nest as does a ground-nesting gull, the ancestral cliff-nesting would have had to defend many separate ledges. Deviant individuals psychologically "satisfied" with defending merely the nesting ledge might have left more offspring because parental care improved with lessening of territorial disputes (see discussion of "aggressive neglect" in Hutchinson and MacArthur, 1959). The converse holds for ground-nesters, which gain an anti-predator advantage by spacing-out.

Some problems arising from this reduced living-space are: (a) reduced space for territorial "fighting" and display; (b) reduced room for display and copulation between mates; (c) undue hostility aroused between mates because of continual propinquity; and (d) fouling of the nest. Some adaptations of these problems are summarized in Table 1, and further explained by the following notes.

In the early part of the nesting season, Kittiwakes display on the water at the foot of the cliff (Tinbergen, 1958). That Swallow-tailed Gulls do not appear to do this might be explained by the fact that they have available other areas for display (see above). Twice I observed copulation on flat land near the nesting-cliffs but never saw it on the cliffs (which were watched for much longer periods of time). E. Curio (pers. comm.) also observed copulation once on the flat surface of a large rock. Allo-preening within the pair (J. M. Cullen, 1962) is probably a "display" evolved to reduce hostility. (This characteristic was not commented upon by E. Cullen, 1957, but was discovered in the cliff-nesting tern (*Anous tenuirostris* by J. M. Cullen and Ashmole: see Postscript to this paper.)

Appeasement Head-flagging exhibited between Kittiwake chicks is absent in *furcatus*, which has only one chick per clutch. (I do not know whether or not strange Swallow-tail adults landing on a nest tend to peck the chick; in this situation the

TABLE 1. Characteristics presumably related to amount of nesting space

Species: Living space:	Typical gulls Large	Swallow-tailed gull Reduced	Kittiwake Small
1. Fighting			
a. frequency	frequent	infrequent(?)	frequent
2. Displays			
*a. chasing and moving displays	several	none (?)	a few
*b. long distance displays	several	none (?)	a few
*c. upright threat	common	absent (?)	absent
d. Long Call ceremony	loud	silent (?)	"Kittiwaking" (?)
*e. pairing displays	on pairing territory	occasionally on flat land	occasionally at base of cliff (and top of)
3. Copulation			
*a. where	on ground	on ground	on ledge
*b. female	stands	stands	sits on tarsi
4. Hostility Reduction			
*a. allo-preening (pairs)	absent (?)	present	present
*b. Head-flagging (adults)	uncommon	uncommon (?)	common (in certain situations)
*c. Head-flagging (chicks)	absent	(never seen)	present
*d. dark neck band	absent	immatures	chicks
5. Prevention of Fouling			
a. older chick defecates (cf. Table 4:2e)	off nest	over ledge	off nest or over ledge
b. eggshells (cf. Table 4:1d)	parents remove or eat	parents remove	not removed

Kittiwake chick Head-flags. The very dark neck-band of *tridactylus* chicks used in appeasement (Bill-hiding—Cullen, 1957: Fig. 1) is likewise absent in *furcatus* chicks. However, such a neck band is found in white-plumaged (prefledged) immature birds, which interact with their parents and possibly birds on other territories. (This band is shown in Hailman, 1964c: Fig. 2.)

Scarcity of nesting sites

The Swallow-tailed Gull may breed at any time of the year (Hailman, 1964a; Leveque, 1964:87), although Snow (Hatch, pers. comm.) has found evidence that individual pairs breed on a 10-month cycle. Furthermore, synchrony of breeding is pronounced only in local areas, not on whole islands or between islands (Hailman, 1964a), although Snow's recent observations and also those of E. Curio indicate a general synchrony within whole colonies as well (Curio, pers. comm.). Therefore the competition for nesting sites might be less acute than in the seasonally breeding Kittiwake.

Furthermore, on Tower Island I noticed many unusual areas that seemed to me capable of supporting *furcatus* nests. However, this situation seems to be true of Kittiwakes as well. The real competition for nest sites may be for nest sites near other pairs. Probably as a result of a reduced competition for nest sites, the territory of the ledge is not guarded (at night) before the egg is laid as strongly as after this time. Adaptations to nesting space are summarized in Table 2.

TABLE 2. Characteristics presumably related to availability of nesting sites

Species: Nesting sites:	Typical gulls Abundant	Swallow-tailed Ample (?)	Kittiwake Scarce
1. Reduction of Competition			
a. breeding	seasonal	probably a 10-month cycle, with islands not in phase	
b. site	stereotyped	varied	stereotyped
2. Territory			
*a. when assumed	after pairing	probably after pairing	before pairing
b. guarding before first egg laid (cf. Table 3:3b)	rare	sometimes	always, but not necessarily continuously

TABLE 3. Characteristics presumably related to availability of nesting materials

Species: Materials:	Typical gulls Abound	Swallow-tailed Abound	Kittiwake Scarce
1. Nest Materials			
a. materials used	vegetation	lava stones	mud and vegetation
*b. place	near nest	near nest	distant areas
*c. collecting	individual	individual	social
2. Building Nest			
*a. timing	individual	individual	synchronized
*b. technique	simple	simple	elaborate
3. Stealing from Other Nests			
*a. frequency	seldom	seldom	often
b. guarding before first egg laid (cf. Table 2:2b)	rare	sometimes	always

Scarcity of nest materials

Kittiwakes compete for nesting material because vegetation does not grow on the breeding cliffs. For such materials, the birds must go to flatter land, which they "fear" (Cullen, 1957). The Swallow-tailed Gull solves the vegetation shortage by

TABLE 4. Characteristics presumably related to amount of nest predation

Species: / Nest predators:	Typical gulls / Aerial and ground	Swallow-tailed / Aerial	Kittiwake / (none)
1. Parental Protection			
*a. alarm call	frequent	frequent	rare
*b. flight distance	far	variable (far to very near)	near
*c. attacks	vigorous	variable (vigorous to very weak)	very weak
d. eggshells (cf. Table 1:5b)	parents remove or eat	parents remove or eat	not dispersed
2. Eggs and Chicks			
a. egg coloration	cryptic	cryptic	cryptic
b. clutch size (cf. Table 5:2a)	three	one	two
*c. plumage	cryptic brown	cryptic grey	not cryptic
*d. behavior	hides in vegetation	hides in cracks	does not hide

using lava stones, and sometimes coral fragments and sea urchin spines, all of which I found abundantly near the nests in which they occurred. This difference in abundance of materials correlates well with the multiple differences between the two species (Table 3).

There is some local synchrony of the general breeding cycle among *furcatus* pairs within sight and sound of one another. However, this synchrony may be an "accidental" extension of the normal responses to displays of the mate (i.e., a sort of "behavioral pleiotropism") and may not have been specifically selected for (Hailman, 1964a). The Swallow-tailed Gull's synchrony is certainly not as remarkable as the specific synchrony of building found in Kittiwakes. The latter's unique building synchrony was thought to be due to the availability of mud only on rainy days and to the social process of inland collecting (Cullen, 1957).

Predation at the nest

In Kittiwakes, nest predation (e.g., by foxes) is virtually eliminated because of the inaccessibility of the nest. However, in the Galapagos there are relatively few potential predators that cannot fly (e.g., two species of native rats which, however, climb readily) so the cliff-nesting habit of *furcatus* has not completely eliminated nest predation. Frigatebirds (*Fregata magnificens* and *F. minor*), which abound in the Galapagos, regularly patrol the nesting-cliffs. Furthermore, the native owl *Asio galapagoensis* (which hunts both by day and night) and the hawk *Buteo galapagoensis* probably prey on *furcatus* nests (Murphy, 1936). In fact, one of the primary selective advantages of nocturnal habits of *furcatus* may be to allow the parents to stand guard at the nest through all the daylight hours when the aerial predators abound (see Hailman, 1964c for a discussion of other possible factors). However, it is not known whether the parent's merely being at the nest actually reduces predation or not.

Table 4 compares anti-predator adaptations in typical gulls, in *furcatus* and in *tridactylus*. Reduction of the clutch size might make a nest less conspicuous to predators, thereby decreasing the probability of destruction of all the eggs (also see below). However, this possible reduction of predation would not seem, *a priori*, to be of such magnitude to offset the approximately 66% reduction in productivity caused by fewer eggs. Table 4 indicates that not only *furcatus* but also *tridactylus* have cryptically colored eggs. Cullen (1957) considers this crypticity to be ancestral, and, although of no benefit to the Kittiwake, retained because it is of no disadvantage.

Danger of falling off the cliff

Most Kittiwakes nest at the very edge of a real precipice; Swallow-tailed Gulls do not, on the average, nest in such a dangerous situation, although certain individuals may (see Fig. 1). Kittiwakes prevent eggs from rolling out of the nest by making an extra deep nest cup, whereas Swallow-tails use lava stones. I tried rolling eggs out of several *furcatus* nests and found it very difficult (much more so than from the nest of *L. atricilla*, the American Laughing Gull, for instance). Gulls themselves might accidently dislodge the egg from the nest, though possibly other natural causes do too. Curio (pers. comm.) recorded one incidence of sea breakers washing an egg from its nest. Possibly the reduction of crowding due to the small clutch size also helps prevent eggs from rolling over the ledge.

TABLE 5. Characteristics presumably related to danger of falling off cliff

Species: Cliff danger:	Typical gulls (none)	Swallow-tailed Some	Kittiwake Great
1. Eggs and Chicks			
a. clutch (cf. Table 4:2b)	three	one	two
*b. nest	shallow cup of vegetation	shallow cup of vegetation	deep cup of vegetation
2. Chicks			
*a. stay in nest	a few days	long period	long period
*b. face toward	any direction	cliff wall	cliff wall
*c. locomotion	frequent	immobile	immobile
*d. when attacked	run	do not run	do not run
*e. flight movements	vigorous	intermediate	weak
f. "visual cliff" behavior	random choice (?)	avoid deep side	?
*g. feed from	ground and parent's bill	parent's bill	parent's throat
*h. parental feeding call	present	present	absent

One trait not appearing in Table 5 requires comment. Cullen (1957:300) notes that the Kittiwake has "strongly developed claws and toe-musculature" compared with ground-nesting gulls. The Swallow-tailed Gull appeared to me to have strong claws as well, but no stronger than those of the Lava Gull, *L. fuliginosus.* I attributed this similarity to the fact that the latter species, although not a cliff-nester, spends its life on the rock substrate of Galapagos shores. However, I later compared a long series of specimens at the U.S. National Museum, and could find no consistent difference between species, of gulls, adults or chicks. Perhaps important differences are obscured in dried skins, so further checking of claws and musculature in the field is desirable.

Emlen (pers. comm.) is attempting to do "visual-cliff" experiments (Walk and Gibson, 1961) on the Kittiwake. Emlen (1963) has already shown that newly hatched chicks of the ground-nesting Herring Gull (*L. argentatus*) may avoid the deep side of an artificial "cliff." However, his apparatus and experimental procedures are sufficiently different from the standard visual-cliff situation that a direct comparison with the usual experiments cannot be made. Cullen (1957) reports that chicks of the ground-nesting Black-headed Gull (*L. ridibundus*) placed in *tridactylus* nests wandered "blindly" off the cliff. However, Shinkman (1963) showed that newly hatched domestic chicks (*Gallus gallus*) do recognize and avoid the deep side of a visual cliff apparatus; this shows that such perceptual organization is possible in a newly hatched precocial bird.

Of the ten newly hatched chicks I tested, six chose the shallow side, one the deep, and three made no choice during the 10-minute test period. The probability that this choice is due to chance is small (binomial of $\frac{1}{7}$ is $p = 0.062$). Of the three immobile chicks, one turned and stepped onto the shallow side when pushed toward the deep; the other two refused to take a step in any direction. It is further of interest that the single "deep-choosing" chick scampered ("without looking") onto the deep side immediately upon removal of the translucent box. Thus, it seems quite likely that *furcatus* chicks (like those of *Gallus gallus*) possess depth perception at hatching.

Cullen (1957) considers that the feeding of Kittiwake chicks is adaptive to cliff-nesting. Most gull species (including *furcatus*) regurgitate food upon the ground or hold it in the bill in response to the chick's pecking at red markings on the parental bill. (The marking is a white tip in *furcatus*, presumably an adaptation to nocturnal feeding: Hailman 1964b, 1964c.) However, Kittiwake chicks take food from the throat of the parent. Lacking red markings on the bill, the parent Kittiwake has a bright red throat, to which the chicks direct pecking-like movements when it is open (although they also peck at the yellow beak; J. M. Cullen, pers. comm.). It could be that the releaser has been moved inside the bill so that Kittiwake chicks will not be tempted to approach adults and topple over the edge of the cliff. At any rate, chicks do not need a "long distance signal" in order to find the parent, nor does the parent require a Pumping display of the chick in order to find its offspring (Cullen, 1957). Since the Swallow-tailed Gull's feeding is additionally influenced by its nocturnal timing, it is not reasonable to expect this species' throat to become white, since this would probably reflect very little light indeed. However, the *furcatus* parent does have a "feeding call" that releases the approach of the chick, as has *atricilla* (Hailman,

1964b) and other ground-nesting species. The Kittiwake lacks this call, presumably to prevent accidentally calling chicks over the cliff.

Since in all other adaptations relating to prevention of falling over the cliff, *furcatus* resembles *tridactylus* (see Table 5), the chick-feeding differences seem to be anomalous. I suggest that in all species bill and throat colors are also under selective pressures relating to displays between adults. I have argued elsewhere, for instance, that the position of the white bill-tip of *furcatus* in relation to the white feathers at the base of the bill indicates the displaying bird's head position in very low light intensities (Hailman, 1964c). Surely the throat color of all gull species is evident during displays in which the mouth is held wide open during vocalizations. It is possible, then, that the chick-feeding method is influenced by display-methods and vice versa.

DISCUSSION AND CONCLUSIONS

Multiple selective pressures

Few characteristics are governed by only one selective pressure during evolution. Thus, the removal or eating of eggshells and the young chick's droppings might serve both to prevent fouling of the nest and to prevent discovery of the nest by predators. Tinbergen and co-workers (1962) have demonstrated by field experiments that nests with broken eggshells are found and destroyed by predators more readily than nests without shells. Fouling has not been studied experimentally. Older chicks of all species defaecate out of the nest. However, Kittiwakes with little nest predation defaecate on the nesting ledge, while Swallow-tailed Gulls with more predation defaecate over the ledge. This difference suggests that predation is important as a selective agent in defaecation habits. Also, guarding of the nesting ledge prior to laying may serve to protect both the site and the nesting materials from being usurped by conspecifics in Kittiwakes.

The clutch size of *furcatus* might be explained by Lack's (1954) proposal that clutch size in birds is determined by the number of young that can be fed successfully, although there seem to be other factors acting as well. The essence of Cullen's (1957:289 ff) interpretation of the reduction of clutch size from three to two in the Kittiwake seems to be a special case of Lack's hypothesis: If a pair of gulls can feed only two young successfully, Kittiwakes need lay only two eggs to have the maximum clutch, while "typical" gulls must lay three since there is a high probability that at least one will die from causes other than starvation (e.g., predation) that do not affect Kittiwake chicks.

Cullen's suggestion probably could not apply to the Swallow-tailed Gull, which has many potential nest predators. Instead, two additional hypotheses were advanced for *furcatus* (above). The first, that clutch reduction makes the nest less conspicuous to predators, is presumably not effective in Kittiwakes because of the lack of predation. However, I think it is unlikely as the major force in reducing clutch size in *furcatus*. The other explanation, lessening of crowding of eggs and chicks to prevent their accidental falling over the cliff, might operate in Kittiwakes as well, although Cullen does not specify this possibility.

However, still a fourth factor may be acting in the Swallow-tailed Gull, one that is a corollary of Lack's hypothesis. The breeding period of seasonally breeding gulls coincides with the abundance of food available for the young and clutch size is expanded to utilize the food maximally. In tropical species for which food is available in moderate supply the year around, the long nocturnal trek at sea for food may severely restrict the number of chicks that can be fed successfully. Although laying but one egg, *furcatus* pairs may actually rear more than one chick per year by breeding more often than annually (Snow's recent evidence, mentioned above, indicates a 10-month cycle).

Test of Cullen's hypothesis

With the data at hand, we are now in a position to test E. Cullen's (1957) hypothesis that the peculiarities shown by Kittiwakes are indirectly the result of selective pressures accompanying cliff-nesting habits. Given the degree of environmental similarity in Kittiwakes and Swallowtails, we can see how closely their characters match. (The following comparison omits (a) characters that cannot be evaluated as being either like Kittiwakes or ground-nesting gulls and (b) characters that cannot be assigned, *a priori*, to a single presumed selective pressure.)

TABLE 6. Summary of the Swallow-tailed Gull's morphological and behavioral characteristics

	Morphological/Behavioral Characteristics*		
Environmental conditions	Like ground-nesting Gull species	Intermediate	Like or equivalent to Kittiwakes
Like Kittiwake			
reduced nesting space	4 } 6	0 } 1	6 } 11
cliff danger	2	1	5
Intermediate			
nest sites scarce	1 } 5	0 } 2	0 } 0
nest predation	4	2	0
Like Ground-Nesters			
nest materials scarce	5	0	0

* Those relatively unambiguous characteristics are marked with an asterisk (*) in Tables 1–5. See text.

Table 6 divides 30 characters of *L. furcatus* into a matrix of the degree of similarity with *tridactylus* versus the degree of similarity of the environmental conditions presumably related to the characters. It is evident that in those respects in which the environmental conditions (i.e., presumed selective pressures) are similar, the morphological and behavioral characters are also similar. Taken as a whole, the data constitute a clear vindication of Cullen's (1957) hypothesis that peculiarities of the Kittiwake are the result of special selective pressures that accompany cliff-nesting.

Why does the Swallow-tailed Gull in some respects resemble ground-nesting gulls when the environmental characteristics are similar to those of Kittiwakes? Several answers are possible.

a) First, *furcatus* does not experience as extreme an environment as does *tridactylus*, even in those respects where the environment is designated as "like Kittiwake" in Table 6. (For instance, Fig. 1 shows that the danger of falling over the cliff is not as great.)

b) Secondly, the independent adaptation of *furcatus* to cliff-dwelling may not yet have proceeded far enough to evolve the full complement of characters possessed by *tridactylus*. That is, in evolutionary time *furcatus* may be a more recent cliff-nesting species; or *furcatus* may have some kind of genetical limitations which have not produced the variation for natural selection to work upon.

c) Lastly, other selective pressures which have escaped the notice of Cullen and me might be acting upon these characters in different ways in the two species. Very probably all of these reasons have some validity. The important thing is, I think, that *furcatus* completely lacks Kittiwake-like traits where its environment resembles that of ground-nesting gulls.

Epistemological status of comparative data concerning natural selection

The most satisfactory method of demonstrating that a morphological or behavioral character is under the influence of a specific selective pressure is to measure that pressure within a population of organisms; deviants from the norm of the character should be more strongly selected against. For instance, Kruuk (1964) has shown that the farther a pair of Black-headed Gulls (*Larus ridibundus*) nests from the center of a colony, the heavier the nest predation by foxes.

Another method, setting up an artificial situation closely resembling the natural one, is often necessary because of the rarity of natural deviants or the difficulty of measuring deviants and differential selection in natural populations. Thus, Tinbergen *et al.* (1962) have shown that Black-headed Gull nests artificially set up and placed near a nesting colony will be preyed upon by both aerial and ground predators. Aerial predators find and destroy such nests more readily when broken shells are placed in or near a nest with chicks or eggs. This demonstrates rather satisfactorily at least one of the selection pressures that maintain the eggshell removal behavior of nesting adults.

Least cogent among methods of demonstrating selective pressures on specific characters is the method of this paper. A population (which may be a species, as in this case) is discovered which shows differences in morphology or behavior from other, presumably genetically related, populations. This discovery, in and of itself, is not a valid demonstration that the characters are under selective pressures due to observed environmental differences between the populations. However, this correlation does function as a prediction as to what characters will be found in another population with the same environment as either the deviant or the "normal" populations already known. This third population, from which the prediction was made, constitutes a valid test of the hypothesis (i.e., environment-character causation) only if its characters were unknown at the time of conception of the hypothesis. (Conversely, if the new population's characters were known—say from museum skins—but its environment was not, prediction of the conditions of its environment would constitute a valid method of approach.)

This indirect, "comparative" method is, however, full of methodological pitfalls. The gene pools of all populations concerned must be similar enough that the same variations would be produced for natural selection to act upon. The populations should have been isolated and living in their present environments for sufficient time for natural selection to work. Furthermore, multiple selective pressures will usually be involved, as well as selective pressures of which the investigator is unaware. There are certainly other problems as well.

In conclusion, the present method for studying natural selection has a rather low reliability. It is, however, a vast improvement over glibly assigning a "selective advantage" to a particular morphological or behavioral character just because to do so seems "reasonable" *a priori.*

A postscript

Shortly after the manuscript of this paper was finished, there appeared a study of the cliff-nesting tern, *Anous tenuirostris* (the Black Noddy). J. M. Cullen and N. P. Ashmole (1963) found many differences between this species and other terns, and these unique characters closely resemble those of the Kittiwake and the Swallow-tailed Gull. The one "new" possible cliff-nesting adaptation reported for *furcatus* (allo-preening), Cullen and Ashmole discovered in the Black Noddy as well. Their study adds a further confirmation of E. Cullen's (1957) hypothesis.

ACKNOWLEDGMENTS

The research trip to the Galapagos was sponsored by NSF Grant No. GB98 to Dr. Peter H. Klopfer. I was accompanied in the field by Jeremy J. Hatch and Robert Risebrough, who were helpful in many ways. Dr. A. Brosset, then director of the Darwin Station, was instrumental in arranging field trips in the Galapagos. Dr. George Watson of the U.S. National Museum made available the gull collections there for study of claws and toe musculature. The manuscript was improved greatly through the critical comments of J. J. Hatch, Dr. E. Curio, Dr. P. H. Klopfer, and especially Dr. J. M. and E. Cullen.

REFERENCES

Bent, A. C. (1921), Life histories of North American gulls and terns, *U.S. Natl. Mus. Bull.*, **113**, 1–345.

Coulson, J. C. (1963), The status of the Kittiwake in the British Isles, *Bird Study*, **10**, 147–179.

Cullen, E. (1957), Adaptations in the Kittiwake to cliff-nesting, *Ibis*, **99**, 275–302.

Cullen, J. M. (1962), Allo-, auto- and hetero-preening, *Ibis*, **105**, 121.

Cullen, J. M., and Ashmole, N. P. (1963), The Black Noddy *Anous tenuirostris* on Ascension Island, II. Behaviour, *Ibis*, **103b**, 423–446.

Emlen, J. T. (1963), Determination of cliff edge and escape responses in Herring Gull chicks in nature, *Behaviour*, **22**, 1–15.

Hailman, J. P. (1963), Why is the Galapagos Lava Gull the color of lava?, *Condor*, **65**, 528.

Hailman, J. P. (1964a), Breeding synchrony in the equatorial Swallow-tailed Gull, *Amer. Nat.*, **98**, 79–83.

Hailman, J. P. (1964b), The ontogeny of an instinct: the pecking response in chicks of the Laughing Gull (*Larus atricilla* L.) and related species, Unpublished Ph.D. Thesis, Duke University, Durham, North Carolina.

Hailman, J. P. (1964c), The Galapagos Swallow-tailed Gull is nocturnal, *Wilson Bull.*, **76**, 347–354.

Hutchinson, G. E., and McArthur, R. H. (1959), On the theoretical significance of aggressive neglect in interspecific competition, *Amer. Nat.*, **93**, 133–134.

Kruuk, H. (1964), Predators and anti-predator behaviour of the Black-headed Gull, (*Larus ridibundus* L), *Behaviour Suppl.*, **XXI**.

Lack, D. (1954), *The Natural Regulation of Animal Numbers.* London: Oxford Univ. Press.

Leveque, R. (1964), Notes sur la reproduction des oiseaux aux Iles Galapagos, *Alauda*, **32**, 5–44.

Moynihan, M. (1959), A revision of the family Laridae (Aves), *Amer. Mus. Novitates*, **1928**, 1–42.

Murphy, R. C. (1936), *Oceanic Birds of South America*, New York: Amer. Mus. Nat. Hist.

Paludan, K. (1955), Some behavior patterns of *Rissa tridactyla*, *Vidensk. Medd. Dansk. Naturh. Foren.*, **117**, 1–21.

Shinkman, P. G. (1963), Visual depth discrimination in day-old chicks, *J. Comp. Physiol. Psychol.*, **56**, 410–414.

Tinbergen, N. (1953), *The Herring Gull's World.* London: Collins.

Tinbergen, N. (1958), *Curious Naturalists.* New York: Basic Books.

Tinbergen, N., Broekhuysen, G. J., Feeks, F., Houghton, J.C.W., Kruuk, H., and Szulc, E. (1962), Egg shell removal by the Black-headed Gull, *Larus ridibundus* L.; A behaviour component of camouflage, *Behaviour*, **19**, 74–117.

Walk, R. D., and Gibson, E. (1961), A comparative and analytical study of visual depth discrimination, *Psychol. Monogr.*, 75 (no. 519).

2. WHAT IS THE HISTORY OF BEHAVIOR IN A POPULATION?

GENETIC AND CULTURAL CHANGES IN BEHAVIOR

Formerly, ethologists were often distinguished from comparative psychologists by the interest of the former in evolutionary problems. Although the distinction is largely obliterated now, it remains true that students of animal behavior trained in zoology retain an active interest in the phylogenetic history of the behavior patterns they study. Indeed, sometimes behavioral differences can be understood only in terms of phylogeny, as has been true with morphological differences. The insect ancestoral to modern flying species could not have evolved a bird's wing any more than the bird's reptilian ancestor could have evolved the insect's wing; yet each wing is functional for its possessor. A key to understanding their differences lies in the recesses of evolutionary history.

There are other reasons for wanting to trace the evolution of behavior patterns. For instance, such studies may help to understand the phylogeny of the animals themselves. The studies may also aid the taxonomist in his efforts at a natural classification of animals. And certainly, evolutionary studies contribute to our understanding of functional, ontogenetic, and physiological aspects of behavior.

There is another kind of history of behavior that is beginning to receive some needed attention: the history of culturally-transmitted patterns. Here the effects of natural selection through time may be minimal because the genetic endowment of individuals in a population is not being greatly altered. However, the actual behavior shown by individuals may have to be understood in the framework of the individual's own experience, and these experiences may be structured differently in different populations of the same species. Washing of sweet potatoes occurs in one population of Japanese monkeys', whereas in others the behavior is absent. White-crowned Sparrows in one location sing similarly, but quite distinctly from otherwise similar birds at another locale. The study of the origin and transmission of culturally-determined behavior was formerly the domain of anthropologists, but is now receiving comparative attention.

The papers collected below give only a test of investigations into the phylogenetic and cultural history of behavior patterns. For further information the reader is referred to Chapter 10 of *An Introduction to Animal Behavior* by Klopfer and Hailman.

A. THE RITUALIZATION OF DISPLAY

"Displays" are favorite behavior patterns among ethologists. These are the patterns that serve a communicative function among members of a species, and may have no other function. The natural question is "how did displays arise in evolution from non-communicative behavior?" The evolutionary process by which this happens has been termed "ritualization" from the non-communicative behavior origin.

The collection of papers below concerning the ritualization of displays brings together two famous names in ethology: those of Konrad Lorenz and Niko Tinbergen, a fitting testimonial of the central role played by display patterns in ethological theory. Lorenz's lead paper summarizes his early attempt at a comparative study of duck courtship displays, an interest derived from his mentor, Oscar Heinroth. Daanje's paper shows in detail how various displays might have arisen from locomotion patterns, and he introduces the important notion of the so-called "intention movement." Tinbergen's paper then attempts an overall summary of "derived activities," as he then called them. The paper by Morris is a natural follow-up to Daanje's paper, taking autonomic rather than motor responses as a place of evolutionary origin. Moynihan then makes some more general comments, and Tinbergen ends with a summary of principles and results to date on an ambitious comparative study of the evolution of displays in gulls.

Comparative Studies on the Behaviour of Anatinae

KONRAD Z. LORENZ

If one wants to formulate a classification of group with any real success one must rid oneself once and for all of the idea that a linear arrangement of the forms can really represent the relationship existing between them. All animals living to-day are growing branch tips of a

Reprinted in part by the kind permission of the editor of *The Avicultural Magazine*; volume 57, pages 157–182, (1951); volume 58, pages 8–17, 86–94, 172–184 (1952); volume 59, pages 24–34, 80–91 (1953). Original numbering of figures is retained. This paper is a translation by C. H. D. Clarke, with new revisions by Lorenz, of a paper that appeared in the *Journal für Ornithologie* (1941).

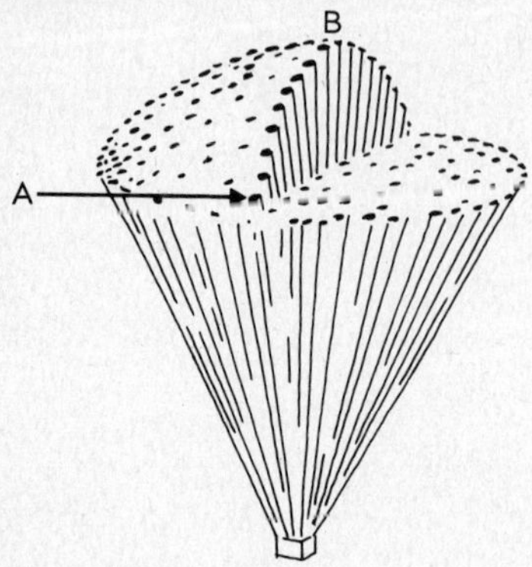

Fig. 51. Diagram of a series of similarities in recent animal forms whose neighbouring members are not joined by close genetic relationship. When part of the lines of descent fall out, the rest of the forms, lying in a row A-B, can be mistaken for phylogenetic series.

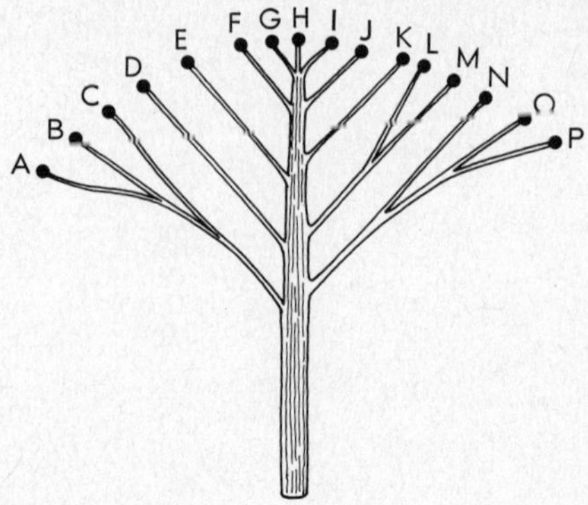

Fig. 52. Diagram of a series of similarities of recent animal forms, resting on true phylogenetic relationship. Every two neighbouring forms in the row A-B owe their similarities to the part of the path of development that is common to them.

"family tree," and can *ipso facto*, not be derived from "one another." The comparison of their characteristics results, therefore, in an arrangement which can be represented freely by the likeness of a family tree with branch tips like those of a small round-chipped beech or yew tree. They lie all together in one surface, which represents a cross-section through a bush growing outwards at one point in time. Just as we can only tell by guessing and judging appearances, because of the thick foliage of the tree which we cannot see through, as to which ends belong to a common branch and how far down they branch from the main stem, so the best systematic arrangement can only give us reasonable speculations about true evolutionary relationships.

I am now going to try to represent graphically, in a tabular diagram what we have gained through that kind of systematic intuition which I tried to describe in the Introduction as a simultaneous review of as many characters as possible. Only by evaluating all accessible characters simultaneously can we achieve a proper judgment as to the relative value of individual characters. First of all we must make a brief deliberation of a fundamental nature. The similarity of a series of forms, even if the series structure arises ever so clearly from a separation according to characters, must not be considered as establishing a series of development stages. Let one imagine that, out of one common root, a number of forms have grown out, all equally old and equally differentiated from the root. We shall represent this family tree structure in Fig. 51 as a kind of shaving brush. Now let us imagine further that, as represented in one half of the brush, part of the hairs have fallen out in such a manner that the rest remain somewhat in the form of a fan. The tips then represent a step ladder leading from A to B, which seem convincing evidence that the forms have descended "from one another," especially if the degree of differentiation is less on one edge of the fan than on the other. Doubtless the end-points of such a "family tree" have already often been taken for phylogenetical series, which unfortunately, places ever more welcome weapons in the hands of the opponents of the theory of descent. On the other hand, we must not fall into the opposite error to this too ready formation of series and generalise in the view, that all similarity of recent organisms might be explained from the principle of fan-shaped classified lines of descent, as an over-generalisation of Kleinschmidt's theory would imply. Doubtless there are many cases in which not only a monophyletic development of large groups of animals has taken place, with many individual forms not splitting off till later, but also the further differentiation of the single forms, at least in respect to single characters, took place at such different rates that similarity lines arose as represented in the diagram in Fig. 52. However in observing these similarities which correspond to a phylogenetical staircase *one must not forget for an instant that the term "primitive" may be used only for one or several characteristics of a recent animal form, never for the form as a whole.* Even *Sphenldon,* or *Ornithorhynchus,* is not a "primitive animal." The circumstance that some or even very many characteristics of such a form are quite certainly phylogenetically primitive, does not justify us in the presumption that all the rest of their characteristics are also primitive. The stopping of further differentiation of one character means nothing in respect to the course of differentiation of the others.

The "intuitive feeling" of the professional systematist, which we discussed in the Introduction, is generally well enough developed to decide between similarity series which rest upon a common descent, as has just been described, and such as arise through the previously discussed phenomenon of classified descent-lines arranged in a fan shape. In order to have a more objective criterion for this separation I suggest the following consideration of probability. If one grants that all the representatives of an animal group come from one source, independent and diverging, without nearer connections with each other, as represented in Fig. 51, then one would expect that the similarities of characters which determine the arrangement as to which lines of descent were to be placed side by side, would be divided rather equally over the whole brush. If, for the sake of simplifying the graphical representation we take a longitudinal section of the brush, getting this a number of fan-shaped, diverging lines of descent, the similarities which bind each form to its systematic neighbour, would have to go through the whole sheaf of lines homogeneously, and especially from every point to both sides, hence, in the three dimensional sheaf diagram, to all sides; binding species with species in like manner.

This type of distribution of characters actually occurs: in all so-called "circles of forms" (*Formenkreise*) this holds true and Kleinschmidt's theory of family-bushes (instead of family trees) is undubitably correct if applied to these cases. A *Formenkreise* is nothing else than a "shaving brush" in which the middle part was worn away, leaving a circle of hairs, respresenting the single forms.

Now if one represents characters held in common as cross connections and arranges the more general, older characters and those common to the larger divisions of the group toward the base and the others more and more towards the periphery in proportion to the narrowness of their distribution and their degree of specialisation and, consequently, their phylogenetic youth, there would be, in the ideal case of the diverging sheaf-like type of species-formation, a classification such as is represented in the diagram in Fig. 53.

Now we shall try to represent graphically, in the manner described, the group Anatinae, using as many of their systematically useful characters as possible, in order thus to form an opinion as to how far their representatives can put together in groups according to true phylogenetic relationship, and to what extent the evolution of their species corresponds to the type of the sheaf-form diagram with diverging straight lines in Fig. 51. Although the sheaf of the lines of descent can be symbolized only in three dimensions we shall have to use several flat projections. He who is very particular about clarity may draw them thus and then glue them together so that they fit. I confess that I myself used, for the arrangement of the

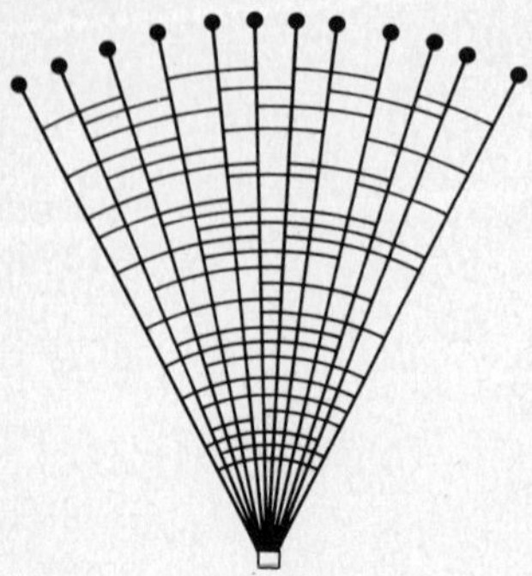

Fig. 53. Diagram of the division of characters one would expect in unbranched lines of descent, diverging in shelf-like form. As the variabilities and similarities are explained only by greater or lesser divergence, the distribution of most characters overlaps.

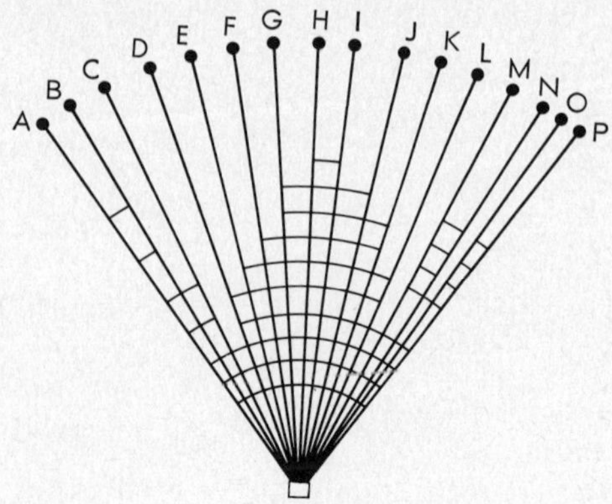

Fig. 54. Diagram of the distribution of characters which is to be expected when lines of descent branch in the form of a tree. As the connecting characteristics are results of common paths of development, they are divided according to their common descent and do not intersect except for convergences.

species, a bundle of stiff wires with thin wires representing "common characters" joining them together into sub-groups.

If one grants that not every species of a group of forms has developed quite by itself, independently from all other, a very different distribution of characters is to be expected. If several forms have only branched from one common ancestor after a long period, we can expect that they have in common such characters as have developed since the time of its branching off. If two branches grew away from each other very far down we are not surprised if they are connected only by very old characters common to larger group categories. Cross connections in character distribution, as we have represented in Fig. 53, are not to be expected if, for the time being, we ignore the possibility of convergence. Fig. 54 shows the type of character distribution which is to be expected in a family tree as opposed to the family "bush" represented in Fig. 53.

In species where there is reason to believe that the lack of a character is not primitive but of secondary occurrence, this has been shown in Table 1 by a little cross at the intersection of the character connection and the line of descent. As is clear from the diagram, the placing together into groups of common origin becomes more probable the nearer we move to the base of the sheaf of lines of descent, while for very many characteristics of more recent date clear cross-connections in the distribution of characters after the manner of Fig. 53 have yet to be found. Notice, for example, the distribution of the grunt-whistle, head-up-tail-up, and turning-of-the-back-of-the-head.

The few morphological characters, interspersed in the diagram are to show how similar their distribution is, in many cases, to that of behaviour traits. After filling in those gaps, which are especially obvious in the list of species investigated, I plan to construct a much larger diagram, built upon the same principle, in which will be entered all or nearly all the available morphological and behaviour characters, such as the fertility of hybrids. The publication of this diagram must above all await the result of comparative studies of the drake's bone-drum, which is so very rich in characters that can be compared. Heinroth had left some unpublished material on this subject, as well as a very complete collection of bone drums.

Even in its preliminary incomplete state our grouping shows clearly the possibility of using the phylogenetic homology concept for characters of inherited behaviour. This fact, which it has been the principal task of my investigation to prove, is of the very greatest significance in comparative psychology.

INTRODUCTION AND STATEMENT OF PURPOSE

The success of the investigator in zoological classification is more dependent than in other branches of biological research upon an instinctive feeling which can, it is true, be learned but not taught. In the introduction to his work "The Birds" in Bronn's *Classes and Orders of the Animal Kingdom,* Gadow made the experiment of a "Thirty Characteristic Systematics." He chose thirty common characteristics, which, from the point of view of classification, are certainly important, and divided the groups of birds in tabular arrangement according to whether the characteristics are present or not. Along with far-reaching harmonies in many bird groups, the systematics thus improved showed, in places, astonishingly gross deviations from the commonly accepted arrangement considered to represent true relationship. This is explained, in the first place, by the fact that the so-called "Classifying instinct" rests upon an unconscious valuation of a very much greater number of characteristics which, not easily accessible to individual observation, are woven into the general impression which such an animal group makes upon the investigator. Such unanalysed complex qualities include very minute individual characteristics which cannot be severed from the general impression at will, although they influence it by determinative qualities. This fact, which is obvious to the psychologist of observation and form, must now be taken into consideration, if one wishes to analyse the instinctive feeling for classification and know the grounds which determine one's own judgment as to the degree of relationship of the different animal forms.

The inadequacy of a classification built upon a limited number of predetermined characteristics, does not rest merely on the fact that their number was too small. Much more disturbing is the circumstance that a definite character does not possess the same weight within the separate parts of a greater classified whole. The rate of differentiation and the variability of a definite character can be decidedly different even in two closely related species. The hypothesis that a character, such as the lack of the fifth secondary, or the form of the furcula, possesses equal taxonomic value throughout the whole class of birds, in all orders and the families, is false to begin with. The weight that can be ascribed to a characteristic as a measurement of phylogenetic relationship must be decided in each case by its relative proportion to other characteristics of the group examined. The statement, as to whether a characteristic is relatively fixed or variable can concern only a narrow selection of closely related

EXPLANATION OF TABLE 1

The vertical lines represent species, the horizontal ones characters which are common to these species. A cross indicates that the character in question is lacking in that particular species. A circle indicates an exceptional differentiation of a character in a species, a question mark ignorance of the author.

List of species 1. Muscovy Duck, *Cairina moschata.* 2. Carolina Duck, *Lampronessa sponsa.* 3. Mandarin Duck, *Aix galericulata.* 4. Chiloe Wigeon, *Mareca sibilatrix.* 5. Wigeon, *Mareca peneolpe.* 6. Gadwall, *Chaulelasmus streperus.* 7. Teal, *Nettion crecca.* 8. Chiloe Teal, *Nettion flavirostre.* 9. Chestnut-breasted Teal, *Virago castanea.* 10. The Genus *Anas,* Mallard, Spotbilled Duck, Meller's Duck, etc. 11. Chiloe Pintail, *Dafila spinicauda.* 12. Pintail, *Dafila acuta.* 13. Bahama Duck, *Poecilonetta bahamensis.* 14. Red-billed Duck, *Poecilonetta* (?) *erythrorhyncha.* 15. Garganey, *Querquedula querquedula.* 16. Shoveler, *Spatula clypeata.* 17. Sheld-Duck, *Tadorna tadorna.* 18. Ruddy Sheld-Duck. 19. *Answe* as a genus. 20. *Branta* as a genus.

Compared characters: *EPV*: monosyllabic piping of lost chick; *ANtr*: Displacement drinking as a signal of peace *Kn Tr*: Bone drum in the trachea of male; *AKk*: Ducklike markings of down chick; *Fs*: Glossy speculum on secondaries; *Ssn*: Bill with lamellae functioning as sieve; 2 *St*: Contact-call of chick consisting of two syllables; *H*: Inciting ceremony in female bird; *Fs*: Displacement shaking as a form of display; *PE*: "Aiming" Intention movements of the head in pre-copulatory display; *Sp*: Displacement preening behind the secondaries in the male; *Ges*: Social display of drakes; *Afs*: Burping; *Skh*: The Duck's inciting movement backwards over the shoulder; *Spf*: Feather differentiations enhancing the function of displacement preening; *Els*: Preliminary shake; *P*: Pumping as pre-copulatory movement; *Dc*: Decrescendo call of the female; *EPf*: Whistling in the drake; *Kh*: Chin lifting; *Hkz*: Turning the back of the head towards the female; *Gp*: Grunt whistle; *Abf*: Down-and-up movement; *Pn*: post-copulatory display with bridling movement; *Izh*: Head-up-tail-up; *GlSp*: Speculum identical in both sexes; *Ar*: Bridling movement; *KrSp*: Speculum parti-coloured black and green; *Trkh*: Chin-lifting in the function of a "Triumph ceremony"; *JA*: Bridling as an independent movement, not coupled with head-up-tail-up; *Kr*: Teal Whistle (of two syllables) coupled with burping; *Kd*: Teal whistle independent of burping; *Ns*: Nod-swimming in the female; *Gg*: "Geee" call of true Pintails; *Spi*: Elongated middle tail feathers; *RR*: Rolling Rrrr call of female while inciting; *HV*: Lifting of anterior part of body while inciting; *Ss*: Markedly wedge-shaped tail; *Sz*: Marking a bill with light spots on the side; *OP*: Drake without a whistle; *Lsf*: Lancetlike feathers on shoulder of drake; *Bfk*: Small feather of wing light blue; *FZ*: Small feathers of wing white, speculum glossy green, elbow feathers ruddy; *SwK*: Down young marked black and white; *MkSt*: Contact call of young with many syllables; *Ef*: Downy young uniformly coloured, without sharp markings; *He*: Dipping of the neck as pre-copulatory display.

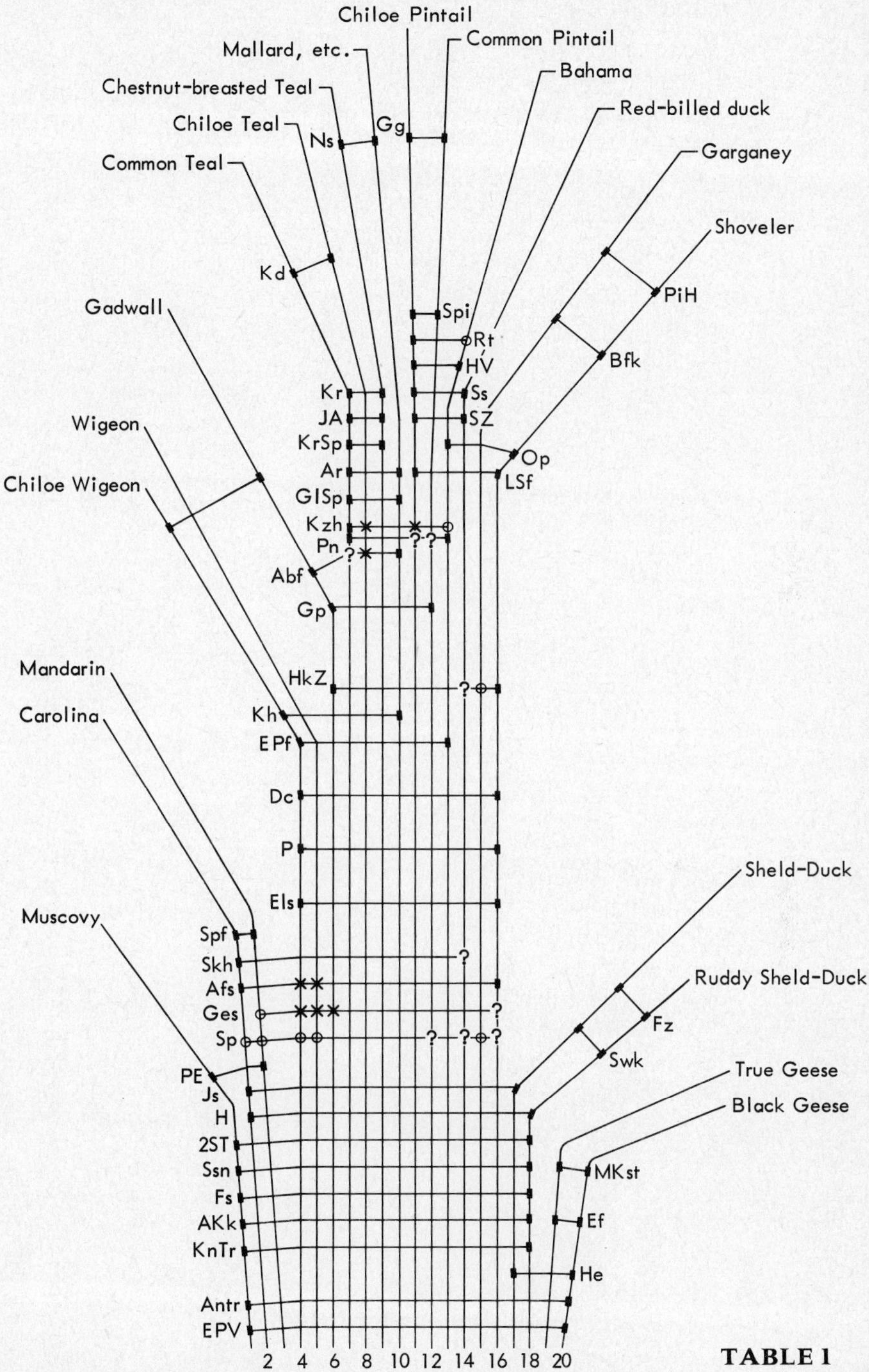
Chiloe Pintail
Common Pintail
Mallard, etc.
Bahama
Chestnut-breasted Teal
Red-billed duck
Chiloe Teal
Garganey
Common Teal
Shoveler
Gadwall
Wigeon
Chiloe Wigeon
Mandarin
Carolina
Muscovy
Sheld-Duck
Ruddy Sheld-Duck
True Geese
Black Geese
Ns
Gg
Kd
Spi
PiH
Rt
HV
Bfk
Kr
Ss
JA
SZ
KrSp
Op
Ar
LSf
GlSp
Kzh
Pn
Abf
Gp
HkZ
Kh
EPf
Dc
P
Els
Spf
Skh
Afs
Ges
Sp
Fz
Swk
PE
Js
H
2ST
Ssn
Fs
AKk
KnTr
Antr
EPV
MKst
Ef
He
2 4 6 8 10 12 14 16 18 20

TABLE 1

forms. This is so, not only for critical characteristics of limited distribution, but very often for characteristics of general distribution. Even the usually fixed ontogenetic characteristics, as, for instance, the immature plumage of so many bird groups, can be affected in a narrowly circumscribed systematic unit by secondary changes in such a way that it would lead to the greatest confusion if one were to ascribe to them that taxonomic value which generally belongs to them. Imagine one considering peculiarities of natal down, which in the Anatidae are so unusually fixed and taxonomically useful, in the taxonomy, in a similar manner, of Rails, in which they are overlaid by secondary differentiations, acting as releasers. The number of characters known to the one who judges the degree of phylogenetic relationship is important not only in itself in the determination of group-specific "complex qualities" but more than that because the more characteristics the examiner can take, consciously or unconsciously, into consideration, the better he can evaluate the relative importance of a single character. The "classifying instinct" does all this without the man who possesses it needing to be able to analyse it himself. However, only when the analysis is accomplished does his performance become science.

In the basis given here briefly, of conscious or instinctive judgment of phylogenetic relationship, one can determine that it is not the man who knows an organ in all its forms in which it appears and where it is represented in a large systematic group, who has the most reliable judgment about relationship, but the man who surveys minutely a small unit with as large a number of characteristics as possible in mind. The possibility of *a priori* conclusions about phylogeny mount not only in arithmetical progression with the number of known characteristics, but decidedly in geometrical progression, because with every additional characteristic examined in all the representatives of the group, our estimate of the importance of the known characteristics gains in correctness. Now we can understand for once why those zoologists who, as keepers of zoological gardens or as fanciers, know a definite animal group through many of its living representatives, attain such peaks in the development of systematic intuition and in the critical understanding of phylogenetic relationships. One thinks of Heinroth's work on Anatidae and the "Study of Equidae" by Antonius. The Zoo director, who is equipped with a knowledge of the anatomy and perhaps even of the paleontology of many representatives of a group of animals has a well of knowledge of inherent specific characters to draw on as compared to the purely museum classifier. This indubitable fact is of significance and value not only for the systematist, but still more so for the psychologist. Since Wundt's time it has been emphasized that the problem of comparative phylogenetic development is just as vital in psychology and the theory of behaviour for the understanding of any given patterns in man and animals as it is in morphology. From the psychological point of view all forms of life are, in a way, phylogenetic attainments whose special objects would have to remain completely obscure without the knowledge of their phylogenetic development. Comparative psychology, which until now, unfortunately, has remained almost entirely a mere syllabus, has the urgent task of carrying on first of all a purely descriptive study of the behaviour of a group of animals suitable for the purpose, in order to combine the characteristics thus learned with every conceivable morphological characteristic, into a critical characterisation of the group. Because of its harmony with the related anatomical characteristics, the applicability of the taxonomic homology concept even fro special behaviour patterns would be established once and for all against all attacks, and thus a working hypothesis for all comparative psychology in the true sense of the word would be set up. Only such a critical characterisation of a sufficiently thoroughly investigated animal group can give us knowledge on the ways and means whereby racial changes in instinctive behaviour, taxes, and inherited patterns and also, in a broader respect, basic mental structure, can come about. The fundamental importance of such knowledge, at present unfortunately still lacking, does not need to be stressed. The path that the investigation has to follow lies clearly before us, even though it be a particularly difficult and thorny one. The present work represents a very incomplete attempt to give a critical characterisation serving the aims mentioned, for one group, including behaviour characteristics.

TECHNICAL

The task of making a purely descriptive inventory of the behaviour of very many species makes very great demands on the examiner's powers of observation. In order to make, even approximately, a sufficiently complete inventory, the observer must live with the animals day after day, year after year. It is really impossible to gain the necessary knowledge by observing the animals in the wild state. Even when one has kept the animals under observation for years and has been most conscientious about keeping entries in a diary, there are still, after the work is done, single dates missing here and there, which would have been of great importance in making comparisons, and concerning which, unfortunately, previous work contributes an all too feeble testimony. The animals investigated must be amenable to captivity. Moreover the group concerned must be rich in comparable species and genera with, if possible, consecutive grades of relationship with the individual forms, also rich in comparable individual behaviour traits, analogous from species to species, yet different enough to serve as a test for the applicability of a common origin concept. All these requirements will be fulfilled ideally by two groups of animals, both of which I have kept to carry on the research of which we have been speaking, namely, in birds the Anatinae and in fishes the Cichlidae. The former will be dealt with here. The Anatinae have been very carefully studied by Heinroth, Delacour, von Boetticher, and others, and in addition they afford the investigator in the field of comparative phylogeny a special advantage by the fact that hybrids can be bred easily. In addition these are in very many cases fertile, so that the inheritance of specific behaviour characters can be studied. Here we have a fruitful field for a synthesis of phylogenetics and genetics. In many cases a special peculiarity of hybrids helps us in making phylogenetic assertions on specific behaviour characters, namely, that they are not intermediates between the parents but, physically and in behaviour, show a third and a more primitive condition. A further value of hybrids lies in the fact that the degree of their fertility can be made a measurement for the degree of the relationship existing between the parents, as Poll (1910) has shown.

Of the family of the Anatinae and the two adjoining families of the Cairininae and Casarcinae, I have been able to examine the following species:

Anatinae Mallard, *Anas platyrhynchos* (L.)
Meller's Duck, *A. melleri* (Scl.)
Japanese Spot-billed Duck, *A. zonorhyncha zonorhyncha* (Swinhoe)
Indian Spot-billed Duck, *A. z. paecilorhyncha* (Forster)
Pintail, *Dafila acuta* (L.)
South American Pintail, *D. spinicauda* (Vieillot)
Bahama Duck, *Paecilonetta bahamenis* (L.)
African Red-billed Duck, *P. erythrorhyncha* (Gm.)
European Teal, *Nettion crecca*
Yellow-billed Teal, *N. flavirostre* (Vieillot)
Gadwall, *Chaulelasmus strepera* (L.)
Wigeon, *Mareca penelope* (L.)
Chilöe Wigeon, *M. sibilatrix* (Poepp.)
Garganey Teal, *Querquedula querquedula* (L.)
Shoveler, *Spatula clypeata* (L.)

Cairininae Muscovy Duck, *Cairina moschata* (L.)
Wood Duck, *Lampronessa sponsa* (L.)
Mandarin Duck, *Aix galericulata* (L.)
Sheld-Duck, *Tadorna tadorna* (L.)
Ruddy Sheldrake, *Casarca ferruginea* (Pallas)
Egyptian Goose, *Alopochen aegyptiaca* (L.)

The Hybrids I shall describe in more detail elsewhere.

GENERAL REMARKS ABOUT EXPRESSION BEHAVIOUR

The taxonomic usefulness of signal movements

Since Whitman (1898) showed that instinctive actions require phylogenetically the same amount of time for their evolution as do body structures, and since Heinroth (1910) published his studies of Anatidae and his work (1930) on the systematic distribution of determined behaviour patterns in vertebrates, the taxonomic usefulness of inherited specific behaviour has gradually become common property of zoological systematics (Stresemann "Aves" in Kükenthal's *Handbuch*). I myself pointed out several years ago that well defined instinctive behaviour is particularly suitable for systematic use and hence, in obvious corollary, for research on the phylogenetical relationship of inherited behaviour patterns, namely those instinctive actions whose significance to the preservation of the species lies in the sending out of stimuli which are answered in a specific manner by other individuals of the same species. Such behaviour patterns, functioning as "releasers" (Lorenz, 1935) are especially characteristic of birds, as in them reaction to optical movement-perception plays a great role. In other animal groups, chemical, acoustic, and tactile releasers are more important. These do not afford man, the eye animal, nearly as good opportunities for experimental and comparative research, but apart from purely observation-technical grounds, the optically functioning signal movement of birds, like those of the bony fishes, is a very fruitful subject for investigation. First, it is very rich in identifying and striking characteristics which greatly aid in phylogenetic comparisons of species with species, and group with group. This wealth of characteristics is very closely related to their function as signals, for just like a human signal, the animal releaser must be incapable of being misunderstood if it is to fulfill its function completely. I have gone into these relationships more closely in a special work (1935) and have explained there that a combination of the greatest possible simplicity and the greatest possible general unreality is the main attribute of a clear and easily understood signal. Both properties make the releaser for purposes of appraisal an especially valuable criterion in a systematic comparison of characters. Even the fact that it can be easily described adds much to its value.

There is a second circumstance that makes the releaser a very valuable aid to an investigation of genetic relationship. Because its form is not directly derived from its function and influenced by it, as is the case in mechanically effective actions, the possibility of convergence can most probably be excluded when the releasing ceremony is the same in two related species. That the wagging of the tail in the dog indicates friendly greeting, while in the cat, hostile feelings, has nothing to do with function. The meaning rests purely on a "convention" between the inherited expression movement in the one, and the equally inherited understanding in the other members of the same species. As far as the special form of the movement is concerned, it could just as well be reversed. In the same way the philologist, who finds the same word with the same meaning, among two different peoples, does not need to bother about the infinitesimally small probability that this is mere chance, but without more ado and quite correctly assumes a single historical root for both words. So also in many cases the comparative psychologist can assume a genetic homology where the releasing ceremonies, which have often many distinguishing features, agree in two related forms. All these circumstances have led to the point where we know far more about the origin, systematically, for instinctive releasing movements than for other inherited behaviour traits. In two ways, both independent of each other, we can trace back many, if not most, of them with high probability to other more primitive and mechanically effective instinctive movements without, naturally, knowing how these, for their part, have arisen. At any rate, in our present poverty of information about the phylogenesis of psychic behaviour traits, even this small provable fragment of phylogenetical developmental history is very welcome.

Symbolic activity

It is a distinguishing characteristic of any centrally co-ordinated autonomic automatism which we designate as instinctive behaviour that it changes into visible movements with the least intensification of the irritation created by it. Thus arise more or less pronounced indications of behaviour which can serve its real purpose in the preservation of the species only by a much more intense response. The weak responses which become noticeable even in the weakest intensities of stimulation are in themselves without meaning and purpose for the preservation of the species. The experienced person, however, can deduce from them the nature of the specific excitation which begins to well up within the animal. He is, therefore, in a position to predict what kind of activities, if any, are to be expected from it. Therefore, these initial activities have been called intention movements. Many instinctive actions, for instance the flying-up of a duck or goose, never happen quite suddenly, except as a result of a very strong external stimulation. When it is the inner urge of an instinct that is gradually cumulating within the organism, then we invariably find a very regular series of intention-movements preceding the final, fully intensive, and biologically adequate activity. Thus there is the possibility that within the species concerned there is developed an "understanding" for the intention-behaviour of the same species, through which inherited patterns are formed which are related to the responses engendered by any given intention-behaviour in the same way as the radio receiver is to the sender. In other animals of the same species these inherited patterns first, for the most part, cause an evoking of the same instinctive action as that which the response-stimulating animal was carrying out. The specific stimulation quality connected with this behaviour is transmitted thus through bringing out the specific intention-behaviour and evoking its corresponding inherited specific behaviour pattern, as between an individual and all others of the same species present. In my work "Der Kumpan in der Umwelt des Vogels" (1935) I have brought together many examples of this form of transference of a mood, which is never better exemplified than in the uniform orientation seen in the behaviour of social animals, falsely called "imitation." Through this procedure, obviously of the very greatest survival value for social animals, the once meaningless intention-behaviour now takes on a great significance, and, like every functioning organ, comes under the influence of those factors which are working towards higher differentiation. Releasing intention-movements, corresponding to their signalling function, are changed from the original form in such a manner that they become visually more pronounced. They become "mimic-exaggerations" like the nod movements of leading mallards, the upward-flying movement of the Casarcinae, and the Anserinae, the swimming movements of nest-leaving Cichlids, etc. The mimic exaggeration can often go so far that the original root of the action, the intention-movement, reduced to a purely mechanical protective behaviour without signal form, becomes practically unrecognisable and can be determined only by bringing in, for systematic comparison, other species in which it is little altered. The

Egyptian Goose's flying-up behaviour, the movement in *Nannacara* indicative of leading young, etc., are certainly derived from intention-behaviour, indicating flying or swimming away, but no one who was ingnorant of the homologous behaviour of their relatives would recognise them as such. Such ceremonies, derived from intention-behaviour but often "formalized" beyond recognition, we designate by the term "symbolic activity."

Displacement activity

Besides these releasing symbol-actions arising out of intention-movements, there are others which have arisen in an entirely different manner, the displacement activity, as N. Tinbergen calls it. At the upper levels of general excitement, there may be released instinctive movements which, so far as any real survival value is concerned, do not belong at all to the immediate biological situation and are, so to speak, performed by mistake. This seems to take place especially when the normal movement, appropriate to the situation, is hindered for some reason in finding its outlet. Then the special stimulus "jumps over," so to speak, into another track and an action as unexpected as inappropriate ensues. Man, too, shows many examples of this procedure. Best known is the scratching of the head in embarrassment. Also, everyone knows the various automatic movements that lecturers make during their speeches. Tinbergen and Kortlandt, who, independently of each other, described this procedure for the first time, have given many examples of it, mostly from the behaviour of bony fishes and of birds. In ornithological literature displacement activity is designated mostly as sham-reactions, and also as mock-preening, mock-feeding, etc. Just as intention behaviour produced by the energy of an "autochthonous" reaction becomes a bizarre symbol of itself through visually acting "mimic" exaggeration of certain phases of behaviour, so also can the movements brought on by displacement activity undergo, in their higher evolutionary development, such formalisation that it is necessary to examine comparatively many related forms in order to unveil their origin.

Most of the display actions, which we shall have to describe for swimming ducks, can be explained as symbolic and as displacement activities, and this interpretation takes on a high degree of probability in view of the self-evident phylogenetic relationships and their agreement with those arrived at by comparing other characteristics. Displacement activities outnumber symbolic activities; most of them are movement which are derived from the behaviour of preening and shaking. Also prevalent among the automatisms in other animals, innervated as displacement activities by allochthonic excitement (Kortlandt), are the common daily automatisms of the care of the body (head scratching in man, mock-preening in pigeons, cranes, etc.), the taking of food (mock-feeding by the threatening barnyard fowl) and others of similar significance. For more detailed treatment of symbol and displacement behaviour, you are referred to my "Der Kumpan in der Umwelt des Vogels" as well as to Tinbergen's Die Uebersprungbewegung."

THE MALLARD *ANAS PLATYRHYNCHOS* (L.)

General remarks

Not only is the Mallard the duck whose behaviour has been known longest and in greatest detail, but it also occupies a central position among them in many respects. As a pronounced "primitive" characteristic of the species, we can use the fact that the drake has, in addition to the courtship whistle, another rather loud note, which can take the form of one-syllabled and drawn-out call ("raehb"), and a two-syllabled conversation note ("raebraeb"). For the same purpose we can use the corresponding note of the duck which is alike in both rhythm and meaning. In all other Anatinae, in the three Cairininae which we are concerned with here, as well as in the Casarcine Tadorna, the male's true voice, produced in the syrinx, is very considerably underdeveloped in favour of the whistle which is produced by a special organ, the drum of the trachea.

The non-sexual reactions and calls

The one-syllabled call-note of both sexes of the Mallard arises ontogenetically in smooth transition from the likewise one-syllabled "whistle of desertion," which is common homologically to all ducklings at present known, and duplicates it fully even to the position of the head and feathers. The conversation-call in all the Anatidae ducklings known to me is similar, as well as in the Anserinae in which it was originally two-syllabled. In the genus *Anser* this certainly original condition lasts only a few hours, after which it develops from the two-syllabled generalized duck-like conversation-call into the many syllabled type characteristic of the genera *Anser, Eulabeia, Branta,* etc. In both sexes of the Mallard, the one-syllabled call-note has also the significance of a warning, the sound, however, being just the same to the human ear. By the appearance of a strange dog the long, drawn-out "raehb" is released in all the drakes. In the latter case perhaps it sounds especially loud and long; at least, I am never able to decide from the call alone whether a drake is calling or warning his mate. Especially from the female, a repeated uttering of the one-syllable call signifies the intention of changing location. Before flying away it is given especially long and loud. It is uttered more softly, but more persistently by the duck when she is wandering around restlessly looking for her nest and when she is going to lead her brood over a rather large area. This form of the call-note which we term "the going-away call" is found in very many Anatinae, and even in *Aix* and *Lampronessa,* where the real call-note no longer exists in its original form but is replaced by a specialized call. In males the going-away call is lacking.

The two-syllabled conversation-call, the accent of which is on the second syllable, is uttered even by the ducklings of a brood, but it is uttered oftener and with special intensity by both mates when they find each other after a rather long separation, or when ease of mind returns after a scare which has made the birds freeze for a long time. In both cases it is the drake especially that utters his two-syllabled "raebraeb" with

Fig. 1. The whistle of desertion of a surface-feeding duckling. **Fig. 2.** One-syllabled call-note of Mallard drake, *Anas platyrhynchos* L.: a drawn-out "raehb." **Fig. 3.** The "raebraeb palaver" of a pair of Mallards with chin raising which is similar to the triumph-cry. Note the position of the feathers of the head.

very great rapidity and loudness, while the duck, in her more intense reaction generally begins to "incite." Such a palaver of a pair, or a brood of ducklings, reminds one very much of the so-called "triumph-cry" of Casarcinae and Anserinae, of which it probably represents the phylogenetic first-step. In high degrees of reaction intensity, corresponding perhaps to the duck turning to incite, the drake particularly has a special way of holding his head with raised chin (Fig. 3); in moments of still greater intensity there appears, but very rarely, an action called the "down-up movement," which we shall discuss later in connection with the antics of courtship. It is established with certainty by phylogenetic links in Anatinae which are more highly specialized with respect to a true triumph-cry, such as *Chaulelasmus, Mareca penelope,* and *M. sibilatrix,* that the "chin-raising" and "down-up" movements undergo a special differentiation. I will return to this when I discuss these species. When these "raebraeb" palavers of the Mallards, analogous to the "triumph-cry," reach exceptionally high degrees of excitement, passing even the threshold of the "up-down" movement, there appears an indication in the "raebraeb" of the drake, and even sometimes in the two-syllabled call of the ducks, of a peculiar tone, which in the Gadwall has led to a further differentiation. It consists in the fact that in the three successive double sounds the middle one is always uttered more strongly than the ones preceding and following; thus "raebrae'b, raeb-RAE'B, raebrae'B."

The sexual reactions and calls of the female

Just as, morphologically, the females of closely related ducks differ much less from one another than do the males, so also do they in respect to the inventory of their behaviour patterns. Therefore a sufficiently detailed description of the female Mallard will save a longer description of other ducks, with the exception of *Mareca.*

Inciting

The most common female action of courtship is the so-called "inciting" that is found in a form fundamentally the same, certainly homologous, in all Anatinae, in *Tadorna* and *Casarca,* and among the Cairininae in *Aix, Lampronessa,* and *Amazonetta.* The duck turns to her mate, or the one to be her mate, swims after him and at the same time threatens another male of the same species over her shoulder. With the Egyptian Goose both these orientation reactions, i.e., the placing of the body in a line with that of the actual or potential mate and the movement of threatening "the symbolical enemy," are quite independent of each other, and in their relationship with each other are quite plastic. The angle between the axis of the body and the direction of the threatening movement of the head is determined solely by the position of the female, the male, and the threatened enemy. It can happen, for example, that the enemy, as seen from the position of the female, is behind the male; so that then the Egyptian Goose, pressed closely against her mate, threatens the symbolical enemy straight ahead. However, most frequently the female in inciting is placed more or less with her tail towards the enemy, because she herself first makes a little dash against him but, not daring to attack, has now to run or swim from the "enemy" to the mate in order to incite at his side. This performance, therefore, determines wholly in itself the particular angle between the head and the axis of the body. Even in *Casarca ferruginea* one often sees an inciting movement of the head directed in this manner backwards over the shoulder, and not motivated by the relative position of female, male, and "enemy." In inciting Anatinae, in which the inciting is formalized to a pure ceremony, and in which the original significance of the instinctive behaviour, namely, bringing the mate into conflict with a real adversary, is completely obscured by the secondary meaning of "avowal of love" for the incited male, the movement of the head backwards over the shoulder is nothing but a simple, taxis-free instinctive movement, and the birds can no longer perform it other than in the one way, fixed for all time. A Mallard female, for example, incites over the shoulder even when the "enemy" is not in the direction in which she is threatening. The direction of the threat can be directed only within a very small angle. I wish to emphasise that I am far from propounding a Lamarckian explanation of this remarkable differentiation—succession of inciting behaviour. I am convinced that it represents a real "phylogenetic order" for, beyond all doubt, the forms of inciting as they are found in Casarcinae and indeed, to a certain extent in Anserinae, represent the original form of the movement. Until one recognises this one cannot understand at all the ceremonies of the Anatinae which, in their significance, are strongly modified. Nevertheless, I reject the proposition that an instinctive movement can arise from an orientation reaction which has become habit, through inheriting acquired characteristics. On the other hand we know of many examples of the rise of firmly fixed centrally co-ordinated forms of movement out of originally oriented movements, for example, the so-called zig-zag dance (Leiner) of the male Stickleback.

When "inciting" the Mallard makes a specific call, a peculiar trembling sound, accented mostly on the third syllable, "queggege'ggeggeggeggegg," through which is detectable a peculiar, querulous accent. This sound is not to be confused with the one that the duck makes when she is being pursued by a strange drake. When the bird is inciting, the very mobile head articulation enables the upper mandible to be bent down ventrally to its limit (Fig. 4); the head and back feathers are

Fig. 4. The inciting of the duck Mallard. Note the angle between forehead and beak and compare with Fig. 5 Backward movement of the head of the drake with chin-raising, compare with Fig. 14.

Fig. 5. Gestures of repulsion of the Mallard.

kept flat. Both cause a peculiar sheep-like expression of the face. Because of this downward bending of the bill the line of the chin presses close to the curve of the rump along which the head moves.

The gestures of repulsion

The note of repulsion that one hears from a duck pursued by a strange drake, and which one hears from ducks already incubating, sounds similar only superficially; it does not consist of a running succession of "quegg" sounds but of a broken series of single "gaeck" sounds, sharply uttered like a cough. While uttering these the upper mandible is bent upwards as far as possible at the so-called frontal-nasal joint, the feathers of the head and back are greatly ruffled, and the head is drawn deeply into the neck (Fig. 5). Female Casarcinae, when disturbed on the nest hold their bodies and feathers in the same way and make a very similar sound, as does also a brooding *Cairina* if a drake wants to tread her. In Mallards, as well as in the species named, mothers leading broods answer with the same sound and movement to every little disturbance, I am inclined to believe that this is the primary significance of the expression. It has been diverted by the Mallard into the more specific meaning of repulsion of a strange drake even for a non-brooding but fast-paired duck. The reactions and sounds, which were originally peculiar to a duck that was unwilling to mate because of brooding, found a new field of application in the Mallard, Pintail, and probably the Teal (*Nettion crecca*) in the dominance reaction of the drake, to be discussed later. When one hears clucking sounds from a Mallard duck in flight, being pursued by two drakes it is practically always the repulsion note just described. Only once in my life did I ever hear a Mallard in flight "incite."

The "decrescendo" call

All female ducks, including the genus *Mareca*, can make a very peculiar sound which, doubtless, with 14 syllables in *Nettion flavirostre* and one syllable in *Mareca*, represents the same homologous instinctive behaviour in all forms. In the Mallard it is generally a six-syllable "quaegaegaegaegaegaegaeg" with the strongest accent on the second syllable and a decreasing sound in those following. This sound is uttered especially by unmated ducks, and by fast-paired ones only when the male has flown away. The sight of another bird of the same species on the wing is the strongest releasing stimulus of

Fig. 6. Nod-swimming of the Mallard.

Fig. 7. The "pumping" prelude to pairing in the Mallard. The pair face each other each taking the opposite extreme positions of the vertical head movement.

the decrescendo call. However, unmated ducks with a very considerable lowering of the threshold to the reaction in question will respond in the same way to any flying duck and even to birds belonging to different orders.

The "nod-swimming"

This instinctive movement, which has been called by Heinroth the "coquette-swimming," corresponds in the female Mallard to the many differentiated behaviour patterns of the social display of the drakes, the "social-play" as Heinroth calls it. We shall see that the drake also possesses the nod-swimming behaviour pattern, which in his case is linked with other behaviour, and is not independent as in the female. The duck brings the behaviour we have described into play only when several drakes are gathered together and have made their desire for courtship known by shaking themselves and by ruffling their head feathers, as we shall describe later. Then, holding her head in a peculiar flat way with a very marked nodding, the duck shoots among the drakes, swimming in short arcs around as many of them as possible. In nodding, the head is held so close to the surface of the water that the duck's chin grazes it (Fig. 6). This nodding has clearly a releasing action for the ensuing display behaviour of the drakes. The phylogenetic origin of nod-swimming is not at all clear, as we do not as yet know any preliminary stages of the ceremony. We know this behaviour so far only in the Mallard and the most closely related members of the genus *Anas* in the old sense, and in the Chestnut-breasted Teal, *Virago castanea*, and the Grey Teal, *Virago gibberifrons* (E. Virchow).

The prelude to mating

This consists of jerking movements of the head, which have great similarity to the intention-behaviour for flying up, but it proceeds in reverse. The head is here not moved slowly downwards and jerked upwards but the other way round. The reaction has obviously arisen out of an intention-behaviour for "ducking-down-flat." While in the flying-up behaviour the head is quickly thrust upwards, the bill being held horizontally and brought back to the starting place with much less speed; in the prelude to mating the downward movement is accelerated with a jerk. Apparently this reaction has arisen from an intention-behaviour for the "ducking-down-flat" which takes place in mating itself. This theory gains in probability by the following chance observation: Once I saw standing on the upper edge of a very steep bank a Mallard drake, which as I thought was performing the "pump" movement of the prelude to mating, but I could see no duck. The next moment the drake flew down over the bank, landing at the base. What I had thought was a prelude to mating was the intention-behaviour for flying downwards—something that hardly ever happens. I am quite sure that an intention movement downwards, i.e., to get flat, is the origin of the "pumping" prelude to mating in most Anatinae. In the Mallard, as well as all other Anatinae, perhaps with the exception of *Mareca,* the movements of the prelude to mating are alike in form in both sexes. Those of the female are much more intense, as she herself mostly gives the occasion for the prelude ceremony and for copulation itself. We call this behaviour "Pumping" (Fig. 7).

The sexual reactions and notes of the drake

The general form of courtship

As we know, the males of most Anatinae as well as those of *Aix galericulata* in the Cairininae and a number of species included in the Fuligulinae gather together for social courtship play in which the females take part only as onlookers, except to provide for the necessary releaser behaviour to stimulate the males to courtship. On the other hand the females of all Anatinae that have a social-play (Heinroth) seem to have a very active role in choosing the mate, a matter we shall come back to, especially with the Mandarin Duck. Mallard drakes, along with Mandarin drakes are probably those in whose social-play the presence or absence of the female plays the least role. As is the case with hens of the Black Game, Capercaillie, Turkey, and Peafowl, their courtship is not the wooing of a special female but a general exhibition that takes place in apparently the same way whether the female is present or not. Within the Anatinae we shall see all the conceivable smooth transitions from such an impersonal mass courtship to the personal courtship of a particular female. In their social courtship the drakes jointly perform a number, varying from species to species, of highly differentiated movements, mostly accompanied by sounds which are produced by means of the drum in the syrinx of the drake. The movement connected with sounds which, as we shall see, have arisen from very different instinctive behaviour of the symbol and displacement activity type, have however this one thing in common, i.e., they lead with very few exceptions to a tensing of the windpipe which is obviously necessary to bring about the courtship note. Preceding the true sound-accompanied courtship movements there is always an instinctive behaviour which, perhaps, serves as a self-stimulus, and whose distribution within the group allows one to conclude a greater phylogenetic antiquity than is required for the real courtship behaviour.

Drinking

I shall first describe one such introductory movement which, perhaps has not an exclusively epigamic character. When two ducks meet in a pond either the one gets out of the way of the other or both drink. One might first consider this drinking as

Fig. 8. Mock-preening of Mallard drake. The rubbing of the bill on the quills of the wing produces a loud noise.

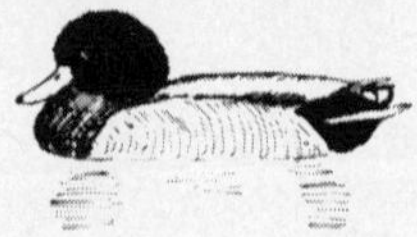

Fig. 9. The starting position of Mallard drakes congregating for "social play." Ruffled head feathers and smooth back feathers.

chance. Heinroth often tells how he observed a long time before it was clear to him, that this drinking at meeting was no chance but a ceremony, with a definite social function and the significance of a "sign of peace." The origin of the significance of "drinking" he explains as a symbolic activity in the sense that two birds that eat or drink close together have no bad intentions toward each other. Thus, originally, drinking was a reaction with a purely social meaning, and is not at all confined to the relationship between the mates of a pair, even though it is very frequently done by them. With greater differentiation it is linked with mock preening behind the wing in many species, and in the Gadwall and in the Mandarin Duck, which belongs to the Cairininae (q.v.), this has become fixed; in the Mallard such a linkage is merely suggested. The drake quite often does mock-preening right after he has drunk opposite to his duck, and then they go through both behaviours several times, and after each other, without any set order. I have already mentioned here that the established linkage between drinking and mock-preening in the Gadwall and the Mandarin is fixed in the reverse sequence.

Drinking is very widely spread in the family. In the Anatinae there is no form of it lacking (see diagram). With the significance of a male courtship act it is linked with other epigamic behaviour even with the Cairininae, *Aix* and *Lampronessa* as well as with the Fuligulinae, *Netta* and *Metopiana*, very close relatives of the Anatinae.

Mock-preening

When a drake has drunk in front of a duck he is wooing, he frequently reaches with his bill behind his slightly lifted wing, just as though he were going to preen himself (Fig. 8), but instead of that, he moves the nail of his bill quickly and roughly over the underside of the keel of his wing so that there is a rather loud "Rrr" sound that can be heard many meters away. The movement is so short that I am still unable to tell whether the bill is rubbed over the secondaries or the primaries. The strength of the tone seems to indicate the latter, but on the other hand, one must remember that in the Wigeon, Garganey Teal, Wood Duck, and Mandarin, the drakes certainly mock-preen on the inner side of the secondaries, the last two on a definite feather especially coloured and differentiated for this purpose. It is an interesting commentary on the psychology of animal observation that mock-preening was first discovered by us only in 1939, after we had devoted many years to the purposeful observation of these birds. What first opened my eyes to the procedure I had so often seen was that I noticed for the first time a homologous movement of the Garganey drake, done a little differently, in which the colours of the small wing feathers, obviously differentiated expressly for this behaviour, made it impossible to overlook. Then I seemed to realise vaguely that I must have seen this ceremony somewhere before, and in the course of special observations to discover this "somewhere" I came to the astonishing conclusion that all the species of ducks I had observed carried out this behaviour (see diagram). Today it seems quite unbelievable that the unusual yet characteristic sound had not drawn my attention to it much sooner.

Mallard drakes begin mock-preening, which almost always begins with drinking (not the other way around!), not only upon meeting the female but, in the same way, and perhaps more intensively, when meeting drakes, when the mood for social courtship begins to rise. With this significance the behaviour can be considered the first prelude to what Heinroth calls "social play."

From the point of view of its phylogenetic origin, mock-preening is certainly to be regarded as displacement activity. There is hardly any behaviour pattern that can take on so many means as the avian displacement activity of preening. Besides its particularly frequent use as courtship behaviour it can be used as a threat, as with cranes, and as a prelude to mating, as among Goosanders and Rock Doves. In the last-named case there are, to be sure, very real doubts as to whether the preening ceremony has arisen from a real transition and not as symbolic activity from an autochthonous act, in about the same way as Heinroth (q.v.) proposes for the origin of drinking.

The preliminary shaking

When several drakes have gathered in the way in which I have described above, the mounting of their specific courtship-impulse is expressed first by a special position of their body and feathers. The head is drawn in firmly between the shoulders so that the white neck ring disappears completely. The under feathers are lightly ruffled so that the bird swims imposingly high on the water, while the back feathers are held very smooth in an "intended" contrast to the usual position of rest. The head feathers are erected as high as possible so that, seen from almost every angle, the green sheen disappears and gives place to a deep velvety black (Fig. 9). This position differs from the one that a Mallard takes just before normal autochthonous action of shaking itself only in the fact that it is often held for several minutes. In ordinary shaking the drawing in of the head and the ruffling of the feathers lasts only a few seconds. The experienced observer can tell from their progression exactly when the shaking will actually happen, just as one can foretell pretty well the moment of sneezing by the amount of preliminary drawing up of another's face. In the Mallard drake's "social-play" one has not only to wait much longer for the shaking but when it does occur it is not followed by relaxation or release from the preparatory position. Indeed the first shaking, in which the head, strangely restrained, is thrust upward timidly and

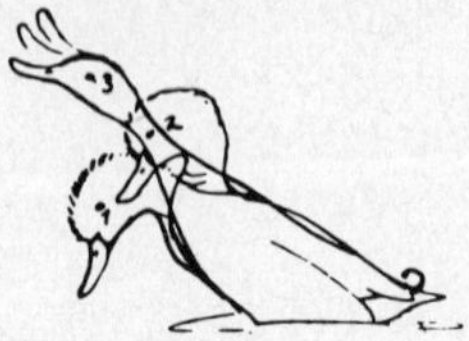

Fig. 10. Diagram of the action of preliminary shaking.

Fig. 11a and b. Two phases of the movement of the grunt-whistle. Note the arch of water-drops thrown up.

with nervous haste (Fig. 10) is followed in a few seconds by a second and third shaking. The intensity of the action then increases very gradually each time until finally the shaking movement seems to hoist the drake high out of the water as though he had a cramp. When this degree of intensity is reached there follow, almost regularly, instead of another shaking one of the three actions of courtship that I am now going to describe, i.e., the grunt-whistle, down-up movement, and the head-and-tail-up, whereupon all the drakes perform at each other. The whole company is relaxed for a while and stop courting, or else they begin again after a short pause with a preliminary shaking of less intensity. The preliminary shaking is certainly a substitute activity. It is found in all ducks except *Spatula*.

The grunt whistle

As in ordinary shaking the bill is first lowered so that the shaking action begins under the water and continues above it (Fig. 11a). This lowering is so decided that the end of the bill, at the first transverse movement, cuts the surface and throws up a shower of little drops in a high arch. This circumstance was not observed by any previous describer of drake courtship and we ourselves discovered it first when we noticed that a line of little black dots always appeared in photographs of drakes performing the grunt-whistle. The erection of the body then runs contrary to an ordinary shaking in that the head is so far forward that the body is standing rather upright in the water while the head is still quite low, with the bill close to the surface. This strange bent-in position (Fig. 11b) has its mechanical explanation obviously in the stretching of the wind-pipe, for just at the moment of its peak there is a loud sharp whistle followed by a deep grunt, while the head straightens up again and the body sinks back to the surface of the water. The grunting sounds as if a body of air, compressed by the whistling, were escaping.

The extent of this grunt-whistle within the order of the Anatinae is peculiar (cf. diagram) in that, on one hand, not all of them have this behaviour or one homologous to it. On the other hand, the Cairininae, *Aix* and *Lampronessa* and *Tadorna tadorna* in the Casarcinae, certainly have a homologous activity even though it is differentiated in a different direction. Other than the species of the genus *Anas* in the narrower sense in which this behaviour pattern is practically the same as the Mallard, the following have almost the same kind of grunt-whilst: The Pintail, *Dafila acuta*, the South American Pintail, *D. spinicauda*, the native and the Yellow-billed Teals, *Nettion crecca* and *N. flavirostre*, and probably all the other members of this genus which are unknown to me as well as the Chestnut-breasted Teal, *Virago castanea*, and in a somewhat changed form, the Gadwall, *Chaulelasmus strepera*. The grunt-whistle is lacking in the Bahama Duck, *Pacilonetta bahamensis*, the African Red-billed Duck, *P. erythrorhyncha*, the various species of Garganey, and the Shoveler.

In view of the intermediate forms between grunt-whistle and shaking still to be discussed in the Wood Duck and Mandarin, as well as the intimation of shaking behaviour which comes at the beginning of the action in the Mallard itself, there can be no doubt that it has had its phylogenetic origin as a mimic exaggeration of what was originally a preliminary shaking. While the Mallard and other Anatinae and the Wood Duck have as reactions, in addition to the usual shaking, two formalisations of this movement, and the Mandarin has no less than four different courtship movements, distinct from each other and from the original shaking, there is a peculiar reverse behaviour in *Tadorna*. The male has a display shaking very much like the corresponding behaviour of the Mandarin drake. At the beginning the head is lowered so as to suggest a grunt-whistle, and then while being shaken and tossed up a trilling whistle is burped out without the body being lifted. This movement, so peculiar to the Sheld-Duck, is in contrast to all other Anatinae that perform homologous movements, its only shaking movement, which is to say that the original, ordinary shaking with its mechanical significances, has merged into this reaction. For example, if one seizes a Sheld-Drake and then lets him go there does not follow, as would be the case among all other drakes, an ordinary shaking, but usually the display-gesture described above. From this one would infer that the feather-straightening effect is much less important in it than the original movement which is contributory only in its mechanical effect. I should like to compare the phylogenetic procedure of this change of a mechanically effective instinctive movement into a releasing courtship-ceremony with that differentiation of morphological structures which is to be seen in the wing of a Goat-sucker and an Argus' wing has so completely merged into a courting organ that its original mechanical function has been considerably encroached upon by its new supplimentary function as a releaser. In the same way with *Anas* and *Aix* the original behaviour of shaking remains in existence alongside the one developed out of it for courtship, while in the breeding *Tadorna* male it has disappeared into it.

Head-up-tail-up (making itself short and high)

Along with the grunt-whistle the Mallard male has two other courting actions that are of equal value, the head-up-tail-up and the down-up movement. Which of the three comes first after the introductory shaking seems to be a matter of chance. Heinroth writes that generally one drake gives the grunt-whistle and the rest of the drakes involved in the social-play answer with the head-up-tail-up or the down-up

Fig. 12. The head-up-tail-up of the Mallard drake; the same phase of movement seen from different sides. Note the prominence of all particularly bright-coloured and morphologically varied parts of the feathers.

Fig. 13. The turning of the head towards the female immediately after the head-up-tail-up. Note the position of the head feathers.

Fig. 14. Mallard drake nod-swimming combined with head-up-tail-up and head-turning.

Fig. 15. The head-turning of the Mallard drake without chin raising, mostly combined with nod-swimming. Note the position of the feathers at the back of the head and their shine.

movement. This is, to be sure, frequent but it is not inevitable. All other possible combinations occur. The head-up-tail-up is perhaps the most striking of all the drake's courtship actions. It is certainly the most complicated. First with a loud whistle the drake thrusts his head, with indrawn chin, backwards and upwards, and at the same time curves his rump upwards with rump feathers very ruffled, so that the whole bird becomes extraordinarily short and high. While doing this the elbows are lifted high so that the curl feathers, sticking up high on the rump, can be seen from the side (Fig. 12). This phase lasts about a second, and then the body sinks back to its normal position. The head alone remains high for a moment, when the bill is pointed to a particular female among those present at the social-play of the drakes. In the case of mated ducks it is pointed at the mate (Fig. 13). In the next moment the drake starts moving and shoots away over the surface, stretched flat in "nod-swimming" (see above) moving mostly in a circle around the duck he is wooing (Fig. 14). Toward the end of the nod-swimming, when the drake is gradually returning to his normal swimming position his head is now lifted high and the back of the head is turned to the same duck to which his bill has been pointing (Fig. 15). Even during the head-up-tail-up and belching out of the whistle his head feathers take on a peculiar "set."

The feathers are held flat on the side of the head but bristled in the medial portion so that the head becomes a narrow but high disk which, seen from the side, becomes very shiny, in sharp contrast to the dull black ball it was in the preliminary shaking. During the "nod-swimming" this set is changed so that all the head feathers are held flat, and only on the neck does a little area of stiffly lifted feathers remain (Fig. 14). This placing of the feathers is not understandable until the next phase, the turning of the head. Then a little black field in a frame of shiny green is presented in a striking manner to the view of the female he is wooing.

This is the place for a word about the comparative morphology of this feather placing, wherein the word "morphology" will denote the concepts of both the form of the movement and the organic structure. The position of the feathers which are purely functional in the Mallard are definitely fixed into an unchangeable form in the Pintail, Teal, Wood Duck, and Mandarin drakes. All of the four species named, and probably many others, possess not only movements homologous to those of the Mallard but also a corresponding feather-placing. With them, however, there are also colour and form characteristics, which show that the same lines that are apparent on the head of the Mallard drake, only in the various arrangements of the feathers, are permanent morphological characters. The disk-like "set," which the Mallard shows at the moment of "burping," has its sharp border exactly at the place where, in the Teal, there runs the fine white line that separates the green of the spectacle around the eye from the dull colour of the top of the head. In the male Wood Duck there is not only the same fine white line, but also a lengthening of the feathers which causes an obvious projection of the border marked in the Teal only by a line. The hoods of *Lampronessa* and *Aix* serve in the first instance as a "disk-set." In the Mandarin drake the same line is marked more clearly by the border of the white temples against the bright colours of the top of the head, and with it the lengthening of the feathers goes so far that the obviously projecting border exists even when the feathers are at rest. In the Pintail drake the "disk-dress" is less clear than in the Mallard but in his case the turning of the back of the head is especially developed in conjunction with erection of the neck feathers. The black velvet cushion that appears in this position in the Mallard drake's neck has become in *Dafila acuta* a permanent morphological characteristic. His corresponding feathers are not only, as we know, a deep black, edged at right and left by a white line, but they are also longer than the surrounding feathers so that in the feather-placing part of the head-turning behaviour a prominent feather bolster appears. There is no doubt that in all these body and feather movements, the movement is older than the organs which has become differentiated in such a way as to increase the optical effect. The wide systematic distribution of the movements, and the much narrower distribution of the specific feather structures and colours, as well as the clear relationship of the latter to all the behaviour patterns underlying them as a common basis, and obviously present in the family a very long time, leaves no other theory possible. It is also proved by the fact that the distinctive body movements are present in quite the same way as in the Mallard, in males of other species of *Anas*

Fig. 16. The down-up movement of the Mallard drake, the right drake at the deepest point of the movement and the other at the highest, which immediately follows. Note the little fountain of water raised on the surface by the great rapidity of the movement.

which have not special display plumage. The placing of the head feathers is also traceable in them but to a lesser extent than in the ducks that have head feathers that are either very shiny or very long.

The distribution of the head-up-tail-up is much less widespread in this family than is the grunt-whistle (see diagram). In addition to the members of the genus *Anas* proper, it occurs only in the Pintail and in the European Teal, as well as in the male Bahama Duck, where it represents a movement of courtship. It is interesting that it is lacking in the drake of *Dafila spinicauda.* In *Nettion flavirostre* parts of the movement exist, i.e., burping and subsequent turning of the head to the female. We can as yet form no theory about the origin of the head-up-tail-up. As with the grunt-whistle the strong bending of the backbone has obviously something to do with the tightening of the windpipe at the burping out of the whistle, but we cannot decide yet whether the movement itself is derived from an instinctive behaviour, originally operating mechanically, or developing through symbol or displacement activities.

The down-up movement

Instead of the grunt-whistle and head-up-tail-up a third movement may follow, in which the drake thrusts his bill into the water as quick as lightning, and in the next movement jerks up his head alone without lifting his breast (Fig. 16), which is still low in the water. At the instant when the head is highest and the breast is deepest there follows the whistle, just when there is the greatest tension on the windpipe. In raising the bill a little fountain is often raised by the quick bill movement which, because of the short reaction time of a bird, may react optically on the duck. Right after this very brief movement the drake says his "raebraeb" quite quickly and with lifted chin. More than in other courtship movements one has the impression from the down-up that the drakes must be influencing each other in some way for, with a frequency far beyond the probability of mere chance, practically all the drakes of a courtship-society perform this movement. Then after the ensuing "raebraeb" palaver the courtship ends for the time being.

The down-up is the only behaviour occurring in social play that is used occasionally also at other times and then only in a quite definite situation as in a "raebraeb" of maximum intensity, which is particularly common after a disturbance, or after a fight between two drakes.

The distribution of the down-up movement is restricted to the closer relatives of the Mallard and is, so to speak, the reverse of the head-up-tail-up. Except for the species of *Anas* in the narrowest sense it is found in a homologous way only in the Gadwall and the different species of Wigeon. An apparently homologous differentiation is found in *Virago castanea.* It is lacking in all kinds of Pintails and Teals, which are otherwise so like the Mallards in courtship behaviour. One possibly homologous courtship movement is extreme lifting of the head, which is the most important courtship movement of the male Rosy-billed Pochard, *Mitopiana peposaca.*

The origin of the down-up movement can be given with some probability. It might be a formalized exaggeration of the "drinking" as found in almost all the Anatinae. If one comprehends the movement of chin lifting done by the Mallard drake after the down-up movement and also, moreover, at great "raebraeb" palavers, as a stage of weaker intensity of the down-up, then we get a somewhat wider distribution of the movement within the family, and at the same time a clearer relationship to the drinking. In the male Wood Duck at times much chin lifting and head turning follows "drinking."

"Gasping"

Often individual drakes without any particular movement, but at the exact instant when the remaining members of the courtship society utter the whistle, make a peculiar hoarse sound. One can imitate it best by saying a three-syllabled "chachacha," while breathing out, then in, then out. One hears the sound especially from weak drakes, or those whose reactions are not yet strong. I have heard it in the same way, but given from a correspondingly different position, from a female Khaki Campbell duck and from a female hybrid of *Virago castanea* and *Paecilonetta bahamensis* which was showing interest in the social-play of the Mallard drakes. The origin and significance of the manifestation are uncertain, as well as its broader distribution.

The combat of the drakes

Fighting Mallard drakes, with necks drawn in, grab each other by the feathers of the crop and shove each other hard, with one trying to shove the other back. Early in the spring one sees round bare places, the result of this specific manner of fighting. Only when the excitement of the fight is very great do the drakes begin to hit each other with the shoulder of the wing. It is then seen that the peculiar position of the head, neck shortened, was predetermined by the proximity of this additional action. Other Anatinae that hit with the shoulder in fighting hold their opponent with the neck drawn in tightly against the breast exactly at the spot where the shoulder hits.

The post-coital play

In the Mallard drake this is marked by a special behaviour. Immediately after treading, while still hanging on, the drake suddenly flings his head and neck far down his back; without lifting them high. Often in this act he is still holding the neck feathers of the female in his bill so that her head is pulled backward (Fig. 17). Then, as if this backward movement had been only the introduction to an exaggerated mimic

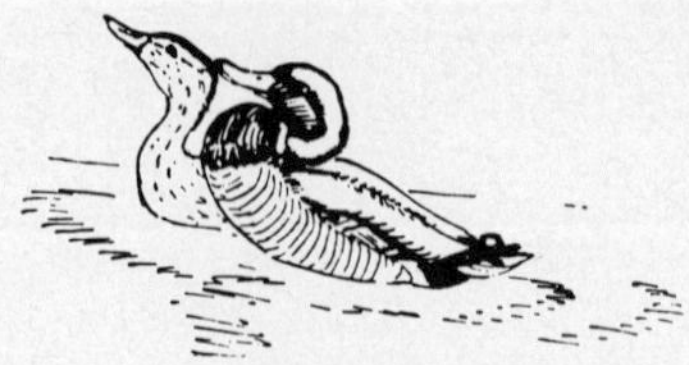

Fig. 17. The flinging back of the head of the Mallard drake after coition. The same action follows the head-up-tail-up (Fig. 12) with head-turning towards the female (Fig. 15) as a preliminary to nod-swimming (Fig. 6).

nod, the drake shoots away with the typical behaviour of nod-swimming, and circles around the duck, just after the head-up-tail-up behaviour. This behaviour of flinging the head upward and backward, for which we shall use the term "bridling," appears often as an insertion in the head-up-tail-up, and it follows the lifting of the head and turning of the bill to the female, with the subsequent nod-swimming, in exactly the same way as it follows the mating.

The same post-coital play is found in all species of *Anas*, in *Nettion flavirostre* (whether in *crecca* or not I do not know) and *Virago castanea*. In both of the last named forms the "bridling" enters in as a quite independent behaviour in social-play (Table 1). In *N. flavirostre* there is no subsequent nod-swimming, and in *V. castanea* it is either quite isolated, or linked up in a single tight chain of actions that begins with the grunt-whistle and closes with nod-swimming and head-turning.

* * *

EDITORS' NOTE

Hereafter follows descriptions of all the species of ducks that Lorenz was able to study, comparing them with the displays of Mallards. These descriptions are too long to reproduce here, but the following list shows the species described and the approximate number of printed pages given to each species; the latter indicates the approximate amount of information that was available to Lorenz for drawing the comparative conclusions that follow the list.

V.	Spot-billed duck, *A. poecilorhyncha*	1 page
VI.	Meller's duck, *A. melleri*	2 pages
VII.	Pintail, *Dafila acuta*	8
VIII.	Chilian pintail, *D. spinicauda*	2
IX.	Bahama duck, *Paecilonetta bahamensis*	4
X.	African red-billed duck, *P. erythrorhyncha*	1
XI.	Garganey teal, *Q. querquedula*	5
XII.	Shoveler, *Spatula clypeata*	3
XIII.	Chestnut-breasted teal, *Virago castanea*	5
XIV.	European teal, *Nettion crecca*	4

XV.	Chilian teal, *N. flavirostre*	4
XVI.	Gadwall, *Chaulelasmus streperus*	5
XVII.	Wigeon, *Mareca penelope* and Chiloe wigeon, *M. sibilatrix*	2
XVIII.	Chiloe Wigeon × Mallard hybrid, *M. sibilatrix* × *Anas platyrhynchos*	2
XIX.	Carolina Duck, *Lampronessa sponsa*	6
XX.	Mandarin duck, *Aix galericulata*	4

REFERENCES

Heinroth, O. (1910), Dir Brautente; *J. Ornithol.*, **58**.

Heinroth, O. (1910), Beitrage zur Biologie, insbesondere Psychologie und Ethologie der Anatiden, *Verh. d. V. Intern. Ornith. Kongr*, Berlin.

Heinroth, O. (1930), Ueber bestimmte Bewegungsweisen bei Wirbeltieren, *Sitzgsber. Naturf. Freunde*, Berlin.

Heinroth, O. and M. (1924–1928), *Die Vogel Mitteleuropas.* Berlin: Lichterfelde.

Kortlandt, A. (1938), De uitdrukkingsbewegingen en geluiden van *Phalacrocorax sinensis*, Shaw and Nodder, *Ardea.*

Leiner, M. Oekologische Studien an Gasterosteus aculeatus, *Zs. Morph*, and *Oek, d, Tiere*, 1929, **14** and 1930, **16**.

Lorenz, K. (1935), Der Kunpan in der Umwelt des Vogels (Der Artgenosse als auslösendes Moment sozialer Verhaltensweisen), *J. Ornithol.*, **83**, H. 2/3.

Poll, H. (1910), Ueber Vogelmischlinge, *Verh. d. V. Intern. Ornith. Kongr*, Berlin.

Tinbergen, N. (1946), Die Uebersprungbewegung, *Z. Tierpsychol.*, **4**, H.1.

Whitman, C. O. (1899), Animal Behavior, Biolog. Lectures of the Marine Biological Lab., Woods Hole, Mass.

Wormald, H. (1910), The Courtship of the Mallard and other Ducks, *British Birds*, **V**.

On Locomotory Movements in Birds and the Intention Movements Derived From Them

A. DAANJE

This paper is intended to show that a great number of movements in birds, the origin of which has not hitherto been understood, are intention movements or movements derived from them.

Reprinted from *Behaviour*, volume 3, pages 49–98 (1950). Original numbering of figures is retained.

Pure intention movements (Heinroth, 1910) being low intensity forms of innate behaviour patterns, are as a rule, the very first parts of an activity. Because many instinctive acts begin with a locomotory movement towards something (food, or a mate, or an enemy) most intention movements are low intensity forms of locomotion.

In order to recognise pure intention movements therefore it is necessary to study the form of locomotory movements first. This is done in the second section, in which special attention is given to hopping and walking. It is shown that hopping involves not only the feet, but also movements of the body, the neck, the wings and the tail. Walking is "one-sided hopping," in which both feet are used alternately, and in which each foot is supported by body, neck, tail and wing movements.

The third section discusses a number of intention movements.

The last section shows that the majority of intention movements are difficult to recognise because they have undergone a secondary evolutionary change (ritualisation) as an adaptation to a newly acquired function, that of acting as a social releaser.

Comparison enables us to trace a number of principles involved in this secondary change, *viz.* (1) exaggeration, (2) a shifting of thresholds of the component elements and (3) loss of coordination between the component rhythms.

Applying these principles, many so-called display, threat and begging movements can be understood as ritualised intention movements.

Owing to the extreme scarcity of accurate comparative studies it is, in most cases, not possible to do more than make a guess as to the origin of a given movement. And although the available facts for the relatively better known groups such as the pigeons, the ducks, and the Gallinaceous birds, support my relatively speculative conclusions. I should like to emphasize the need of accurate comparative studies aimed at a tracing of the origin of derived movements.

INTRODUCTORY

As all observers of animal behaviour know, many activities may occasionally be performed incompletely. This is especially true of locomotory movements. Walking, for instance, in a bird, may be seen in numerous degrees of intensity, arrangeable in a sliding scale from full intensity down to an almost imperceptible indication of it. Heinroth seems to have been the first to have called attention to these low intensity movements; he called them "intention movements" because the trained behaviour student can derive from their study a knowledge of what the animal is intending to do in the next few moments.

It has become clear to me that intention movements are, at least in birds, of much more common occurrence than is usually realised, and that a better knowledge of them could promote our insight into the factors underlying behaviour to a considerable degree. In order to recognise intention movements however it is imperative that the full intensity of the corresponding movements should be well known. A close study of locomotory movements, therefore, is the first step in a study of intention movements.

In this paper I want first to give a description of various types of locomotion in birds, especially of the start, as it is these first phases that are usually found in the intention movement. Most attention will be given to the movements of the legs, and less to those of the wings, because I am better acquainted with the phenomena of jumping and walking than with bird flight. When I have had to consider particulars of flight, I have mostly drawn from the extensive observations of Lorenz (1933) and Stolpe & Zimmer (1939).

Following upon the chapter on locomotion I will present a number of descriptions of more or less easily recognisable instances of intention movements. This will be followed by a chapter tending to show how such intention movements may have undergone certain changes or may have merged with other movements as a result of which it may be difficult to recognise them as derived from intention movements.

Although I have tried to study as great a variety of species as possible, I am quite aware of the limited scope of my study. Circumstances prevented me from studying birds more closely in Zoological gardens. Further, my observations are concerned with the mechanics of locomotory movements, and many anatomical and physiological aspects are ignored, being irrelevant to the present problem. Further, my use of the literature has been limited to papers in which full attention is given to this special aspect of locomotion. Lastly, I must stress the tentative nature of many of my interpretations. An intention movement can only be recognised as such with certainty after a thorough study of the behaviour pattern of the species concerned. I naturally desired to draw my examples from a great variety of species: therefore I was led to include observations of which still contain elements of uncertainty.

In spite of these limitations I think it justified to present this paper as a sketch of the probable origin of a number of hitherto little understood behaviour elements.

I am much indebted to Prof. Dr. E. J. Slijper and Dr. N. Tinbergen for their valuable criticism and advice, and to Mr. P. Sevenster for making most of the drawings.

LOCOMOTORY MOVEMENTS OF THE HIND EXTREMITIES

A bird on the ground or in a tree may move in two ways: (1) hopping on both feet at the same time or (2) walking or striding with alternately moving feet. Many species only hop, others again hop or walk according to circumstances (see Stresemann 1927–1934). According to Stresemann the hopping takes considerably more energy than walking or running. Many kinds of hopping birds get over this difficulty to a certain extent by avoiding using both feet at exactly the same time, instead placing them one after another. One and the same foot remains in advance of the other in a series of successive leaps. The hind foot gives the impetus and the jump is developed to its full extent by the foremost foot so that the movement becomes supple and less jerky, as can be seen, e.g., in the Magpie. When going slowly a Magpie walks; at higher speeds it hops with the feet laterally behind each other; at still higher speed it proceeds with both feet side by side and pushing off simultaneously. This last method is never used for more than a short time.

The hopping type of locomotion, considered by Boker (1935) as the primitive type characteristic of the tree living ancestors of birds, will be treated first.

The hop

A bird, for instance a Blackbird, about to take a hop to a higher branch, partially folds its legs (chiefly by bending the heel joint), lowers its breast, so that the hip joint is strongly bent, and draws in its head and neck, so that the back part of the head is almost on and between the shoulders with the neck forming a more strongly curved S than in the attitude of rest. In taking the hop it begins with stretching the legs, raising the breast relative to the haunches (the hipjoint) and the neck rapidly changes from the S-form into an almost straight line. Shortly before and during alighting the opposite bending movement follows. Finally, after landing, it stretches itself from this "doubled up" position into the position of rest. Mechanically seen, this movement is best compared to the movement of a loosely coiled spring, that is pushed in and then relaxed, so that the spring jumps up. The maximum effect is reached when the stretching is performed by the whole spring. In much the same way, a leaping bird does not confine the stretching to the legs but uses the muscles of legs, body and neck.

When a bird makes a still larger leap, the wing is also moved. At first sight this may make the impression of incipient flight, but I believe that it is really something very different. When a man jumps with both legs at the same time, he not merely huddles up partly with his legs, bends his back and shortens his neck, but he also bends his elbows and even clenches his fists. When he jumps the arms go upwards and laterally forwards, the clenched fists are relaxed. Then the arms move backwards. This has nothing to do with flying, swimming or rowing in the air. I think the most important function is that of supporting the energy of the jump, the arms acting as additional springs, laterally fixed to the main spring. When the automatic movement of the arms is hindered, a much smaller leap will be the result, even in spite of a higher expenditure of energy.

The movement of the arms of a leaping bird is much the same. The relative lightness of the bird arm renders it of less importance in this respect than the arms of man. Moreover, the bird arm with its large carrying planes supports the bird during the leap. Although this latter function now probably overshadows the "spring" function, I nevertheless think the spring function is not to be neglected.

Finally in treating the jump we have to consider the movements of the tail. Many birds erect the tail just before jumping. During the start it is gradually lowered until it is almost in line with the body axis. At the same time it is spread to a certain extent. One function of the lowering of the tail might be the following. When the bird does not merely jump but actually flies up, the turning down and spreading of the tail may help in directing the flight. According to Lorenz (1933), a bird starts with fluttering wing movements (Rutteln). In this type of flight, the pressure centre of the wings usually lies a little in front of the centre of gravity of the bird, so that the back part of the body would be inclined to sink down if the tail were not spread out and pushed down a little.

It may be that the spreading of the tail in a jumping bird has something to do with flight. The initial erecting and the subsequent gradual lowering, however, has a

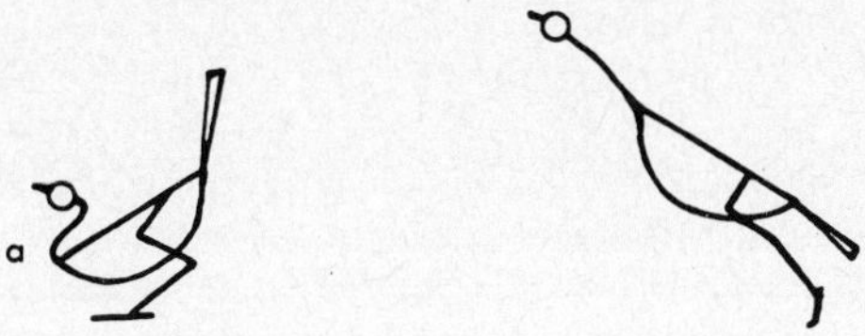

Fig. 1. Scheme of (a) the first and (b) second phase of the jump by a bird.

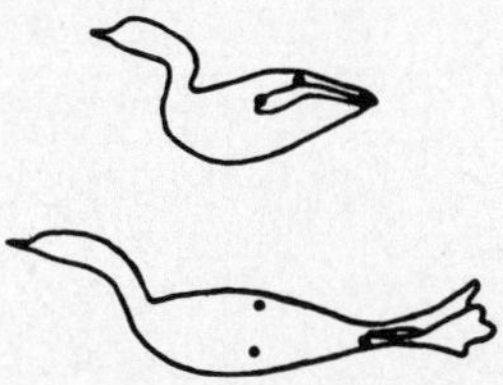

Fig. 2. The huddled and the stretched phase in under water simming of *Podiceps nigricollis* Brehm. From a film by Frank and Neu in Stresemann, 1927–1934.

function in the jumping. If the tail were kept in line with the body axis, the full weight of the tail would have to be carried right at the start. By first erecting and then gradually stretching the tail, its drag is coming into play gradually.

The parts of the body behind the hip joints carry out a similar movement, as they cannot help executing a downward movement as a result of the upward movement of the forepart of the body. The same line of argument could be applied here.

The tail movement just described is only found in few birds; the majority of them simply drag their tails along. This suggests that the movement of the tail is of little importance—which can easily be understood since the mass of the tail is, as a rule, very slight. But even in species that show no discernible tail movement in a normal hop, the movement of the tail does occur in a hop of high intensity.

The preparation for the jump may therefore be described as follows: The legs are bent, the breast is lowered, the neck is drawn in, the hind part of the body and the tail are pushed up and the wings are slightly expanded (Fig. 1a). When the actual jump follows, legs and neck are stretched, the breast moves upward, the abdomen and tail are pushed down and the wings may make a single stroke (Fig. 1b).

All these movements can best be seen in larger, more slowly moving species, especially when executing large jumps. The smaller the bird, the smaller the amplitude of movements. In a hopping House Sparrow for instance, the legs seem to be the chief or may be the only parts of the body that make the bird move. Yet one should not conclude too hastily that the other movements are absent, for I know by experience that absence of movements may be only apparent, the movement being visible under exceptionally favourable conditions only. I once observed a House Sparrow which had some feathers stuck to its neck. These feathers, enlarging so to speak the neck's movements, clearly indicated that at each hop the neck was stretched and refolded. I also saw the movements in a hopping Chaffinch, a species in which the neck movements are very conspicuous during walking.

Figure 2 shows that there is a type of under water swimming that closely resembles hopping.

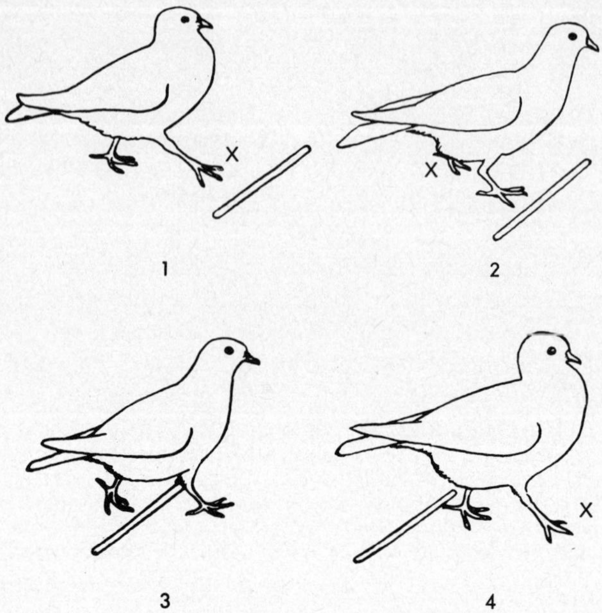

Fig. 3. Sagittal horizontal head movement of a walking Pigeon in connection with the movements of the legs. The feet in the figures, indicated with a cross, were not sharp in the original photographs, which proves that they were still moving as "air leg." For further explanation, see text. From a film by Thauer and Peters, 1938.

Walking

For a discussion of the movements during walking it will be necessary to select three types, *viz.* (1) Domestic Hen, (2) Wagtail, (3) Duck.

Domestic hen. Except for the movements of the legs the most distinct movement in a hen when walking is that of head and neck. With each step the head is brought forward and backward again. This head movement is linked up with the leg movement as follows. At the moment the leg which moves forward through the air (abbreviated: the air leg) almost reaches the ground and the most distal part of the leg on the ground (ground leg) just passes backward under the body's gravity centre, the head reaches its most backward position. The stretching of the neck now begins, simultaneously with the stretching of the ground leg, which is now pushing backwards. The head reaches its most forward position when the ground leg is stretched and is about to leave the ground. Now that this leg, which now has become the air leg, moves forward, the head is withdrawn, to be pushed forward again when the other leg is stretched. Four phases of this movement are given in Fig. 3.

It will be clear that the head is brought forward each time one of the legs is stretched. In other words, each step is a hop with one leg, and walking is alternate hopping. At each step the head gives mechanical assistance in exactly the same way as in each hop.

The movements of the head in walking have often been noticed and interpreted in another way. In some birds the head, when being drawn back, is kept at the same place in relation to the environment. It is possible to take photographic pictures of such birds in which the whole part of the body is blurred owing to movement but in which the head is absolutely sharp. This is undoubtedly an adaptation, which enables the eye to fix the environment during a longer time than would be otherwise possible. I believe, however, that this adaptive function is secondary and has, in some species, evolved on the basis of the already existing movement, the primary function of which is to give mechanical support. I base this belief on the fact that the movement is found in many species, but that only in some of them it results in the head being kept in place in relation to the environment. A hen or Stork, as long as it is not walking too quickly, keeps its head in place; but an Oystercatcher, while withdrawing the head at each step, nevertheless moves it forward in relation to the surroundings. Even when a walking hen increases its speed it reduces the amplitude of its head movement and consequently the head is no longer kept at one place; in the withdrawal phase it is merely retarded. With increasing speed the head movement's amplitude diminishes until, in the running bird, the neck remains stretched and the head's movement is gliding like that of body and neck.

Wagtail. A second element of locomotion is seen, in a rather extreme form, in the Wagtails. A walking Wagtail shows, besides the movements of legs and head, a distinct whipping up and down of the hind part of the body and the tail, accompanied by less distinct counter movements of the forebody. This tail wagging and its counter movements is very similar to the whipping movement of a bird's body in jumping. The only difference, again, is that the Wagtail "jumps" with one leg at a time.

Ducks. While the two elements of locomotion described above are taking place in a sagittal plane, the most striking movements of walking Ducks (waddling) do not. In Ducks, the pushing motion of the hind leg, pushes the body forward, and at the same time lifts the body at its side and turns it in a horizontal plane around the foremost leg. This lifting and turning element of walking is so conspicuous in Ducks because of (1) the relative shortness of their legs, (2) the wide distance between the legs and (3) the nearness of the center of gravity to the ground.

These three elements, *viz.*, the rhythmic movement of head and neck, the tail wagging and the waddling are by no means confined to these types because they are "specialists" which each show one of the elements in a very conspicuous form. Most, if not all, birds, show all these elements. Withdrawal of the head for instance is seen in slowly walking Pigeons, Rails, Cranes, Tringa species, Common Curlew, Ibis, Oystercatcher, Wagtails, Larks, Starling, Crows, Magpie, White Stork, Herons and foraging Black-headed Gull. It is even present in Ducks, although it is difficult to detect and shown only under favourable circumstances. It is difficult to see in the Lapwing because it runs too fast; but one can detect a slight forward movement of the head at the beginning of the run which is not withdrawn until the end.

Birds, such as the Lapwing, the Black-headed Gull and Terns, which take very short steps, i.e., whose legs execute only a small angle displacement, show either very small head movements or none at all. A Heron, when striding almost erect and taking small steps, does not make the head movement; should however it take large steps with the body almost horizontal, as in foraging, then the head movements become very noticeable.

While in the Wagtail the tail moves even in relation to the abdomen, in Crows and other Song birds, in the Flamingo and possibly in many other birds one gets the impression that the tail moves merely as a consequence of the body movements.

Waddling, while most distinct in Ducks, Geese and Swans, is also seen in Pelicans, Cormorants, Parrots, and Crows. The Swallows and the Nightjar, though also short-legged, hardly show the waddling, because they always walk with such small steps. The Woodpeckers, also short-legged, always hop, even when on the ground and therefore do no waddle at all.

It will be clear therefore, that most birds show all the movements. Depending on the proportions of various parts of the body, on speed of motion, on the angular-displacement of the legs, etc. now one of the elements, now another is more prominent. Absence of an element can only be established after very careful observation, because a movement may be hardly detectable, as is especially the case in species that have "specialized" into one of the types of locomotion. Other species, like, e.g., the Crows, are non-specialists; showing all types of movement more or less equally developed.

THE INTENTION MOVEMENT

As has been pointed out by Heinroth (1910) and by Lorenz (1935, 1941), an animal may under the influence of weak internal or external stimulation, perform incomplete movements. Because it is often possible to judge from these fragmentary or incipient movements what part of the behaviour pattern is activated at the moment, or in other words what the animal intends to do, Heinroth called such incipient movements intention movements ("*Intentionshandlungen*"). Apart from the instances of intention movements in Ducks mentioned by Heinroth and Lorenz, relatively few intention movements have been described in literature. Yet my studies of locomotion have convinced me that intention movements are not at all rare in birds. For various reasons however they are not always readily recognisable. I will begin my description with the more obvious cases.

Intention movements of walking or hopping

The most striking case is the initial phase, the bow. An alarmed Robin will perform one or more bows before it actually flies away. From the more or less erect resting posture, it rapidly lowers its breast, at the same time moving the hind part of the body and the tail upward. With a pronounced bow the wingtips are also moved upward and probably the legs are somewhat folded and the head drawn back. After this bow, the bird stretches itself again, raising the foremost part of the body and lowering the hind part. After what has been said about the jump it will be clear that the bow is nothing but the first phase of the jump, and that the subsequent stretching is the incipient form of the next phase of the jump, performed on the spot. The identity of these intention movements with the actual jump is obvious when with growing alarm, the bird finally jumps or flies away.

Another example of this same type of bow is provided by the Magpie. When relatively strongly motivated, the Magpie also makes a very deep bow, raises the tail and the back part of the body, and spreading the tail while doing so, but folding it again at the highest point. Also, it bends the heel joints and flaps the wing tips. From this highest intensity performance down to the merest indication there is a whole scale of increasingly incomplete movements. With decreasing intensity the wing movements are dropped first, then the actual bowing becomes less pronounced and finally only the tail flapping remains. This sequence of "fading" is clearly seen in the series of movements made after perching on a branch, when the locomotion impulses under the influence of an after-discharge, are gradually dying down. Besides this type of bow, which is very frequent, the Magpie, sometimes shows another type in which the bird bows slowly downward, putting up the tail and bending the legs, but remains in this position for rather a long time, sometimes fluttering its wings. The springing up follows after some time. This type of the bow is executed by one bird of a pair in front of the other. It is followed by flying away, the other bird following.

The Lapwing and many other waders when alarmed perform a movement which I believe to be derived from the bow, and which I propose calling the plovers' bow (following German usage, where it is called the "*Regenpfeiferknicks*"). An alarmed Lapwing raises the forepart of the body and stretches the neck a little, resuming the original posture immediately afterwards. This movement is very similar to the second phase of the robin's bow. The omission of the first phase of the bow may be connected with its omission in most cases of actual flying-away, when the Lapwing, having a rather horizontal normal position, does not bother to lower its forebody but jumps right up. Only when strongly alarmed does a Lapwing execute a deep bow. I observed this only once, in a bird with a broken wing which was being attacked by a Kestrel. This Lapwing bowed, and springing up made counter attacks.

This omission of the first phase can be observed in many birds that can fly up without jumping. I even saw it in a tame Duck that was reacting to food in my hand. Even when standing in a quite upright position it was able to jump up for more than 10 cm, using its toes only.

The intention movement of a Kingfisher, a bird that, when searching for fish, has a quite erect bearing, does not include the downward phase either but it raises the body sometimes before it flies off to the sighted prey to catch it.

The Little Owl may execute either one of the two phases. Upon the approach of a human it may duck, and remain ducked for some time (Fig. 4). When the enemy moves, it may keep up the erect posture for some time again and, upon the next movement of the enemy, duck down again.

Fig. 4a and b. The ducked and erected attitude of a frightened Little Owl.

In the case of the plovers' bow and of the Little Owl's movements, one again meets with a different interpretation in literature in which quite another function is emphasized. It is generally supposed that these movements, by enlarging the parallax, help the bird in three dimensional analyses of that visual field. I believe, in both cases, but more especially in the owls, that this function certainly is not primary but that it may have evolved secondarily from the intention movement. My main argument is the fact that the intention movements are so widely distributed and occur in so many cases where nobody would attach any function in three dimensional vision to them. Also, I believe that the movements that help in three dimensional vision are, even in owls, different from the mere intention movements. Raber (1949) in his extremely thorough study of prey catching in owls, describes and figures in the Tawny Owl movements preceding the actual swooping down on the prey. The head is not only moved in vertical but also in a horizontal direction. According to Raber, these movements are, superficially, similar to intention movements, but fundamentally different. I think it quite possible, however, that these fixation movements have evolved from the intention movement, in much the same way as the fixation of the head in some birds when walking may be a secondary development on top of the primary function of withdrawing the head at each step.

Such secondary developments may also be suspected in other, quite different cases. The crouching movement of an alarmed Lark, for instance, is quite obviously the first phase of the bow. It has evolved into an adaptation correlated with the cryptic colouration, and together they have the function of concealing the bird from visual feeders like predatory birds. This type of adaptive colouration and behaviour has evolved in many species.

In the Bittern, an inhabitant of reed beds, the second phase of the intention movement has evolved into a cryptic posture, the so called pole posture (Fig. 5). Both birds, Lark and Bittern (to choose only two examples) are forced, upon approaching

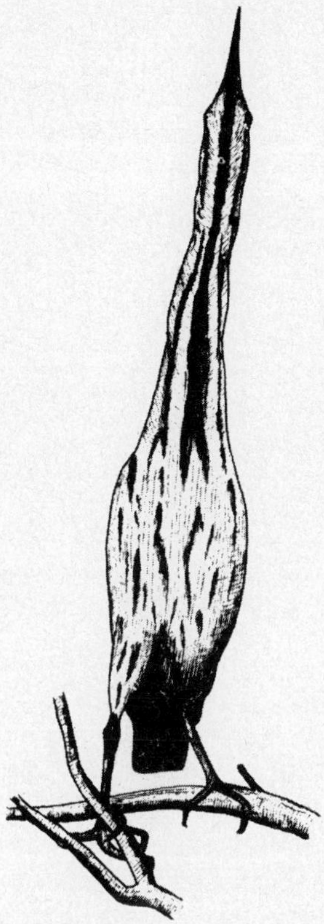

Fig. 5. The pole attitude of the Bittern. From a photograph by Schut in Portielje and Schut, 1938.

danger, to do something in the way of escape, but they freeze in one of the phases of the intention movement, *viz.*, in the phase that has the greatest selective value in relation to the given environment and its own colouration. In the Bittern, adaptation has even gone so far as to make the bird take up its position very gradually.

The first phase of the bow is displayed by the Bittern under different conditions, *viz.*, when the enemy approaches it very closely, which, from the viewpoint of selection, means that the chances of being noticed are greater than those of escaping unnoticed. The bird bends its legs, withdraws its head, and lifts its wings, On top of this, and certainly connected with its secondary value as a threat, the feathers are fluffed and the wings are spread, so as to make the bird appear as large as possible to the enemy. From this posture, the actual jump is carried out when the enemy comes within striking distance. This jump again has a secondary function as a means of defense: The long and sharp bill is thrusted towards the opponent's head, as has been shown experimentally by Portielje (1926).

Fig. 6. Attitude of an alarmed Blond Ring Dove. From a photograph by Craig, 1909.

Another example of the upright posture being adopted in mild alarm is the Blond Ring Dove (Fig. 6) as described and figured by Craig (1909). Unlike the Bittern, the Blond Ring Dove does not point the bill upwards.

A very remarkable type of bow is executed by the Great Titmouse and the Blue Titmouse. These species combine the bow with a lateral movement, executed during the second phase. When stretching two times in succession, the horizontal movements alternate in direction. The head thus performs a movement in the shape of a U. The sideward component is not confined to these species; it can be seen in, e.g., the Robin, though in a less marked degree. These lateral movements may be preparation for a side-ward jump, in which the length of the swing and the phase difference between the legs may increase the effectiveness of the actual jump.

In this type of bow, the movement of the tail is not, as one would expect, the reverse of the movements of the forepart of the body. It has taken me some time to understand the tail movements in the Titmice's bow, but the slower moving Magpie showed me the solution. Unlike most other birds, the Magpie and still more so the Titmice execute the bow with great rapidity. The tail, partly by its inertia and partly by the air resistance, lags behind. The result is a more subtle type of movement than a perfect see-saw would show. Now in the Titmice, the tail lags behind in the vertical direction but not in the horizontal direction. The result is that the tip of the tail describes a horizontal 8. Fig. 7 gives a graphic representation. I have represented the varying speed of the horizontal and the vertical components of the movement as identical with those of a cylinder piston, that is to say maximal speed in the centre and decreasing to zero at the turning points. One phase of the vertical movement coincides with a half phase of the horizontal component. Arbi-

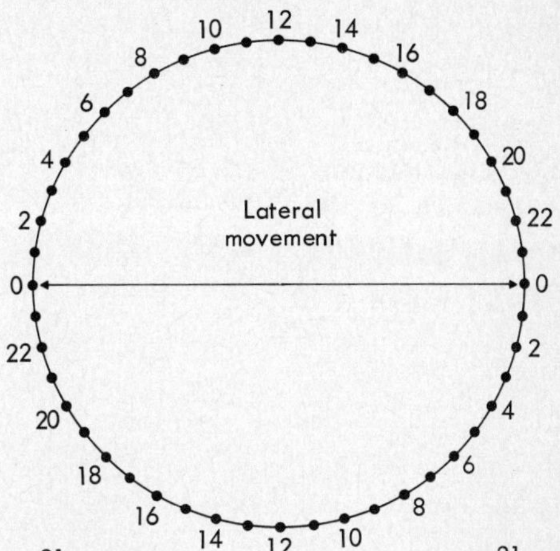

Fig. 7. Scheme of the co-operating lateral and dorso-ventral movement of the tail of the bowing Great Titmouse. Against the circles of the lateral and the dorso-ventral movement and against the curve of the resulting movement are placed the corresponding times. The dorso-ventral movement is shifted 9 time units in regard to the lateral movement. For further explanation, see text.

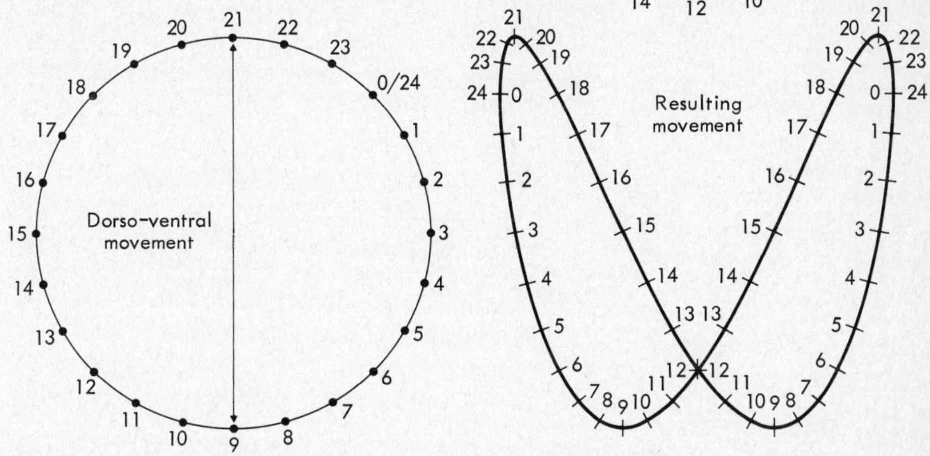

trarily, I made the vertical movement lag 9 time units behind the movement of the breast. In order to simplify the figure I did not shift the horizontal movement, nor did I take into account that the upward movement is as a rule a little faster than the downward movement.

The same lateral movement of the body is executed by the Nightjar when it is disturbed in the daytime. By night it omits these intention movements and flies away immediately. This species, having very short legs, does not execute the vertical bow.

Now there are many birds in which the intention movement of the jump is incomplete in another sense. Instead of a complete bending or stretching of the whole body, neck, legs, and tail included, the movement may be performed by only part of the body. Depending on the species and on circumstances this part may be the neck and head, or the tail.

As mentioned above, the Magpie may move the tail exclusively. In this species it is obvious that this is nothing else but an intention movement of very low intensity, and it seems probable that this holds good also for species who rarely or seldom add head and leg movements to those of the tail. This exclusive use of the tail is seen in many song birds such as House Sparrow, Tree Sparrow, Chaffinch, Buntings, Blackbird, also in Moorhen, Partridge and Water Rail. It seems to be quite possible that many or even all instances of tail whipping in "nervous" birds are derived from the bow.

The tail movement may differ from the original pattern in various ways. In most birds just mentioned the up stroke is quicker than the down stroke. Further, the frequency may increase with increasing nervousness, and in the highest intensity may result in a strained, motionless erected position, as if the tail is kept prepared to help in the jump for a considerable time. Some species, such as the domestic hen, even keep up their tails permanently, as long as they are active. Only when hens are ill, or in absolute rest, or feel quite safe and at ease while feeding at the manger is the tail relaxed.

On the other hand there are birds that only put up their tails under exceptional circumstances. A Jackdaw, whose wings and tail were clipped, being obviously more on the alert than a normal one, increased its push off by erecting its tail, so that in jumping, it could be lowered with maximum effect. The tail whipping is dropped entirely in the Bittern and Little Bittern. Their relative *Tigrisoma lineatum* however does whip the tail (Portielje, 1926).

In some species the tail movement seems the opposite of that described above. Chiffchaff, Willow Warbler and Stonechat for instance do not move their tails up but down, and the downward movement is quicker than the subsequent return into the normal position. In the two firstnamed species I never saw the upward movement at all, the Stonechat performs it when it is highly agitated. The down stroke is, in actual jumping, merely a later phase of the jump; it is executed when the bird actually jumps off.

Other species, such as Blackbird, Nightingale, House Sparrow and Chaffinch usually show the up movement, but sometimes tip the tail down. The Blackbird, of which I have the best observations, first moves its tail up and from this position it tips the tail violently downward, then brings it up more slowly again. With the Nightingale, the downward movement may result in a drooping and spread tail.

The Wagtail keeps the mean between these two types in that the up and down stroke are equally quick and extensive. In the next section we shall see that this movement may reach a climax in two ways: The tail may be kept up in a vertical position, or it may be spread and pointed down. In the Wagtail the tail wagging usually appears after walking when the locomotion impulse ebbs down; in this case the amplitude of the movement gradually decreases. Sometimes the tail wagging appears as a real intention movement.

Whether or not the tail trembling of the Redstarts is homologous with the tail movement of the Wagtail, I am unable to say.

The sideward component of the tail movements, as described in the Titmice, has become prominent in some species, for instance in the Shrikes. In the weaverbirds *Sporaceginthis melpoda* and *Spermophila albigularis*, the vertical movement is even absent and the tail is only moved laterally. The Wren, which puts its tail up at the least excitement, then moves it up and down and to the left and right. The vertical and the horizontal movements do not occur simultaneously as in the Titmice but are kept apart. According to Kluyver and coll. (1940) the lateral movements are shown by the male only. The Red-backed Shrike may execute them simultaneously but often shows them after each other.

The other birds the tail movements are entirely absent; instead, the head movement is prominent. The head movements have three components, *viz.*: (1) a vertical movement, (2) a horizontal-sagittal movement, and (3) a horizontal-lateral movement.

The vertical movement is shown in pure form by the Kestrel, which before jumping always bobs up its head several times, with an amplitude of about 0.5 to 1 cm. This movement again is generally considered a help in three dimensional vision. After what I have said above it will be clear that I do not believe this to be primary. Indeed in this case I believe it has no connection with three dimensional vision at all. My main argument is that a hovering Kestrel, far from moving its head, takes amazing pains to keep it at exactly the same place. As a professional surveyor, I have watched a hovering Kestrel through a theodolite. To my intense astonishment, I found that the head was kept at exactly the same place, even in gusts of wind! Yet if a sitting Kestrel would use head movements as a means for increasing parallax, one would expect a hovering Kestrel at least to take no excessive precautions to prevent head movements.

This is not to imply that I do not believe in the alleged function of parallax increase in the head movements of other types; quite the contrary. For these movements, as e.g., described by Raber (1948) in the Tawny Owl, are relatively slow and occur in more than one direction.

The horizontal sagittal movement is found in Pigeons, Fowl and Ducks. Often, but not always, this movement is preceded by a vertical movement. The quickest motion is found during the forward thrust, so that the head seems to make an aiming movement. With Ducks we can distinguish three forms: (1) one a little upward, and (2) one horizontal. The former type indicates a jump upward, in most cases flight, the latter one indicates forward walking or swimming. (3) Finally Ducks have a downward head movement, which according to Lorenz (1941) precedes flying down, and according to my own observations may also indicate the intention to walk down a slope, or into a pond, and further to dive. I saw the latter case in a one day old duckling. It first performed a "vacuum" performance (*Leerlaufreaktion*, Lorenz, 1935) of the flying dive and repeated this several times without the other duckling of the clutch taking any notice. It then made the downward intention movement and this caused all the others to dive at once.

Just as with the tail movements this aiming head movement may also turn into its reverse. The aiming movement belongs to the actual jump; the reverse one which originally precedes it, belongs to the ducking before the jump. The ducking phase

Fig. 8. Domestic Duck preparing to jump from the water onto a bank emerging vertically out of the water.

can be observed in Ducks wishing to jump up onto a bank emerging vertically out of the water—an action which costs them some considerable trouble. The birds are so "stupid" to swim to the bank as close as possible so that they force themselves to make an almost vertical jump. They then begin by making an aiming head movement, then withdraw it and move the head more and more backward, especially after some misses! Whereas the withdrawal is at first repeated in quick succession, the head is finally kept in a fully withdrawn position for sometime after which the jump is eventually executed (Fig. 8).

Apart from the two types of head movements discussed, there is, also perhaps, a lateral movement. I have no personal knowledge of this type but Lorenz (1935, p. 353) describes it in Geese as a "short lateral shaking of the bill." Judging from what is seen with tail movements one would expect lateral movements especially in the waddling Geese, Ducks, Cormorants, etc.

Special mention should be made of the peculiar head movements seen in Vultures. When intending to fly away, a Vulture puts its head far under the insertion of the neck in the breast; the body does not make a bow, and the wings are lifted a little (Fig. 9). In taking a strong leap the head is thrust forward and upward. I believe that this type of head movement is part of the "spring action"; the only difference from that of all the other birds being that the head is not withdrawn above the body but below the body, which seems to be just as good a method.

Lastly, leg movements may be the most prominent part of the intention movement. I have observed instances in the Jay and in the Spotted Flycatcher. The legs are withdrawn from the branch for a moment, then stretched until they touch the branch again and the actual jump is executed at once.

Intention movements of flight

Apart from the tail movements common to both the jump and the take-off for flight, the tipping up of the wing as an expression of the intention to fly up should be briefly mentioned. This movement is seen very conspicuously in the Blackbird, which usually

Fig. 9. Griffon Vulture intending to fly away.

executes it synchronously with the raising of the tail. It can also be seen in the House Sparrow, the Tree Sparrow, the Willow Warbler, the Chiffchaff, Titmice, Goldcrests, *Spermophila albigularis* and *Sporaeginthis melpoda.* Most remarkable is the fact that I once saw a Teal doing it before it flew up from the water; it seems to be rare in Ducks because neither Heinroth (1910) nor Lorenz (1935, 1941) mentions it.

Terns, and several Waders (Godwit, Plovers) stretch the wings and point them upward before taking off. In the actual take off the upward impetus is chiefly effected by the forceful down stroke of the wings. A similar stretching upward of the wings can be seen in the foremost partner of a pair of Swifts at the moment the hindmost partner, the male, swoops down on it in an effort to copulate (Daanje, 1944).

RITUALIZED FORMS OF INTENTION MOVEMENTS

Just as in other types of derived movements, e.g., displacement activities (Tinbergen, 1940, Kortlandt, 1940b) intention movements may be ritualized, that is, secondarily adapted to the function of social releaser. So in ritualized displacement activities, the deformation may be of such an extent that it is impossible to recognise the origin of the movement unless it is compared with homologous, less markedly ritualized movements in related species. For a satisfactory study of the origin of derived and secondarily changed movements a thorough comparative description is necessary. Our descriptive knowledge being still in a highly fragmentary state, any interpretation of movements as I am going to give in this chapter must necessarily remain highly speculative. Yet I think it necessary to put forward such ideas as I have developed in the course of my study; for I am convinced that some of them will prove fertile for future research.

Principles of secondary change

In the foregoing chapter I have not confined myself to the pure, unchanged, original types of intention movements, for the simple reason that these seem to be relatively scarce. This has led me to touch several times upon the principles of secondary charges involved. Before proceeding to a more extensive treatment of the more highly ritualized types I will discuss the principles of secondary change which I have encountered thus far. These principles are:

1. Exaggeration. ("*Mimische Ubertreibung*", Lorenz, 1941).
2. Shifting thresholds of component elements.
3. Loss of coordination.

Exaggeration. This principle has already been discovered by previous observers, and has been discussed by Lorenz (1941). A good example is the tail whipping of the Magpie, and that of the Wagtails.

I am not quite certain whether exaggeration always springs from the same source. In the case of the Magpie's tail I believe it is connected with a function as a social releaser. I think however that exaggeration may in other cases be the result of the suppression of a movement by another drive. I have observed Willow Warblers continuously fluttering their wings while on the ground. One never sees this to such an intensity in Willow Warblers foraging in the tree. I am inclined to think that the inclination to flee to a safer tree was suppressed by the inclination to feed on some special food on the ground, and that this was the cause of the exaggerated wing fluttering.

Shifting thresholds of component elements. Problems of threshold concern us here in two respects. First, the intention movement as a whole is built up of various elements (movements of head, legs, wings, tail) that lead a more or less independent life within the whole. As I mentioned above, not all the elements have the same threshold for the general excitation evoking the movement as a whole. Thus in the Magpie the first sign of motivation is movement of the tail, while the other components, having a higher threshold, appear with increasing motivation.

As a general rule it seems that usually the most distal parts have the lowest threshold. This is the case in the tail wagging of dogs. I believe tail wagging in dogs to be the intention movement of walking. It is not, as is often supposed a sign of joy. Fierce dogs, growling or barking while on a chain or in the kennel wag their tails just as intensely as any dog let out by its master. The tail wagging is the lowest intensity of walking; with increasing motivation, as every dog owner knows the wagging movement will extend forwards, until at first the hind body is wagging, then the fore-legs beat the ground alternately and finally the hind legs also "walk" on the spot.

Corresponding tail movements are shown by cats, when they prepare to jump, not only to catch prey, but also to reach a higher window. Often the whole tail moves, but sometimes only the extreme tip.

Now the relation between the thresholds of the components appears to vary considerably from species to species. As will be clear from many of the instances cited in the foregoing section, some species use mainly the tail in the intention movements, others the head, others the legs. This shift in the thresholds may even go so far that some of the elements are completely "lost." To complicate the matter still further, the thresholds may even change in one individual in the course of time—an instance of which is provided by the Teal, which as a rule bobs the head before flying up, but occasionally lifts the wingtips. Even such an "element" as tail wagging is a system containing separate elements, and these elements, as we have seen, lead their own life too. Thus in some species the downward movement in others the upward movement, in others again the lateral movement has become prominent.

Loss of coordination. As was set forth in the first chapter, there is, during locomotion as well as in the original intention movements, a very rigid coordination in tempo between the separate elements; for example, rhythm of the head, the legs, the tail and the wings. This synchronisation may break down in some cases, and in order to understand this phenomenon fully it is necessary to study the mechanism of coordination as far as it is known.

Coordination between limbs has been studied by von Holst, mainly in fishes, but also in other animals, including man.

Von Holst found that in many fishes the oscillations of the pectoral fins, the dorsal fin and the caudal fin are synchronous. In the *Labrus sargus* type for instance the pectorals alternate, and the dorsal, caudal and anal fins are synchronised both with each other and with the pectoral rhythm.

Under certain circumstances, this coordination may break down, that is to say, whereas each fin still oscillates regularly, their rhythms are no longer synchronised but are entirely independent of each other. Both complete (absolute) synchronisation and complete lack of synchronisation are observable in fish after severance of the connections between spinal cord and the brain. Also, external stimuli, while apparently necessary to initiate the movements, are not necessary for the synchronisation. The coordination between the various fin rhythms, therefore, is entirely an intrinsic faculty of the spinal cord. This faculty has been studied more closely by von Holst, who showed that synchronisation was due to an influence exerted by the spinal mechanism responsible for the rhythm of another fin. This influence responsible for synchronisation, the exact nature of which is still unknown, was called by von Holst the magnet effect. The magnet effect, exerted by one rhythm upon the other, tends to slow down or to speed up the frequency of the dependent rhythm until it coincides with that of the dominant rhythm.

Von Holst discovered further that the magnet effect did not always secure complete synchronization but often influenced the dependent rhythm in a "half-hearted" way, in that it succeeds in forcing its rhythm upon the dependent fin when the phase difference is not too great, but fails to do so in case of greater phase differences (Fig. 10). He called this relative coordination.

Von Holst also found a type of relative coordination which does not affect the tempo but the amplitude of the dependent rhythm. A dominant fin may increase the movements of a dependent fin when they are moving in a certain phase relationship (e.g., synchronously) but suppress the dependent rhythm in the contrary phase

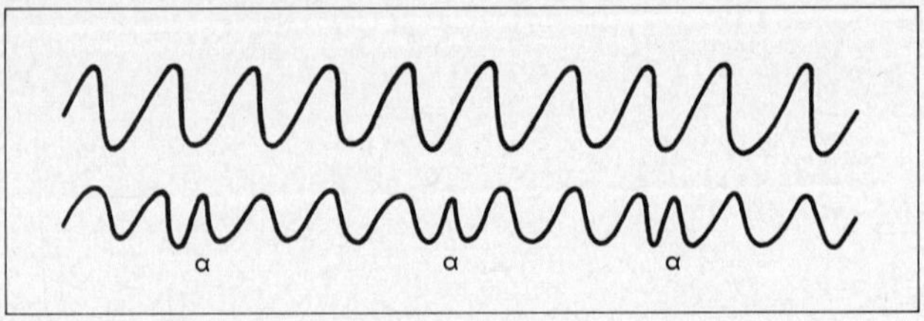

Fig. 10. The magnet effect in *Labrus.* Above: rhythm of the pectorals (dominant). Below: rhythm of the dorsal fin (dependent) influenced by the pectorals' rhythm. After von Holst, 1937.

relationship. This effect upon the amplitude was called the superposition effect (Fig. 11).

The phenomena of relative coordination seem to be much more frequent in fish than in land Vertebrates, although von Holst found some instances in the latter, even in Man. Yet I think the phenomenon is more common in birds than von Holst seems to think. The coordination between the two legs, it is true, is always absolute, but the movements of head and (or) tail are often very loosely coordinated with those of the legs.

Owing to the differences in thresholds of the elements involved in locomotion one element (1) may, when locomotion begins very gradually, come in action well before the element with the next higher threshold (2) steps in. As long as (2) does not act, its possible influence on (1) does not show itself. Now if (2) is dominant to (1), the rhythm and amplitude of (1) suddenly change when (2) comes in action. The following may serve as an example. In a walking domestic hen the movements of the head are absolutely synchronised with those of the legs. The head makes one oscillation at each oscillation of each leg, so that each leg makes one oscillation in the time the head makes two oscillations (phase relation 1:2). Now when a hen starts walking slowly, the head movement is the first to set in, and follows its own "local" rhythm. After some head oscillations the legs may begin to step, but in a very low tempo, so that the phase relation may be 1:6. After a while the leg movement may be speeded up and the head movement slowed down, the relation will become 1:4 and finally 1:2.

When the movements during the 1:4 phase relationship are studied more closely it will be seen that not all head movements are exactly alike. Those which coincide with the leg movement as in normal locomotion are reinforced, those in between are weakened. This is relative coordination of quite the same type as described in fish by von Holst.

The influence of the magnet effect on the tempo of the oscillations can be observed for instance in the Moorhen. When the bird is a little alarmed it tips up its tail; then follows a rather slow movement with the legs. The tempo of the leg movement gradually increases while the rhythm of the tail becomes retarded until at a certain

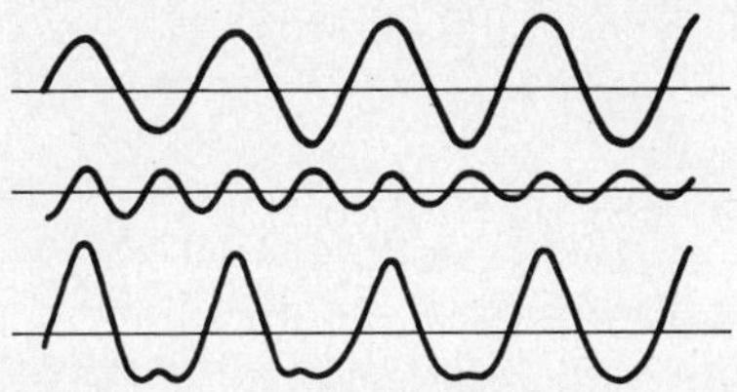

Fig. 11. Schematic representation of the super-position effect. Above: rhythm 1, of a sinusoid character, with increasing amplitude. Center: rhythm 2, of a sinusoid character but with smaller and decreasing amplitude. Below: result of super-position of the rhythms. After von Holst, 1937.

moment the phase relation 1:2 of the oscillations of legs and tail is almost reached. The rhythm of the tail however is still a little too fast; the magnet effect is too feeble to achieve absolute coordination and the tail movement gets more and more ahead in relation to the leg movements until at a certain moment the difference in time between the two oscillations becomes so great and the influence of the magnet effect so feeble that again the tail moves at its own higher frequency, inserting one quick oscillation; after which both movements of tail and legs coincide again in about the relation 1:2. Such insertions of a quick oscillation may be repeated more often till finally absolute coordination is reached.

Review of instances

I will now proceed to describe and interpret a number of secondarily changed intention movements. Most of these form parts of the prenuptial behaviour pattern. They may have releaser function as threat or as sexual display or as begging movements and their ritualisation probably is a consequence of this releaser function.

Because sexual display and threat are often difficult to tell from each other, I will treat them together.

Sexual display and threat postures derived from the bow

The male Turkey displaying in front of a female, withdraws the head to such an extent that it almost touches the raised back feathers. The tail is spread and raised. The wings are partly spread downward and kept very rigid, the tips of the primaries touching the ground. The hindpart of the abdomen is somewhat raised. This posture is kept up for a considerable time, while the cock walks either towards or in front of the hen. The tail is tilted sideward so as to offer a broad view to the hen. She, if interested, picks at the ground as if feeding.

This male display posture resembles the bow in many respects. The legs however are not bent, the tail and wing postures are exaggerated, and the bird is frozen into one of the extreme postures of the bow, *viz.* the crouched position. Whether the origin of this movement expresses the intention to jump towards the female for the purpose of

Fig. 18. Displaying male House Sparrow.

Fig. 19. Display with stretched neck of the male Tree Sparrow.

mating, or the intention to make a show of force before the female I do not venture to say.

This type of display, the ritualised first phase of the jump, is found, with highly varying types of ritualisation, in many other Gallinaceous birds, such as *Centrocerus urophasianus, Dendragapus obscurus richardsoni, Tympanachus cupido* L., *Bonasa umbellus* L., the Peacock, and the Blackcock. *Tympanachus* has a stretched neck, presumably for the sake of blowing up the gullet; *Bonasa* hardly lowers the breast; the Peacock lowers the breast very deeply but scarcely withdraws the neck at all, the Blackcock, though lowering the breast, may stretch or withdraw the neck or may keep it in the position of rest.

The bow, while utilised to such an enormous extent in the Gallinaceous birds, is by no means confined to them. Among the song birds the House Sparrow-male displays in much the same way before the female (Daanje, 1941) (Fig. 18). The Sparrow's display is, however, not a "frozen" posture. The male bows slowly, and quivers the wings now and then. In the beginning the tail is raised, sometimes until nearly in the vertical position and it may be spread a little. Further the male strives to attain a position as much as possible in front of the female. The fact that the female adopts a defensive attitude suggests that the display has threat-character.

The male Tree Sparrow's display is almost identical. In my paper on the House Sparrow (1941) I described a slightly different display in the Tree Sparrow in which the neck was stretched (Fig. 19). Later I found that the display with withdrawn neck is really the more common of the two. Yet this stretched attitude is of considerable interest. The stretched neck belongs to the second phase of the jump, the raised tail and dropping wings form part of the first phase. This is just another instance of a time shift in the elements in relation to each other; while the legs, tail and wings are still in the first phase, the head is in the next phase.

The same type of display is seen in male Bluethroats, White Wagtails, Yellow Buntings, Crested Larks and sometimes in Blackbirds.

Another type of display derived from the jump is found in the Blond Ring Dove (Craig, 1909). This Dove first makes the deep bow (Fig. 20b), remains in this posture for a short time and then raises and stretches itself (Fig. 20a). When in the raised position it makes some greatly exaggerated marking-time movements.

Fig. 20a. The up phase of the bow of the Blond Ring Dove. From photographs by Craig, 1909.

Fig. 20b. The down phase of the bow of the Blond Ring Dove. From photographs by Craig, 1909.

Displaying Carrier Pigeons perform a bow accompanied by head movements; in the erect posture they execute little jumps.

The Wood Pigeon's display is much like that of the Blond Ring Dove, but the bow is deeper, bringing the breast down below the branch upon which it is sitting; the tail is sometimes raised.

Whitman (1919) who made a very extensive comparative study of pigeon behaviour, shows that many species have a similar display. *Ocyphabs* bows and moves its legs alternately erecting and spreading the tail and lifting the wings a little, "and all these movements are combined in a very effective way." *Geopelia numeralis* and *G. striata* show the same display. *Phabs* exaggerates the leg movements to such an extent that the half of the body over the lifted leg is also lifted, as in a walking Duck; it further makes the same horizontal head movements as Carrier Pigeons. Our Turtledove makes similar bows and leg movements.

The Gallinaceous birds mentioned above hardly move at all during this type of display. The Sparrows show the relation of the display to the bow slightly better; bowing now and then, though relatively slowly. The doves show more action, and the relation to the bow is obvious. Other birds do not slow down or "freeze" but they speed up the bow. The Ducks have done so to such an extent that only a cinematographic picture enables us to see the movement clearly. Lorenz (1941) has filmed and described a number of these movements. The Mallard drake withdraws the head, erects and at

Fig. 21. The huddled phase of the "*Kurzhochwerden*" of the male Mallard. From a film by Lorenz, 1941.

Fig. 22. The huddled phase of the "*Kurzhochwerden*" of the male Pintail. From a film by Lorenz, 1941.

Fig. 23. "*Aufstossen*" of the male Pintail. From a film by Lorenz, 1941.

Fig. 24. "*Kopfzurücklegen*" of the male Garganey. From a film by Lorenz, 1941.

the same time spreads its tail and raises the elbows (Fig. 21). Immediately afterwards the neck is stretched upward, and tail and wings are lowered. This type of display ("*Kurzhochwerden*") takes place on the water, as do the displays described below. It is remarkable that the primaries are not lifted but only the secondaries. Possibly the Teal behaves in the same way but the movement is too quick to be certain. The homologous display in the Pintail differs in that the tail is raised even more, the heads less (Fig. 22). The Bahama Duck's display is similar to that of the Pintail; the movements of tail and body are still more extreme and by trampling the feet the male manages to stand in the withdrawn position for about $\frac{3}{4}$ of a second. *Virago castanea*, Teal, Gadwall and others have about the same type of display as the Mallard.

Lorenz has not expressed any opinion about the origin of this display. It will be clear that I believe them to be derived from the bow.

Another type of display found in Ducks is the "*Aufstossen*." It consists of stretching and subsequently withdrawing the neck. It is shown by the Pintail (Fig. 23), the South American Pintail, the Bahama Duck, *Paecilonetta erythrorhyncha*, the Garganey, the Mandarin Duck and others. I quite agree with Lorenz that this movement is nothing but the second part of the "*Kurzhochwerden*." While "*Kurzhochwerden*" is a combina-

Fig. 25. Clappering White Stork. From a photograph by Siewert, 1932.

Fig. 26. "*Aufreissen*" of the male of *Nettion flavirostre*. From a film by Lorenz, 1941.

tion of first and second phase of the jump,"*Aufstossen*" is the isolated second phase. I am inclined to believe that it represents a lower intensity than the "*Kurzhochwerden.*"

As discussed above, ritualisation of an intension movement may, in part, consist of exaggeration of one component, usually at the cost of other components. A fine example is provided by the Garganey, as described by Lorenz. The "*Aufstossen*" of the Garganey is the low intensity form of another movement, called by Lorenz "*Kopfzurucklegen*" (Fig. 24), both forms being connected by intermediates. In the highest intensity of "*Kopfzurucklegen*" the head is swung backward to such an extent that the neck is stretched and the occiput touches the back. This is followed by a quick forward movement of the head, which in my opinion is nothing but an extremely exaggerated second phase of the jump, also exaggerated by bending the head backwards so as to make the "springing" as forceful and conspicuous as possible. A similar development has taken place in the clappering posture of the White Stork (Fig. 25) and in one of the display postures of the Cormorant, as I will show below. Another type of exaggeration of the same intention movement is the "*Aufreissen.*" Here the backward swing is not emphasised by the head, but rather by including the breast in the upward swing. This type is shown by *Nettion flavirostre* (Fig. 26) and, in a lesser degree, by the Mallard. A similar movement, in the form of a frozen posture, is shown by the domesticated Croppers among the pigeons, which overdo it sometimes to such an extent that they would tumble on their backs if the wings did not support them in time. The head is moved but little.

Finally, the strongly overdone horizontal head movement of the male *Virago castanea* (Fig. 27), which is shown in a less extreme form by the Mallard drake is, in my opinion, an intention movement indicating rapid swimming, which as a matter of fact usually follows it. The same stretched swimming posture is seen in the female Mallard when she is swimming among displaying males.

Fig. 27. Strongly exaggerated horizontal head movement of the male of *Virago castanea.* From a film by Lorenz, 1941.

In all these movements of displaying Ducks, again the problem of primary versus secondary function has to be raised. Lorenz explains the stretching of the neck in "*Kurzhochwerden,*" in "*Aufstossen*" and in "*Kopfzurucklegen*" as an adaptation to the mechanism of vocalisation. The stretching is necessary to stretch the trachea which alone enables the drakes to produce the calls involved. In my opinion, again, this is a secondary development, superimposed upon the original intention movement. Here again my main argument in favor of such a hypothesis is the widespread occurrence of the same intention movement without such a secondarily superposed element.

It may be that Lorenz is right in assuming that in the "*Grunzpfiff*" the stretching movement of the neck might be any adaptation to the vocal apparatus (Lorenz, 1941, pp. 219–220). There are however slight indications that this movement may be derived from the copulation act.

These various display movements are, in most species, executed in entirely irregular sequence. In some species there are rigid series; in the Mallard, e.g., "*Kurzhochwerden*" is followed by the horizontal stretching of the neck and subsequent rapid swimming.

The Goldeneye shows another type of the jump (Stresemann, 1927–1934, Bernhardt, 1940). First the neck is stretched forward and upward, then it is swung far backwards, the occiput touching the back, and at the same time the tail is raised and spread. The spreading seems to be dropped occasionally—while A. Brooks pictures it, Bernhardt's photograph does not show it. After this, the head is again swung forward, the tail, still spread, is pushed down, and the legs simultaneously thrown back, causing the water behind the bird to splash up high (Fig. 28).

Among Ducks there is further the widespread "pumping" display, which initiates the copulation. The head of male and female makes the vertical, downward movement as described earlier. After some time this pumping movement gradually changes into the flat horizontal stretched neck posture of the female, whereupon the male ascends her to copulate.

The display of the Cranes, the so-called dance, is less static than most displays discussed so far, as it is combined with walking. The whole performance can be easily understood as a combination of walking, bowing and actual jumping. The elegance is partly due to the great length of neck and legs, which cause the movements to be slower than the homologous movements of smaller birds. Wings and tail are kept spread, and sometimes the wings make flapping movements. Spreading the tail is an ordinary feature in fast running birds, e.g., the domestic fowl. Wing flapping may be understood as incipient flight. The bow shows affinity to that of the Vultures, in that the head is

Fig. 28a. The huddled phase of the displaying male Goldeneye. From a picture by A. Brooks in Stresemann, 1927–1934.

Fig. 28b. The stretching phase in the display of the same bird. From a photograph by Bernhardt, 1940.

Fig. 29a and b. First and second phase of the "*Reckbewegung*" of the Blue Heron. From drawings by Verwey, 1930.

lowered far below the plane of the shoulders. Exaggeration seems sufficient to account for the special traits of the Cranes' dance.

One of the display postures of the male Blue Heron, as described by Verwey (1930) resembles, as the author himself remarks, the pole attitude of the Bittern. The Heron, standing upon the nest or the nest site, stretches its neck upward to its full length and brings the bill into line with it (Fig. 20a). Unlike the Bittern, the Heron does not stretch the whole body nor the legs. After stretching, the neck is withdrawn and the legs are bent while the bill still points upward (Fig. 20b). The second phase, which I regard as the deep phase of the jump, corresponds to the defence posture of the Bittern. According to Verwey the Purple Heron shows the same stretching movement. The

Fig. 30a and b. Positions of the "rhythmic wing flapping" of the "sitting partner" of the Cormorant. From photographs by Kortlandt, 1940.

Fig. 31. The "gurgling" of the "sitting partner" of the Cormorant. The top of the bill moves along the dotted line. From a photograph by Kortlandt, 1940.

female of the Blue Heron shows the two movements especially at the relief at the nest. Portielje (1926, 1938) reports that in *Tigrisoma lineatum* both male and female show the stretching, but here the whole body takes part in it. The second phase seems to be absent in this species.

The same display is shown by male and female Cormorants. According to Kortlandt (1940a) male and female have the same types of display at their disposal, and can only be distinguished by differences of voice. When indulging in one of their ceremonies one of the partners—either male or female—adopts the "sitting" posture, the other the "standing" posture. The sitting partner is on the nest or on the future nest site. It displays by raising the head until it is pointed vertically upward, the tail is spread and raised a little, the wings are lifted about twice a second, the body is horizontal, the legs are bent (Fig. 30). This movement, called by Kortlandt "rhythmic wing flapping," is chiefly but not exclusively shown by the male. Another movement often following wing flapping, is "gurgling," in which the head is jerked backward and swung forward slowly (Fig. 31). Gurgling also takes place on the nest. It closely re-

Fig. 32. The "copulation attitude" of the "sitting partner" of the Cormorant. From a photograph by Kortlandt, 1940.

sembles the backward swing of the Garganey. The quickness of the backward swing in the Cormorant is remarkable; it might be understood in the same way as suggested in the case of the domestic duck jumping out of the water.

A third movement, executed after gurgling, is a stretching of the neck; after which head and neck are slantingly moved forward and upward, while the body remains in the same position as before (Fig. 32). This movement, called the coition posture by Kortlandt, can be executed by both male and female. This coition posture is, in my opinion, nothing but an incipient jump. We are strengthened in this opinion by the fact that the birds make rhythmic sideward movements with the head—components possibly derived from the Duck—like the gait of the Cormorants. It is also possible that the lateral head movement sets in before the body movement, just as in displaying Great Tits the lateral body movements occur somewhat in advance of the tail movements. A phase between head and body would be difficult to detect in a small bird like a Tit, but in tall birds the phase differences might be greater and more easily observed. In the "dancing" Cranes, it may be remembered, these phase differences between head, neck and body are very conspicuous.

The lateral neck movements shown by displaying Nandu cocks (Portielje, 1925) have another origin again; they are intention movements derived from the type of escape behaviour typical of more species of the steppe, *viz.*, making hooks.

The standing partner of the Cormorant executes a small jump. Before and during this the neck remains stretched and is, therefore, in disharmony with the jump. The standing partner shows its locomotion urge by making short flying excursions in the neighbourhood, each time returning to the nest.

The White Wagtail, apart from a display much the same as that described in the House Sparrow, sometimes shows another type, which I, having observed it only rarely, cannot describe very accurately. The male runs straight at the female with a spread and lowered tail, sometimes even dragging it over the ground, while the body is horizontal, neck and head directed straight forward, and the wings are kept apart from the body and lowered a little (Fig. 33). It is in accordance with the high speed that head, neck and body are stretched forward. Wings and tail are kept ready for flight.

Fig. 33. Attacking attitude in the display of the male White Wagtail.

As soon as the male has reached the female, it jumps up or flies up and moves back to the starting place to make another dash towards the female. This display is simply a charge at the female, stopped at the last moment.

Charges are also carried out by part of the displaying House Sparrow males that often gather around a female. These attacks may even be carried through, aimed at the female's abdomen. The female reacts by the same type of charge and makes violent assaults at the males (Daanje, 1941).

A charge similar to that of the male Wagtail may be executed by the cock Nightingale. The male in the bush, is on the lookout for his mate, which is foraging on the ground. His tail is spread and lowered, so that the reddish brown colour of the tail stands out beautifully. Suddenly he rushes down at her but stops dead close before her and flies back to some other spot on the ground, from where he repeats his charge, sometimes as many as five times in quick succession.

The Song Sparrow pounces down upon his female, turning back and swooping over her just before he reaches her (Nice, 1937, 1943), much after the manner of the Nightingale.

Whereas I have thus far been speaking of locomotion in general, in these last instances I mention a charge or attack. This is no contradiction of any kind. I have left open the question of the further intentions of any animal showing incipient locomotion because it is, usually, impossible to say into which type of behaviour the introductory locomotory movements might evolve. Directive locomotion may be the introduction (appetitive behaviour) to various activities. In the reproductive seasons we usually have to do with either an intention to mate or an intention to attack. The last mentioned "charges" are presumably attacks, and that is why I abandoned the more cautious and neutral expression "locomotion." As a rule however, the boundary between attack and mating might be vague. Who shall say whether the cock Golden Pheasant, when displaying in front of the female is carrying out an incipient attack or an incipient mating? The Wagtail and the Song Sparrow just "miss" the female, in that they move past her instead of carrying through the charge. The Golden Pheasant may have emphasised the "passing" and—secondarily—evolved this phase into the brilliant lateral display that is so well-known. The display of the Lady Amherst Pheasants and probably of many relatives is of much the same type. Whether all these displays are considered incipient attack or incipient mating may be a matter of discussion; but it seems to be safe to consider them incipient locomotion.

Fig. 34. (a) Robin posturing at intruder only slightly above it. The position of the legs shows the extent of the lateral swaying. (b) Robin about to launch direct attack on intruder. (c) Robin swaying in the stretched attitude at intruder, well above the agressor. From photographs by Lack, 1939.

The Magpie, in defending its territory against other Magpies or Crows performs deep bows as described previously. The Great Tit threatens an adversary by bowing accompanied by lateral body movements also described previously. Tinbergen (1937) noticed an additional element; when standing in the stretched position with the head pointed upward, two fighting males make swinging movements with the heads, showing off the white cheeks in alternation. According to Tinbergen the head moves in respect to the body.

The Jay makes a similar, but very slow, bow. It is also accompanied by lateral head and body movements and the wings droop a little. I cannot say whether these bows are directed against sex partners or against rivals. It is dubious whether the Jay knows it himself. Before or after the bows there may be a fight.

The Robin threatens intruders into its territory by withdrawing or stretching the neck and by making large, relatively slow, lateral swinging movements with the forebody. The tail is lifted (Fig. 34) (Lack, 1939). The Robin, when thus threatening is always facing the intruder, while the Great Tit may, in care of encounters, face in any direction.

The Robin sometimes threatens without lateral swings but with vertical movements, during which it may change the position of its legs, "sometimes so rapidly that it appears to dance round the intruder. Other Robins maintain a stretched neck with little movement" (Lack, 1939, p. 179). Robin and Tit may both flutter their wings rapidly upwards.

A Mute Swan, in defending its territory, adopts the well-known "proud" attitude in which the elbows are lifted, and the head is bent in S-form. It pushes itself forward with strong strokes of both feet at the same time. Solitary females show the same carriage in the reproductive season and with such a female I saw that this movement may turn into flying up at any moment.

A Cormorant chasing another Cormorant from a resting-perch in the water shows the same movement (Fig. 35) (Kortlandt, 1940a). The legs, however, probably alternate.

Fig. 35. Threatening Cormorant, intending to chase another Cormorant from a perching pole in the water. From a photograph by Kortlandt, 1940.

Fig. 36. Female Mallard, threatening at a male that wants to copulate. From a film by Lorenz, 1941.

Fig. 37. Female of the Great Crested Grebe, assuming a hostile attitude towards the male. From a photograph by Berg, 1929.

Much the same posture is taken by the female of the Mallard (Fig. 36) and by that of the Great Crested Grebe (Fig. 37) when they are unwilling to mate and threaten the obtrusive male.

The Moorhen, especially the female, but sometimes the male, receives its partner in much the same way (see Fig. 38 of the same attitude of the Coot). The neck is drawn in, the wingtips are pointed upward, and the tail is spread and raised. When the other partner comes from behind, the tail is kept symmetrical; when it comes from the side the tail's underside is turned towards it. Whether this tilting of the tail has evolved purely as a visual display or whether it formed originally a mechanically useful part of the movement, facilitating the jump sidewards, is difficult to decide.

Fig. 38. Defensive attitude of the Coot. From a photograph by Berg, 1929.

Fig. 39. Defensive attitude of the White Stork. From a photograph by Siewert, 1932.

It is a remarkable fact that all the water birds mentioned, such as Swan, Great Crested Grebe, Mallard, Moorhen, Coot, and Cormorant, raise their wings during all these intention movements, while most land birds lower them. This must be connected with the fact that land birds, standing upon the substratum, have their wings well above it while water birds would touch the water and even submerge part of the wings when they lowered them.

According to the excellent description by Horst Siewert (1932) the movements of the Black Stork when defending its nest against a rival are very similar to those of the "dancing" Cranes. The defensive posture of the White Stork (Fig. 39) is slightly different and shows some similarity to the display of the House Sparrow. Also, the White Stork clappers.

A Domestic Hen leading chicks adopts a posture resembling the Turkey cock's display. In particular, when in the presence of a predator she draws the neck in, spreads and erects the tail, lowers the wings and lifts them away from the body.

Characteristic of all these movements is a certain stiffness and muscle tension. In my opinion the explanation of this must be found in the fact that we are not concerned with pure intention movements but with inhibited intention movements. In all these cases the bird is in a state of high motivation, it is prepared either to attack or to mate, but at the same time it is forced to refrain from actually doing so, either by fear, or by the absence of further releasing stimuli, or by the urge to stay and show off on the spot. The result is the simultaneous contraction of antagonistic muscle groups, which causes the obvious tension. Sometimes, as for instance in the Gander (Heinroth, 1910) the two patterns are so accurately balanced that there is only the slightest indication of an intention movement and the only characteristic of the display is the tension and the resulting rigidity of movement.

Display flights

The intention movements treated thus far may be derived from walking or jumping or from flying up. Actually, there does not seem to be a sharp boundary between jumping away and flying away, and therefore there is no need to discuss separately the intention movements of flying away.

During actual flight however, and especially during the start, we may observe intention movements indicating a tendency to increase the speed and these intention movements may have developed into signals by exaggeration. This can be done in two different ways, *viz.*, (1) by speeding up the rhythm or (2) by increasing the amplitude. The two principles may be illustrated by a human example: When playing with a child and trying to make it believe that we want to overtake it we may either take very long but very slow strides, or we may make very frequent but exceedingly small steps. Both procedures carry the impression of rapid locomotion.

As a rule smaller birds seem to use increase of tempo rather than amplitude, while taller forms increase the amplitude at the expense of tempo. We saw the two methods applied in the exaggeration of walking. A disturbed Hen makes frequent but small head movements, whereas displaying Ducks may overdo the amplitude of the head movements. The Robin may meet an intruder now with frequent small bows now with large and slow bows (Lack, 1939).

House Pigeons taking to flight at the appearance of a Sparrow Hawk start with some extraordinarily large wing strokes in which the wings touch above the back. This has evolved into an auditory signal, and the movements and sounds of this wing clapping often causes all the inhabitants of the pigeon loft to fly up. It is difficult to state with certainty whether the increased amplitude is proportionate to an increase in speed of locomotion or whether there is an excess due to exaggeration. From the fact that the wings are actually clapped together with force one would conclude that there is much exaggeration that is mechanically worthless.

The male of the Black-tailed Godwit executes his display flight (ceremonial flight, Huxley and Montague, 1926) with large slow strokes of the wings, turning its body from side to side and spreading the tail. No increase in speed is effected.

Increase in amplitude with resulting wing clapping is also practised by the male Nightjar (Stulcken and Brull, 1938). Again the speed is not at all increased.

Increased frequency, resulting in what is commonly called wing trembling, is seen in a number of Passerine birds in the beginning of the reproductive period. In the House Sparrow the male does it during display, and the female when "saluting" her husband. While this is usually done while sitting on the substratum both male and female may fly around with trembling wings, when they are preparing for a second brood, a fact not mentioned in my paper on the House Sparrow (Daanje, 1941). During this trembling flight the actual speed of flight is less than during normal flight. The same type of trembling flight is seen at the mating time in both sexes of the Chiffchaff, in the male Great Spotted Woodpecker, and presumably in many other small birds. The Magpie which shows the trembling when sitting, does not do it in flight.

Swifts show the trembling flight just before copulation takes place in the air. As a rule the trembling is only shown by the female but occasionally the male also shows it. Sometimes a pursuit with both birds trembling their wings may be observed without a copulation following. While in normal flight the wings beat at a rate of about 280 a minute; this is increased till about 360 a minute in the trembling flight. The speed of flight is notably less than normal (Daanje, 1944).

Among many Song birds the wing trembling is an invitation to mating.

Begging movements

The young of many species of Passerine birds show the same movements when begging for food. As I pointed out in my paper on the House Sparrow (1941) this type of wing trembling has something to do with flight and, in young Swallows for instance, it may become true flight when begging increases in intensity, the birds really lifting themselves up into the air. It is remarkable however that the movement has no forward component. Failure of a forward component is fairly common in intention movements of locomotion, as for instance in the pole attitude of the Bittern (Fig. 5), the alarmed posture of the Blond Ring Dove (Fig. 6), the bow of the Little Owl (Fig. 4), the vertical head movement of the Kestrel, Lapwing, and presumably in the bow of Magpie and Robin, etc.

Young Wagtails do not only flutter their wings but also wag the tail, another indication of locomotory intention. Fledged Wrens show, in addition to the wing trembling, a very quick rhythmic moving up and down of the head and forepart of the body; doubtless a symbol of hopping. Quieter bows are performed by fledglings of the Song Sparrow (Nice, 1943) and by the begging young of the Little Owl (Haverschmidt, 1946).

A very remarkable begging movement is shown by young Moorhens. Very young chicks in particular may run around, flapping the "wing" alternately. Moreover the wings move in coordination with the running legs in the same way as the arm movements of Man are coordinated with his leg movements, which suggests that the wing movements support the leg movements. If its primary task were to help in maintaining equilibrium the wing movement would be irregular. This coordination of leg and wing movements of the Moorhen can best be studied in very young chicks when they swim away as fast as possible from an enemy; for instance the observer. This is very difficult to detect in running individuals, as their movements are usually too quick. Once I saw a halfgrown young one change over from the alternating wing movement into the synergistic movement. It was fleeing over a flat bank, waving its wings alternately. However, when it reached a slope, which had to be climbed, the alternating movement was (gradually?) replaced by the synergistic movement. In running, the alternating movement apparently gave the best support, in climbing a lift with the wings was of more help.

Very young Moorhens, waving the wings alternately, when running towards the parent to be fed, continue these wing movements till they are fed. This can be observed very well when the young are stopped by reaching the water, where the parent is approaching with food. The wing movements get visual support from the red colour at the tips of the wings.

In the Moorhen and in many Songbirds, begging is also accompanied by a stretching of body and neck towards the parent, and of the tail backward. This too might belong to the intention of rapid locomotion. With decreasing intensity the stretching of the body is first abandoned, then the head is withdrawn and finally the trembling of the wings is diminished. The forward stretching component is also found in the pre-mating posture of female Passerines, and in the pre-mating posture of a female Mallard. The mating posture of the Cormorant is also a clear instance of a locomotory intention movement, as I showed above. The invitatory movements of the female Robin are similar to those of the Cormorant. Apart from the normal Passerine invitation posture the female of the Robin will, if the male does not respond immediately, perform a lateral swaying movement of the neck or a head movement up and down (Lack, 1939).

Finally I want to mention the remarkable movements of some displaying Song birds, such as the Hedge Sparrow, in which only one wing is flapped. Now one wing flicks out, now another. These movements may be alternated with quite synchronised synergistic movements of both wings or with movements of both wings, that are approximately synchronised but shifted in time. Although these movements correspond in many respects with the fin movements of fish, studied by von Holst, it would be too speculative to make an attempt at any interpretation until more comparative data are available.

REFERENCES

Allen, A. A. (1934), Sex rhythm in the Ruffed Grouse (*Bonasa umbellus* Linn.) and other Birds, *The Auk*, **51**.

Berg, B. (1929), *Tookern.* Berlin.

Bernhardt, P. (1940), Beitrag, zur Biologie der Schellente (*Bucephala clangula*) *J. Ornithol.*, **87**.

Boker, H. (1935), Vergleichende biologische Anatomie der Wirbeltiere, **1**. Jena.

Craig, W. (1909), The Expressions of Emotion in the Pigeon I. The Blond Ring Dove (*Turtur risorius*), *The Journal of Comparative Neurology and Psychology*, **19**.

Craig, W. (1918), Appetites and Aversions as Constituents of Instincts, *Biol. Bull.*, **34**.

Daanje, A. (1941), Uber das Verhalten des Haussperlings (*Passer d. domesticus* L.), *Ardea*, **30**.

Daanje, A. (1944), De vliegende Gierzwaluw (*Apus a. apus* L.) *Adrea*, **33**.

Haverschmidt, F. (1940), Observations on the breeding habits of the Little Owl, *Ardea*, **34**.

Heinroth, O. (1910), Beitrage zur Biologie, insbesondere Psychologie und Ethologie der Anatiden, *Verhand. d. as V. Intern. Ornithol. Kongr.*, Berlin.

Holst, E. von (1937), Vom Wesen der Ordnung im Zentralnervensystem, *Naturwiss.*, **25**.

Holst, E. von (1938a), Neue Versuche zur Deutung der relativen Koordination bei Fischen, *Pflugers Archiv.*, **240**.

Holst, E. von (1938b), Uber relative Koordination bei Saugern und beim Menschen, *Pflugers Archiv.*, **240**.

Holst, E. von (1939), Entwurf eines Systems der locomotorischen Perioden-bildungen bei Fischen, *Z. vergl. Physiologie*, **27**.

Huxley, J. S. and Montague, F. A. (1926), Studies on the Courtship and Sexual Life of Birds VI. The Black-tailed Godwit. (*Limosa limosa* L.), *The Ibis*, **2**

Kluyver, H. N., Ligtoet, J., Ouwelandt, C. v.d. and Zegwaard, F. (1940), De Levenswijze van den Winterkoning, *Troglodytes tr. troglodytes* L. *Limosa*, **13**.

Kortlandt, A. (1940a), Eine Ubersicht der angeborenen Verhaltensweisen des mitteleuropaichen Kormorans (*Phalacrocorax carbo sinensis*, Shaw and Nodder), ihre Funktion, ontogenetische Entwicklung und phylogenetische Herkunft, *Archives Néerl. Zool.*, **4**.

Kortlandt, A. (1940b), Wechselwirkung zwischen Instinkten, *Archives Néerl. Zool.*, **4**.

Lack, D. (1939), The Behaviour of the Robin, I and II, *Proceedings of the Zoological Soc. of London*, **109**.

Lorenz, K. (1933), Beobachtetes uber das Fliegen der Vogel und uber die Beziehungen der Flugel-und Steuerform zur Art des Fluges, *J. Ornithol.*, **81**.

Lorenz, K. (1935), Der Kumpan in der Umwelt des Vogels, *J. Ornithol.*, **83**.

Lorenz, K. (1941), Vergleichende Bewegungsstudien an Anatinen, *J. Ornithol.*, Erganzungsband III.

Nice, M. M. (1937), Studies in the life history of the Song Sparrow, I, *Transactions of the Linnaean Soc. of New York*, **4**.

Nice, M. M. (1943), Studies in the life history of the Song Sparrow, II, *Transactions of the Linnaean Soc. of New York*, **6**.

Portielje, A. F. J. (1925), Zur Ethologie bezw. Psychologie der *Rhea americana* I, *Ardea*, **14**.

Portielje, A. F. J. (1926), Zur Ethologie bezw. Psychologie von *Botaurus stellaris* L, *Ardea*, **15**.

Portielje, A. F. J. and Schut, W. F. H. (1938), *Dieren Zien en Leeren Kennen*. Amsterdam.

Raber, H. (1949), Das Verhalten gefangener Waldohreulen (*Asio o. otus*) und Wald kauze (*Strix a. aluco*) zur Beute, *Behaviour*, **2**.

Siewert, H. (1932), *Storche*. Berlin.

Stolpe, M. und Zimmer, K. (1939), *Der Vogelflug*. Leipzig.

Stresemann, E. (1927–1934), Aves in Kukenthal-Krumbach, *Handbuch der Zoologie*, **7**, Berlin und Leipzig.

Stulcken, K. und Brull, H. (1938), Vom Nestleben der Nachtschwalbe (*Caprimulgus e. europaeus*), *J. Ornithol.*, **86**.

Thauer, R. und Peters, G. (1938), Sensibilitat und Motorik bei lange "uberlebender Zwischen-Mittelhirntauben," *Pflugers Archiv.*, **240**.

Tinbergen, N. (1937), Uber das Verhalten kampfender Kohlmeisen (*Parus m. major* L.), *Ardea*, **26**.

Tinbergen, N. (1940), Die Ubersprungbewegung, *Z. Tierpsychol.*, **4**.

Tinbergen, N. (1942), An objectivistic study of the innate behaviour of animals. *Bibliothea Biotheor.*, **1**, pars 2, Leiden.

Verwey, J. (1930), Die Paarungsbiologie des Fischreihers, *Zool. Jahrbucher. Abf. F. allg. Zool. u. Physiol.*, **48**.

Whitman, C. O. (1919), *The Behavior of Pigeons*, Vol. III. Washington.

Yeates, G. K. (1936), On the fighting of the Blackcock, *Brit. Birds.*, **30**.

The Feather Postures of Birds and the Problem of the Origin of Social Signals

DESMOND MORRIS

The primary thermoregulatory function of body feathers and their movements is discussed and four general feather postures are distinguished: (1) Sleeked: The feathers are fully depressed against the body. This characteristic of active birds, or birds ready for action, and streamlines them in flight and also reduces the insulating effect of the plumage. (2) Relaxed: The feathers are neither depressed nor erected. This is the usual state of birds when only moderately active, or inactive. (3) Fluffed: The feathers are partially erected giving the body a rounded appearance with smooth outlines. This increases the thickness of the insulating layer and helps to keep the bird warm. (4) Ruffled: The feathers are fully erected giving the bird a round but ragged appearance. The separating out of the individual feathers tends to destroy the insulating layer and a cooling effect results. It is observed in very hot birds which are inactive.

The secondary signal function of body feathers is investigated. It is shown that in non-thwarting situations the mood of a bird may under certain conditions be conveyed to its fellows by its unspecialised feather postures, which therefore act as simple signals.

Reprinted from *Behaviour*, volume 9, pages 75–113 (1956). Original numbering of figures is retained.

In thwarting situations, the autonomic discharge, which accompanies the primary somatic response, is known to produce dramatic pilomotoric effects. It is suggested that this physiological relationship between thwarting and feather erection is the basis of all feather-posture displays (with the exception of the simple ones mentioned above which occur in non-thwarting situations). Not only generalised body fluffing or ruffling, but also the highly specialised body-feather displays are explicable in this way. Examples given include the evolution of crests, ruffs, chin-growths, throat-plumes, ear-tufts, flank-plumes, etc. It is stressed that a detailed study of the physio-ethology of thwarting would greatly assist in advancing the study of social signals in animals.

The general problem of the evolution as signals of responses to thwarting situations is next considered. Three conditions of thwarting are recognised, namely: The absence of indispensable stimuli, following arousal; simple physical obstruction; and simultaneous arousal of incompatible tendencies.

The Primary Response to thwarting is subdivided into a Somatic Response, which tends to adjust the animal to its external environment, and an Autonomic Response, which tends to adjust the internal environment of the animal to the requirements of this Somatic Response. Five types of Somatic Response are conceived: "Perseverance," "Snap Decision," Thwarted Intention Movements, Ambivalent Posturing and Alternating Ambivalent Movements. Five types of Autonomic Response are also mentioned as being relevant to ethology, and the signals which have probably evolved from them are discussed.

The most important are as follows: Alimentary changes leading to urination and defecation, which have evolved into territorial marking systems. Circulatory changes leading flushings which have evolved into bare-skin-flushing displays. Respiratory changes leading to alterations in the breathing rate, amplitude and regularity, which have evolved into vocalisation on the one hand, and inflation displays, on the other. Thermoregulatory changes leading to sweating and pilomotoric activity, which have evolved into scent signals and hair and feather erection displays, respectively.

The Secondary Responses which appear as alternative somatic responses, when the unsuccessful Primary Somatic Responses are abandoned, are categorised as follows: Displacement Activities, Redirection Activities, Regressive Activities, and "Neurotic Inactivity." Only the former is discussed in any detail and it is pointed out that the autonomic changes accompanying thwarting may be important in determining the nature of the Displacement Activities which follow. The question of whether or not Displacement Activities arising in this way can really be considered as irrelevant activities is left open.

In conclusion it is pointed out that three, and not two, fundamental types of signal must be considered as arising from thwarting situations. Not only are there Primary Somatic Signals arising from Thwarted Intention Movements and Ambivalent Activities and Secondary Somatic Signals arising from Displacement Activities, but there are also many Autonomic Signals, the origin of which has hitherto been ignored.

INTRODUCTION

From a functional point of view there are basically two kinds of feathers: those with which a bird flies and those which keep it warm. In the course of evolution either type may acquire additional functions in connection with concealment or display and, in the latter case, the new function may be such that it even obliterates the older one.

The origin of the secondary display-function of feathers is an important problem for the ethologist in interpreting the social behaviour of birds and much progress has been made in certain directions. The work of Daanje (1950), in particular, has clarified the situation in connection with locomotory intention movements. This author has shown how the actions involved in incipient locomotion have become modified in different ways in various bird species, as social signals. The modifications which accompany the development of signal function involve a number of kinds of change such as component exaggeration, the shifting of component thresholds and component re-co ordination. In addition, bright markings often occur on the most conspicuously displayed regions of the bird.

In the case of the locomotory intention movements, the feathers especially involved are only those of the wings and tail. The body feathers, whose primary function is temperature control, are not concerned. Nevertheless, these body feathers are often seen to be strangely depressed or erected as components of display patterns.

However, despite the fact that body-feather posturing, like wing and tail posturing, is quite accepted as a common and widespread bird display component, and despite the existence of the derivational analyses of the latter type of posturing, the question of the origin and evolution of the former appears to have attracted little or no attention. The present paper is an attempt to correct this omission and to focus attention onto this special category of bird signal.

II. THE PRIMARY FUNCTION OF BODY FEATHERS

First, it is essential to discuss in some detail the primary function of body feathers, namely that of insulation against heat loss and the control of the degree of this insulation.

The general problem of temperature regulation in birds has recently been reviewed by Sturkie (1954, Ch. 8). He points out that heat loss can be brought about by increased respiration and panting, and heat gain by shivering. In addition, he states that "Heat loss is also regulated by the pilomotor and vasomotor nervous mechanisms. Erection of hair or feathers (pilomotor system) tends to conserve heat; in the non-erected state, heat loss is facilitated." Concerning the vasomotor system, he points out that "When the air temperature is high, the blood vessels in the skin dilate, thus increasing heat loss, and when the temperature is low, the vessels constrict, which tends to conserve heat."

Wetmore, however, in an earlier paper (1921) on the subject of bird temperatures, after discussing the excellent insulatory properties of the plumage, proceeds to consider it as a non-regulatory heat-conserving system. He regards it as a system which retains heat for the bird, but which does not control how much heat it retains at any given moment, and goes on to say that "This lack of heat regulation by means of the skin (including feathers) would throw the vital work of temperature control directly upon the respiratory system." He then devotes considerable space to the question of the heat-reducing properties of the pulmonary air-sacs which, he postulates, are of fundamental importance in the temperature regulation of birds, and which compensate for the absence of sweat glands. He concludes by adding that, during extremely cold weather, special adjustment of the pulmonary system may result in the air-sacs being used as heat reservoirs, thus enabling certain species to withstand the bitter cold of winter weather.

Although Wetmore failed to recognize the regulatory significance of the pilomotor system, which can bring short-term influences to bear on the rate of heat loss, he nevertheless noted the existence of the long-term effect of plumage change: *e.g.* "Birds that remain in regions where they are exposed to cold, become more heavily feathered before the winter season so that there is less radiation of heat externally. Correspondingly, in summer the feathered covering is thinner, and the feathers themselves often become worn so they are less burdensome."

However, Wetmore is not alone in ignoring the short-term possibilities of the pilomotor system. Certain other authors have also investigated differences in the insulatory efficiency of the plumage without referring to changes in feather postures. Hutt and Ball (1938), for example, using Wetmore's (1936) data on the body weights and numbers of feathers per individual in 97 species of Passerine birds, were able to show that, although smaller birds have fewer feathers than larger ones (*e.g.* Grebe, fifteen thousand; Humming Bird, fifteen hundred), they nevertheless have more relative to their body weight. These authors conclude that "While the increased metabolism of smaller birds is instrumental in their maintenance of high temperatures, the rapid increase in the number of feathers per unit of body weight with decreasing size of bird is an adaptation for retention of the heat produced."

Neither Wetmore, nor Hutt and Ball, appear to have appreciated the importance of changes in body-feather postures in connection with rapid adjustments to fluctuations in the organism-environment temperature ratio. Even Sturkie's comment concerning the increase in insulation obtained from feather erection is only partially correct. The work of Moore (1945) gives a more complete picture. This latter author discusses the physics of heat loss from a bird's body in a cold environment. Concerning the loss of heat by conduction, he states that "The coat of a bird, consisting of air entrapped by overlapping feathers, is an excellent insulator; that is, a poor conductor. Nevertheless it does conduct heat. When the feathers are "normal" (neither pressed down, nor fluffed out) the air is almost perfectly entrapped. The heat is then conducted from warm skin to colder outer plumage both by means of the entrapped air and the feather material itself. Fluffing of the feathers is the very efficient means whereby the bird can secure a large increase in thickness of insulating coat. With the feathers erected, the coat becomes several times as thick as in the normal state." Thus far Moore and Sturkie agree, but the important point which the latter missed is contained in Moore's next statement, to the effect that "When the feathers are completely fluffed, they cannot do a perfect job of entrapping air among them, and movement of air through the feathers would reduce the insulating effect."

The significance of this last point, in which Moore distinguishes between "fluffed" and "completely fluffed," will become clear below. It is surprising that, in his paper on the subject, Moore gives no supporting evidence for the above statements and indeed one gains the impression that his discussion is mainly hypothetical. If this is in fact the case, then his predictions are most accurate. However, before giving data to support them it is necessary to establish a simple but rigid terminology for feather posturings.

A survey of the literature reveals that many names have been used in the past to describe the state of the plumage; for example: depressed, flattened, sleeked, erected, raised, fluffed, ruffled, fluffled, etc. Since one name has been used for more than one feather posture and, conversely, more than one name has been given to one feather posture, in different cases, it is desirable to attempt to standardise the situation. To do this it is necessary to consider the way in which the feathers move.

Langley (1904), in his analysis of the physiology of feather movements, showed that there are distinct muscles for erecting and for depressing the plumage, the penna-erector and penna-depressor muscles, respectively, and that "Both erector and depressor muscles are supplied with sympathetic nerve fibres and can be caused to contract separately." There are therefore, two distinct processes at work in feather posturing, and at first sight it might seem sufficient on this basis to consider only three basic feather postures, namely depressed, normal and erected. However, although erection and depression are terms which describe adequately the feather movements, the feather postures resulting from these movements require more specific terms. This is because a gradual increase in the degree of erection results in two quite distinct functional stages of "erectedness." As the feathers are raised from the normal, relaxed, position, so the entrapped-air spaces between them increase in size and greater insulation is achieved. This is made possible by the slight curvature of the feathers (see diagram in Fig. 1). But as they are raised even further, there comes a point when they cease to touch one another and they then stand out singly from the bird's body, giving it a ragged appearance. Immediately the feather-to-feather contact is broken, the entrapped-air spaces are lost and insulation with them.

Feather erection can, therefore, either help to keep a bird warm, or cool, according to its extent.

Feather depression, on the other hand, is simpler and can only help to keep a bird cool. The flattening of the feathers against the bird's body decreases the thickness of the insulating layer but, as it does not eliminate it completely, the cooling effect will be less drastic than that of complete erection. The primary function of feather depression, however, is that of streamlining. With its feathers depressed, the surface area of the bird is considerably reduced and it offers less resistance to the air when in flight. A bird with its feathers in the normal relaxed position is one which is inactive, or only moderately active; the same applies to birds with erect plumage, although the demands of temperature control may be so great that an extremely cold or hot bird may be observed to fly with its feathers erect.

Feather posture:	Sleeked	Relaxed	Fluffed	Ruffled
State of bird:	Active	Inactive	Cold	Hot
Feather movement:	Depression	—	Partial erection	Full erection
Schematic feather section	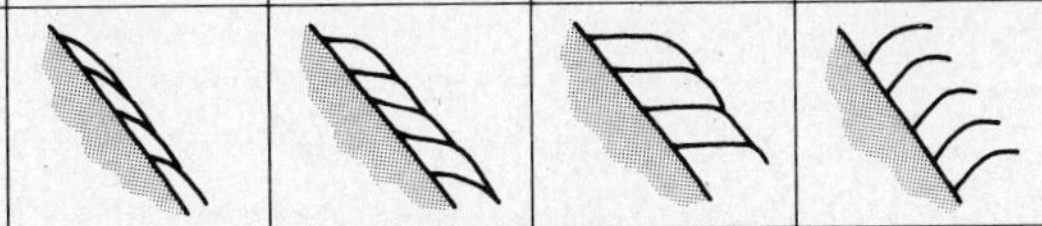			
Supposed degree of insulation	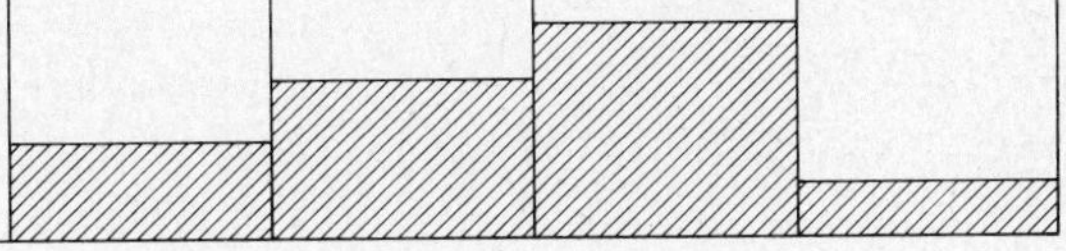			

Fig. 1. The four basic feather postures.

There are, therefore, four basic feather postures that must be distinguished and for which I propose the following names. (It has been unnecessary to introduce new names, but only to standardise existing ones.)

Sleeked. The feathers are fully depressed against the body, giving it a slim appearance. Characteristic of very active birds, or birds holding themselves in readiness for immediate action. The extra heat resulting from the high metabolic rate of the active bird is compensated for, to some extent, by the reduction of insulation produced by sleeking.

Relaxed. The feathers are neither depressed nor erected. The body shape is intermediate between slim and rounded. This is the typical relaxed state of a bird, and is observed in individuals which are only moderately active or are inactive, and which are neither very hot nor very cold.

Fluffed. The feathers are erected, but only partially, giving the body a very rounded appearance, with a smooth, unbroken, outline. Characteristic of birds which are cold. Typically, fluffed birds are inactive; the increased heat production resulting from vigorous activity is usually sufficient to permit sleeking in flight, but in extremes of cold birds may be observed to be fluffed actually during flight.

Ruffled. The feathers are fully erected and the body, although rounded, has a ragged appearance, with a broken outline. Characteristic of birds which are very hot and which are inactive.

Figure 1. summarises the foregoing diagrammatically.

The correlation between these four feather postures and the four sets of conditions can be confirmed not only by simple observations at different times of the year, but also experimentally by controlling the temperature and the degree of activity artificially. This was done accidentally by the author in early attempts to photograph captive birds. The powerful electric lamps employed to illuminate the birds in question produced such a high temperature that the birds began to ruffle and even sunbathe. If they were particularly active in the photographic cage, the birds only responded to the heat by very intense sleeking, but as soon as they settled down and became inactive, the ruffling began and persisted. Naturally, when photographing courtship behaviour, these special temperature responses interfered with results and in order to avoid this, the hot illumination was abandoned and all photographs (including the ones illustrating this paper) were taken with electronic flash apparatus.

In addition to the above observations, experiments have been carried out which show the loss in temperature control which occurs when the feathers are clipped short (Kendeigh and Baldwin, 1928), and, also, Davson (1951, p. 144) quotes investigations by Baldwin and Kendeigh, and Giaja, which reveal that the experimental restriction of feather movements interferes with the temperature regulatory system of birds.

Before passing on to consider the question of the secondary signal-function of the feather postures, it is necessary to mention briefly under what other non-signal conditions feather movements may occur.

A special secondary non-signal function of feather-posturing is, according to Madsen (1941), that of the waterproofing of aquatic birds. This author believes that "the secretion of the oil gland can hardly be the principle cause of the feathers repelling water" and considers that the main factor involved is the air entrapped in the plumage. He states that "Notably swimming birds have a vigorous musculature under the skin, particularly on the neck, belly and sides. This skin musculature has the very important task of keeping the feathers at such a distance from one another that water cannot penetrate." From this it appears that fluffing the submerged plumage is essential to prevent waterlogging, and it is interesting to note that in such a situation the feather posture has the double function of insulating against the cold of the water as well as the water itself.

A general ruffling of the feathers is observed in sleeping birds and birds which are ill. Both are conditions in which the increased insulation thus achieved is obviously an advantage.

Several kinds of feather movement can be seen during cleaning activities. The feathers are fully ruffled during bathing. Just as ruffling, when very hot, opens up the air spaces between the feathers and lets the outside air in, so does ruffling, when bathing, let the water in. In this way the skin and the plumage become efficiently soaked. Then, when the bird is drying and cleaning itself following the bath, the feathers are again ruffled during shaking and during preening. The ruffling that accompanies head and body shaking enables the bird to fling off the water from the places the latter reached as a result of the ruffling during bathing. The ruffling that occurs during preening facilitates the grasping of the individual feathers by the beak.

Ruffling of the feathers also occurs during dust bathing, sand bathing, and smoke bathing, apparently for similar reasons to those given above for water bathing. The ruffling which occurs as a component of so-called sunbathing is undoubtedly primarily nothing more than a simple temperature response.

A localised erection of the feathers in the cloacal region occurs during defecation. If this response is performed incompletely, the vent feathers become fouled with excreta. A similar localised feather erection occurs immediately before and during copulation; this appears to facilitate cloacal contact.

Finally, brooding behaviour also involves a rather specialised type of erection of the ventral feathers. This is most clearly seen as a bird settles down on its eggs at the beginning of a period of incubation.

The above conditions under which various forms of feather movement occur in non-signal situations (other than those concerned with temperature regulation) have only been mentioned briefly here. This is because they are not of main importance to the present study. It would be a mistake to ignore them completely, however, since in any discussion of derivational problems, it is essential to begin by surveying all possible derivational sources.

THE SECONDARY SIGNAL-FUNCTION OF BODY FEATHERS

Having surveyed in some detail the non-signal aspects of feather posturing, it is now possible to consider the way in which such responses have come to act as social signals.

It is valuable to make a broad distinction between specialised and unspecialised signals, according to whether or not the original motor patterns concerned have been modified in any way in connection with their new secondary function. This process of modification was termed "ritualisation" by Huxley (1923), and much has been written on the subject in recent years by ethologists. (See, in particular, Lorenz, 1935, 1937, 1941, 1954; Tinbergen, 1940, 1948a, 1951, 1952, 1953; Baerends, 1950; Daanje, 1950.)

Briefly, this process involves modifications which render the original response more conspicuous. Thus, movements and postures may become exaggerated in some ways and simplified in others. Not only is the response itself "schematised" (Tinbergen, 1952), but also the organs involved in the response may be rendered more obvious, e.g., by the addition of bright markings. Also, the response may become more frequent, as a result of the development of a "Typical Intensity" (Morris, in preparation). In this latter process, the causal range (see Morris, 1954c) of a particular form of a response spreads as the signal function develops, so that although initially the reaction only occurred at a special stimulus-intensity level, it now occurs as a reaction to a whole range of stimulus intensities. It therefore becomes the typical form of the response, or, to put it in another way, the reaction develops a "Typical-Response-Intensity." I consider that this is the fundamental change that occurs during ritualisation and that it is the basis of the so-called "fixation" of a behaviour pattern during evolution (and, for that matter, during ontogeny). However, this whole subject will be dealt with more fully in a later paper. In the present study, we are not primarily concerned with the nature of the changes that take place as a response acquires signal function, but rather with the causal problem of the nature of the responses that become used in this way. These two aspects are so closely linked, however, that the above brief comments on ritualisation are relevant here.

In a later section the general question of the origin of all types of social signals will be discussed. In the present section, the discussion is limited to the problem of the use of feather posturing as social signals.

FEATHER SIGNALS IN NON-THWARTING SITUATIONS

Social signals can be divided up for discussion into two groups, according to the causal conditions involved. Most of the more elaborate signals arise in situations of considerable stress, resulting from the strong activation and subsequent thwarting of a response. The thwarting may be due to the absence of indispensible stimuli, or to the presence of another simultaneously activated, but incompatible, tendency (see Tinbergen, 1952). (Thus, I am including both thwarting proper and conflict under the single heading of thwarting.) I shall deal with this, the major category, second; first it is necessary to describe the (generally simpler) signals which arise in situations where there is no frustration involved.

In many species, there is a marked tendency to "do likewise." As soon as one member of a group starts to perform an activity, the others quickly follow suit and group synchronisation ensues. This is particularly noticeable with the very sociable Estrildine Finches which I am at present studying. They not only move as a group, but also flee, eat, drink, bathe, clean, sleep and rest as a group. Most of the visual signals involved are of the unspecialised type. For example, the bird that starts eating does not do so in an exaggerated manner; but nevertheless the other birds respond to his normal feeding movements by joining him. This inadvertent signalling, which occurs under normal, non-thwarting conditions, requires little further mention as far as the actor is concerned, but the behaviour of the reactors has been the cause of much controversy in the past and must be discussed briefly here.

This controversy has centred around the question of whether there is a "social drive" or not. Tinbergen (1951) has stressed that, contrary to previous contentions, there is no evidence for one and that the "social" responses are always connected with and are a part of some special activity or other. Moynihan and Hall (1954) disagree with this and state that their Spice Finches (*Lonchura punctulata*) "certainly look as if they just want to be together." They conclude from their observations that "This seems to imply the existence of some (partially) independent general or social gregarious motivation," the consummatory act of which is "being near other birds of the same species."

Extending their argument, it would follow that birds eat as a group primarily because they have a strong tendency to stay together and secondarily because this puts them all at the same time into a situation where there are strong stimuli for feeding. In other words, the signal from the actor, which has the ultimate result of group feeding, would, according to them, be "I am moving to the feeding area" and not "I am feeding." When one thinks of the birds in the wild state, this does seem to be much more reasonable.

It is not, however, the whole story. I have observed that, in the laboratory, when birds in one aviary have started bathing, birds in other aviaries some distance away (which have been intently watching the bathers), will rush down to their own water trays and "join in" the bathing. Clearly, there is no question of such synchronisation being the result of a primary tendency to move about together. It seems as if there is simply a tendency on the part of individuals of communal groups to "follow suit," regardless of whether this involves generalised locomotion or specialised activities. It would probably be more valuable in future to investigate the way in which this tendency develops in ontogeny, rather than to argue as to whether there is a "social drive" or not. For the purpose of the present paper the matter must rest there, except to point out that the mechanism underlying the "following suit" responsiveness of sociable species is most unlikely to be similar to the mechanisms normally referred to as drives or instincts, which underlie other groups of activities. As far as the present study is concerned, it is sufficient simply to accept the existence of these responses to "non-thwarting," inadvertent, signals.

The feather posturing which is most frequently observed to play such a role is the fluffing which occurs when a bird is at rest. As already pointed out, the resting bird compensates for the lowering of heat production which results from its inactivity, by fluffing its plumage. This gives it a characteristic rounded appearance, which is further enhanced by the fact that the bird squats on its perch and retracts its neck. In this hunched position, it often approaches a spheroid shape and this contrasts strongly with its active posture in which, not only are the feathers sleeked, but the legs and neck are stretched.

Amongst the Estrildine Finches, it is typical for a semi-reproductive or non-reproductive group to rest in a dense row, or clump (see Moynihan and Hall, 1954). Also, if no roosting platforms, holes, or nests are available, a group will roost and sleep in such a clump. The signal which stimulates clumping is the "spheroid posture," of which fluffing is the principle component. The actor, however, does no more than it would do if it were in complete isolation, and this signal is quite unspecialised. The reactors respond to the sight of a spheroid bird by moving up to it and pressing against its side, settling into a spheroid posture themselves as they do so. If only two birds are involved, they can be clearly seen to lean up against one another. As more birds join the clump, so the warming effect of the dense grouping increases. This undoubtedly helps these very small finches to withstand the colder nights and, in captivity, it is frequently also observed during day-time resting periods, unless the weather is particularly warm or unless the birds are in full reproductive condition.

At first sight, it seems remarkable that the normal Individual-Distance (see Hediger, 1950), which these birds usually maintain, should be eliminated so easily by this spheroid posture. The ease with which this is achieved is probably due to the fact that it is such a complete opposite of the thin "activity posture." Adoption of the spheroid posture, therefore, results in the resting bird giving out none of the usual activity-signals which might interfere with clumping by warding off the reactors. For, to approach too closely to an active bird is to risk being attacked.

The power of these signals is demonstrated by the following observations. When individuals of a number of species of Estrildine Finches have just been purchased, all the birds are sometimes placed together in one aviary as a preliminary to being introduced into experimental units. When the birds are in a mixed group of this kind, it is possible to observe interspecific reactions which are often most illuminating. Occasionally certain individuals of newly imported batches of finches are sick and, as mentioned earlier, sick birds insulate themselves by fluffing their plumage. They sit around hunched up in this manner even at times when all the other birds are very active. Now healthy birds of other species under these conditions are presented with, on the one hand, active members of their own kind, and on the other hand, members of a different species in the full spheroid posture. If the healthy individual is going to "clump," it has to do so either with a bird which has the right markings, but the wrong posture, or with one which has the wrong markings, but the right posture. In a number of cases I have seen the latter choice made.

One example of this may be cited here. A mixed group of finches contained, amongst other things, one sick Long-tailed Grassfinch (*Poephila acuticauda*) and two healthy Star Finches (*Poephila ruficauda*)*. The markings are very different in the two species. The former has a large black throat spot, whereas the latter has a bright red face mask and white chest and flank spots, to mention only a few of the characters. Despite the presence of a second (sleeked) member of its own species, one of the Star Finches clumped with the fluffed Long-tailed Grassfinch and repeatedly preened its head. This illustrates very clearly the power of this particular type of feather signal.

I mentioned above that the Star Finch preened the head of the Long tailed Grassfinch, and this requires some explanation. The preening of the head of one bird by another is a typical subsidiary resting response. The birds, having come very close together during clumping, are often stimulated, by the close proximity of the plumage of the other individual, to preen its feathers. The feathers that offer the greatest stimulus for preening are those of the head and neck, because they cannot be kept preened by their owner, and it is here that the preener concentrates its efforts. It was pointed out earlier that, during preening, the feathers are ruffled and, considering the biological advantage of mutual head preening, it is not surprising that a specialised signal has arisen in this connection. This consists of ruffling the plumage of the head and neck, but not the rest of the body. This helps to increase the stimulating effect of the head feathers and to concentrate the mutual preening in this region. The response has apparently become very distinct during evolution and now may be given as a special initiating signal itself. The full response consists of a more or

*The scientific names of the Estrildine Finches mentioned in this paper are based on Delacour's (1943) revision of the group.

less spheroid posture with the addition of a ruffling of the head and neck feathers, a rolling of the head to one side, away from the preener, and a half or full closing of the eyes. This response to being preened may now be seen to be performed by an individual as it approaches a resting bird, before the latter has made any attempt to preen it.

Whereas body fluffing was acting as an unspecialised signal in a non-thwarting situation, head ruffling has become a specialized signal, in a similar situation. It is worth noting that although the former occurs in isolated birds, the latter never does so. (The only exception to this last point is that parrots and certain other pet birds, which are kept in isolation, may attempt to stimulate their human companions to perform mutual preening by adopting the head ruffle and roll posture towards them.)

Moynihan and Hall (1954), in their study of the Spice Finch (*Lonchura punctulata*), have put forward the suggestion that this head-ruffling signal itself may have evolved into a more elaborate form of ruffling in this species in agonistic situations. They suggest that the ruffling of the head, neck, scapular and upper back in hostile encounters in this species may have evolved from the simpler social preening invitation of head ruffling because "it may thus help to release or stimulate a relatively friendly social reaction instead of an extremely unfriendly one." As far as the evolutionary trend is concerned, I cannot agree with them. Ruffling in agonistic situations is of extremely widespread occurrence and many varieties exist as regards the motivational conditions of the performers and the areas of the plumage erected. Some much more general explanation is required and this is given in the following sections of the present paper. However, once evolved, agonistic ruffling might well function in the manner suggested by Moynihan and Hall in special cases such as the Spice Finch. Briefly, then, they conceive of the similarity in the Spice Finch between social-preening headruffling, on the one hand, and agonistic ruffling, on the other, as a divergence, whereas I contend that it is more probably a case of convergence. I will only give one example to support my view here because the later sections of this paper make it unnecessary to consider the question in any more detail. The male Crimson Finch (*Poephila phaeton*) possesses a specialised threat display, in which the head feathers are ruffled in a way almost identical with that seen in the social-preening invitation. Yet this display is given when the bird is extremely aggressive and may be observed to occur even whilst the performer is pecking or plucking other birds—hardly a situation in which to be "stimulating friendly reactions."

One further group of observations must now be mentioned to further illustrate the power of these clumping feather-signals. These observations concern a bizarre relationship that was observed to develop between some Java Sparrows (*Padda oryzivora*) and some Necklace Doves (*Streptopelia chinensis*), which were kept together in the same aviary during the winter of 1954–1955.

Goodwin (1952) had reported observing that Java Sparrows kept in captivity with Turtle Doves (*Streptopelia turtur*) repeatedly roosted under incubating individuals of the latter species. This author found that "On shining a torch on the nest one would see the two white eggs pushed to one side and, lifting up the Turtle Dove, would find the two Java Sparrows side by side beneath her."

Fig. 2. The Java Sparrow—Dove relationship. The finches were strongly attracted to the doves and not only sat with them, as shown here, but also performed various other social activities with them. For explanation, see text.

Fig. 3. A Java Sparrow clumping alongside a Dove.

The Necklace Doves, in my own observations, were not incubating and had no nest, but the Java Sparrows were nevertheless seen to be attracted to them, and the following interspecific patterns were observed to occur:

1. A Java Sparrow was frequently seen to sit next to a Dove in preference for a member of its own species (Fig. 2).
2. On a number of occasions a Java Sparrow was seen to push up against, or to lean against, a Dove (Fig. 3).
3. A Java Sparrow was seen to push up between the legs of a Dove and plunge itself into the ventral plumage of the latter (Fig. 4). Once, at dusk, two Doves were seen sitting near one another, with no Java Sparrows in sight. The slight disturbance caused by the presence of the observer resulted in the appearance from beneath each Dove of the head of a Java Sparrow.
4. The Java Sparrows were occasionally seen to preen the Doves in preference for preening one another (Fig. 5).

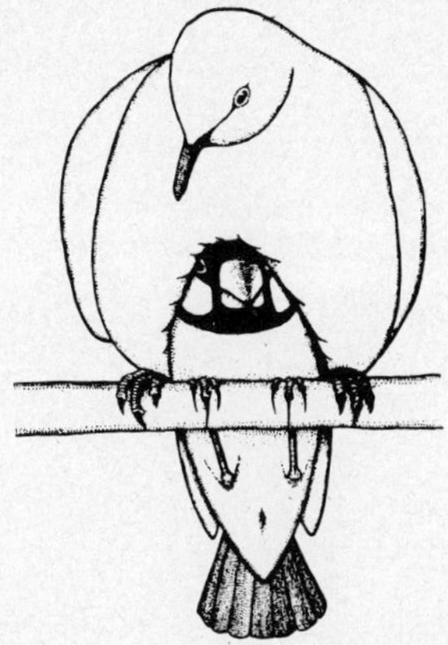

Fig. 4. A Java Sparrow attempting to clump underneath a Dove.

Fig. 5. A Java Sparrow preening a Dove.

Fig. 6. A Java Sparrow defending its Dove against a rival.

5. When the number of Java Sparrows was increased above that of the Doves, the dominant Java Sparrows defended against rivals the Doves they had selected to clump with (Fig. 6).

Fig. 7. A Java Sparrow "playing leap-frog" over the back of a Dove.

6. On a few occasions, a Java Sparrow was seen to "play leap-frog" over the back of a Dove. The latter remained in its place on the branch, whilst the Java Sparrow hopped from one side of it, onto its back, and then off again. Then it turned, hopped again onto the back of the Dove, and again hopped off the other side. This performance was then repeated several times. It should be stressed that this was not a copulation attempt since, when the Java Sparrow was mounted on the Dove, their long axes were at right angles to one another (Fig. 7).

These then are the observations which require explanation. At first sight the behaviour recorded appears to be extremely aberrant, to say the least. However, if the signal effect of the fluffed resting posture referred to earlier is called to mind, the situation becomes considerably simplified.

It has already been pointed out that one species of finch will clump with and preen an individual of another species if the latter fluffs out into the spheroid posture. It is only one step further to the Dove case. Doves, even when active, present an extremely rounded appearance. When Doves assume the resting posture, it is a much larger and much rounder "spheroid" that results, when compared with that of finches. It provides, in fact, a super-normal* clumping stimulus which is more attractive than even that given by other Java Sparrows. It will be remembered that in the previous case the Star Finches had to decide between a fluffed Long-tailed Grassfinch and a sleeked Star Finch. Here, however, the Doves provided such an intense stimulus that a Java Sparrow would prefer them to one of its own kind even though the latter was also in the fluffed spheroid posture.

It was noticed that the colour of the plumage of the two species was rather similar in certain respects and in order to test whether this was a relevant stimulus, the Doves were removed and a group of Domestic Pigeons was put in their place. These were, of course, much larger than the Doves and completely dwarfed the Java Sparrows†, but even so the latter were attracted to them. Each Pigeon was of a

---

*For discussion of supernormal stimuli, see Tinbergen, 1951.
†A Pigeon weighs at least 20 times as much as a Java Sparrow.

different colour and it was hoped that it would be possible to observe if there was a colour preference shown by the Java Sparrows in selecting their companions. Unfortunately, the Pigeons were so active in establishing a peck order amongst themselves that every time a Java Sparrow began to creep up on one at dusk, it was frightened away, before it could make contact, by the sudden aggressive jerks and jumps of the large birds.

Although this made it impossible to obtain any detailed results, it is nevertheless interesting in itself that such comparatively huge birds should still stimulate the small finches to clump. (One cannot help thinking paradoxically of Tinbergen's (1948b) Oystercatcher trying to sit on a huge egg.)

Although the above explanation clarifies observations 1, 2 and 4 listed above, the others require further elucidation. The roosting underneath the perched Doves which I observed, and the roosting under incubating birds recorded by Goodwin, can be taken together and explained as follows. Estrildines not only roost together in a tightly packed row along branches at night, but they may also roost on platforms or inside nests. In the latter instances they may roost tightly together, not only side to side, but also even in layer upon layer. Field observations on the roosting behaviour of African Estrildines to be found in Bannerman (1949, vol. 7, especially p. 227) reveal that as many as thirty of these small finches may pack together into a small roosting cavity. Getting underneath the Doves then, is a perfectly normal roosting pattern, once one has accepted the Doves as supernormal Java Sparrows.

A similar explanation applies to the "leap-frog" phenomenon. When a group of Estrildines is clumped along a branch, the end birds of the clump occasionally attempt to get into a warmer position in the middle. To do this they hop along the backs of the row and push down in between two of the more centrally placed birds. In the case of the Java Sparrows, the clump which consisted of one Dove alongside one finch provided the stimuli for just such a manoeuvre. The Java Sparrow, finding itself at the end of a "row," tried to improve its position by moving to the middle of the "row." Its failure to find this mythical middle led to repeated trips back and forth over the "backs of the row." It should be noted that when two finches clump together they appear to know that there is no middle position and no attempts are then made to leap-frog to it. The Java Sparrows, however, sitting on one side of the fluffed Dove, could see nothing beyond it and appears to have been under the illusion that there must be a whole row of birds there.

This now explains all the listed observations on this interspecific relationship, with the exception of number 5, and this I cannot explain. Estrildine Finches normally defend only the following objects: their nests, nest sites, roosting sites, mates, and young. It is impossible to say into which, if any, of these categories the Doves should be placed when they were being defended by the Java Sparrows against one another.

These few examples of signal-functions of feather postures in non-thwarting situations must suffice for the present. Although they do not cover a wide range of either signals or species, they nevertheless serve to illustrate clearly the significance of this category of feather signals.

FEATHER SIGNALS IN THWARTING SITUATIONS

When a bird is aroused in some way and the aroused tendency is then prevented from expressing itself fully, one of several things may happen. (For the most recent discussion on this subject, see Bastock, Morris and Moynihan, 1953.) The types of response occurring in such a situation, which have received most attention from ethologists in the past, are Thwarted Intention Movements, Ambivalent Responses, Displacement Activities, and Redirection Activities. These will be discussed further in the last section of the present paper.

Although ethologists have recognised that, under such circumstances, other types of response may occur, they appear to have almost completely overlooked the behavioural significance of one whole group of effects, namely the dramatic autonomic changes which inevitably accompany intense thwarting.*

Physiologists have long been aware of the complexity and importance of these autonomic changes which accompany what they have termed "emotional disturbances," and the physiological literature on the subject runs into many volumes. It will be convenient, at this point, to briefly summarise the physiological data, as far as they impinge on the ethological problem.

It is a little difficult to give an objective definition of the causal factor known as "emotional disturbance." In physiological writings many different words are used to describe such disturbances, e.g.: anxiety, tenseness, excitement, stress, startle, fear, rage, resentment, shame, and so on. However, whatever confusions may exist about the scientific meanings of these terms, one thing is quite clear; namely that, whatever else may or may not cause emotional disturbances, one of the best established causal factors, in a general sense, is intense thwarting. And this is sufficient for the present purpose. (I am still, of course, thinking of conflict as a double, or mutual, thwarting.)

The thwarting situation can be thought of as having two major primary effects: somatic† and autonomic. The somatic changes result in the performance of the motor patterns of the stimulated activity, or activities, and are concerned with the adjustment of the organism to its external environment. The autonomic changes, on the other hand, result in the adjustment of the internal environment of the animal to the requirements of these somatic responses which it has to perform.

Briefly, this adjustment of the internal environment, which mobilises the animal for action, involves the following changes: The sympathetico-adrenal system is activated and this results in: (1) The cessation of processes of storing and digesting food. Thus, salivation is restrained; movements of the stomach, the secretion of

---

*This statement already requires modification. Since the present paper was completed, it has come to my notice that Dr. R. Andrew (at Cambridge) has also been investigating the autonomic aspects of feather posturing.

†The term somatic is employed with differing emphasis by geneticists, psychologists and physiologists. As is clear from the following discussion, it is used here in the sense employed by the physiologists.

gastric juices and the peristaltic movements of the intestine are all inhibited. Also, the rectum and bladder do not empty as easily as normally. (2) There are profound changes in the circulation of the blood. The heart beats faster and blood is transferred from the skin and viscera to the muscles and brain. This gives rise to an increase in blood pressure. In addition the number of red blood cells increases as a result of the excitation of the splenic nerves. (3) There are alterations in the rate of respiration; breathing is both quicker and deeper. (4) The temperature regulation mechanisms are violently activated and pilomotor and sudomotor responses are observed.

There are certain other physiological changes which also operate under the influence of the sympathetico-adrenal system, but all those which seem likely to have any bearing on ethological problems have been included in the above list. For further details the reader is referred to the reviews in Cannon (1929, 1932), Morgan and Stellar (1950), and Gellhorn (1943, 1953).

This is not the whole of the autonomic story, however. There are still the parasympathetic responses to be considered. As is well known, the parasympathetic system works in opposition to the sympathetic system. Wherever a sympathetic influence causes a change in one direction, the parasympathetic causes one in the opposite direction. Under usual conditions, a balance is maintained between the two systems, but, under emergency conditions, when there is a sudden emotional disturbance, there is an immediate and powerful activation of the sympathetic system, which upsets this balance. The job of the parasympathetic system being to preserve and restore bodily reserves, it is quickly set in action to counteract the "extravagant" effects of the sympathetic system mentioned above.

Knowing this, it might be expected that, when observing an emotional disturbance, all the above sympathetic symptoms would be observed first, then quickly all the opposite parasympathetic symptoms, and finally a "calming down," or restoration of the autonomic balance. But this is not the case. Exactly what happens depends on the type of emotional disturbance, the particular individual concerned, and also the species involved. In some cases, the sympathetic system dominates the scene, and few or no para sympathetic symptoms are observed. In other cases, the reverse is the case. Gellhorn (1953) mentions that certain studies of the problem "suggest that fear causes reactions predominantly sympathetic, and feelings of hostility and anxiety predominantly parasympathetic discharges." I must confess that I find it strange to see fear and anxiety contrasted in this way, but it is nevertheless worthwhile to note that fear and aggression may have opposite effects in this connection and we shall be returning to this point later. It must be added that individual variations may occur in a quite striking way and interfere with any completely clear-cut association of one type of autonomic balance with any one type of emotional disturbance.

These variations which occur under different circumstances, or with different individuals, are not necessarily all-or-none phenomena. Of the various changes that were listed above, some may go in a sympathetic direction, others in a parasympathetic direction, with the result that one may observe certain symptoms of both systems simultaneously. For example, when a bird is suddenly startled, its heart beats harder (sympathetic) and at the same time the bird may defecate (parasympathetic). Many such examples could be given, but from an ethological point of view they are not yet of primary importance. What really is important is to know the total range of easily observed symptoms which may occur as the result of the activation of the autonomic system during an emotional disturbance such as intense thwarting. (It was nevertheless relevant to point out that various combinations of these symptoms may be seen in different situations.)

It is valuable to conclude this brief survey of the physiological data, therefore, with a list of these symptoms, both sympathetic and parasympathetic, as follows: Where pairs of alternatives are given, the sympathetic effect is given first and then the parasympathetic—dryness in the mouth, or excessive salivation; defecation; urination; extreme pallor, or blushing and flushing; vasodilation of sex organs, including erection of penis; rapid deep breathing, or slow shallow breathing; gasping; sighing; panting; weeping; fainting; sweating; and finally, the most important from the present point of view, namely pilomotor activity. (It will be appreciated that not all the above responses occur in birds, but, in view of the general discussion to follow, it was thought advisable to include them here.)

It is my intention to show below how a number of these different symptoms have, in the course of evolution, become secondarily modified as social signals. The intimate association that exists between the various thwarting situations and these autonomic symptoms lends itself admirably to the selection of the latter as indicators of the mood of the performer.

However, the pilomotor responses must be considered separately first. When a bird prepares itself for action, it naturally streamlines itself in readiness for flight. Thus, the mechanically effective pilomotor response in an emergency situation is the sleeking of the plumage. This will also have the effect of reducing insulation and will thus help to reduce the overheating that will tend to occur with vigorous activity. This sleeking, rapid breathing and, in mammals, sweating, are particularly important as cooling devices here because vasodilation, which is normally a powerful cooling device, cannot occur in this context, the blood from the skin being needed elsewhere.

It is not surprising therefore that Langley (1904) found that "The ordinary effect of stimulating any part of the sympathetic system or of stimulating a cutaneous nerve is depression of the feathers." But it must also be pointed out that he goes on to say that (with sympathetic stimulation) "At times a strong erection of feathers occurs instead of depression"; also, "Frequently the feathers in one part are erected and in another depressed." As far as I know, the exact causal differences which result in these different symptoms are still not understood, but it is nevertheless clear that in birds, the typical symptom of sympathetic stimulation is sleeking.

Despite this, in any general review of the subject, pilo-erection is said to occur as the result of sympathetic stimulation. At first sight this apparent contradiction is difficult to understand. It seems to be due to the fact that most physiologists are "mammal-oriented." (Even the term pilomotor itself bears evidence to this, for it was originally coined in connection with the erection and depression of hair, but is now used to include the movements of both hair and feathers.) It would appear that the basic organisation of the erection and depression responses under sympathetic influences is different in mammals and birds. A full erection can occur in both groups, but only in birds, it seems, is the ordinary effect one of depression. In their general reviews, physiologists do not make any allowance for this difference. Also, it is unfortunate that they do not make any clear statements concerning the exact form of the pilo-erection under such conditions. Most physiologists refer to pilo-erection as a warming device, and then make no comment as to why an animal should set into action both cooling and warming systems at once as a result of sympathetic stimulation.

Of course, if the response was one of full pilo-erection it would give rise to ruffling in birds, which would have a cooling rather than a heating effect and would therefore be more in line with the general syndrome. Perhaps full erection of hair, in mammals, has the same effect. As was stressed earlier, few authors have appreciated the functional differences between fluffing and ruffling in birds, but it is significant that Langley refers at various points to feathers being "more or less erected," which contrasts with the "strong erection" he mentions in connection with sympathetic stimulation. It seems likely that the latter form was indeed ruffling.

Further research is badly needed here. On the present evidence, I would predict that sleeking or ruffling would be expected as the usual or special sympathetic responses respectively, and that fluffing would be correlated with parasympathetic responses. It is important that the exact reasons should be found for the occurrence of sympathetic ruffling in birds, as opposed to sympathetic sleeking, since, as will be shown below, it is this form of the response that is of the greatest ethological interest.

Setting these details on one side for the moment, the general conclusion can be stated as follows: If a bird is intensely aroused and is then thwarted in some way, the autonomic changes which will accompany the somatic reactions to this situation will involve marked pilomotoric activity, which sometimes takes the form of pilo-erection. From this, it can be further stated that there is every reason to suppose that striking feather postures will become intimately associated with thwarting situations. As the latter circumstances are just those where social signals are most important, the use of feather postures as such signals is inevitable. Furthermore, the physiological data suggest that, in all probability, different types of thwarting will give rise to different basic types of feather posturing, although more physiological evidence is needed on this point.

If we now examine the feather-posture behaviour of birds under thwarting conditions, it soon becomes clear that, not only have the autonomic pilomotor responses been used in this way, but also they have, in many instances, undergone varying degrees of secondary modification during evolution. Many highly complex and specialised forms of feather posturing can be observed in displaying birds of many species.

Plate 5. Above: The Bicheno Finch; the pair immediately before a bout of courtship activity. Below: The courtship of the Bicheno Finch. The courting male (on the left) has assumed the posture, characteristic of this species, in which there is a marked fluffing of the body feathers.

This specialisation has taken the form of: (1) The restriction of the regions of the plumage, where pilo-erection occurs, to special areas of the body. There is typically a correlation between the regions thus differentially erected and those which are visible to the reactor. (2) The addition of bright markings and colors to the erected regions. (3) The enlargement of the body feathers on the displayed regions of the plumage. This enlargement may render the feathers useless as regards their original functions. (4) The loss of pilomotor control in special displayed regions. Thus, tufts or crests of feathers may become fixed in a permanently erected position. Here, as in number 3 above, the feathers concerned may lose their original functions.

The examples of the above phenomena, of the ritualisation of autonomic pilomotor responses, are so abundant that it is not possible to give more than a selection here. In selecting, I shall attempt to give one or a few examples of each type of display feather-signal, beginning with the less specialised and ending with the most highly specialised cases.

Firstly, the following examples of unspecialised cases may be mentioned: Morris (1954a) has noted that, in the Zebra Finch (*Poephila guttata*), thwarted escape may lead to a fluffing of the plumage and that this may tend to inhibit the attacks of the dominant individuals. This appeasement function of fluffing appears to operate successfully because of the "inactivity" signals given out by the fluffed spheroid posture of the bird. Hinde (1953) records that subordinate Chaffinches (*Fringilla coelebs*), which are very low in the peck order, sit around most of the time in the fluffed posture and he too has suggested that "This latter posture may have the effect of reducing aggressive behaviour in other individuals."

A rather generalised fluffing of the feathers occurs during the courtship of males of various species of Grassfinch (Erythrurae) (see Pl. V). This appears to be quite unspecialised in the case of the Bicheno Finch (*Poephila bichenovi*) except that it is accompanied by an expansion of the black and white markings of the species. In the case of the Zebra Finch, there are further minor indications of specialisation. The unmarked crown region is sleeked, rather than fluffed like the rest of the body feathers. The fluffing of the flank regions exposes the brightly spotted flank feathers.

Amongst another group of the Estrildine Finches, namely the Mannikins (Amadinae), the courting males often show a more or less generalised ruffling of all the body feathers, e.g., The Bronze Mannikin (*Lonchura cucullata*). The male Cutthroat Finch (*Amadina fasciata*) (see Morris 1945b) ruffles all its feathers during courtship, but some specialisation does occur in the belly region. Here there is a large brown patch flanked with white, and there is a sudden sharp increase in the extent of piloerection just at the edges of this patch (see Pl. VI). Certain other Mannikins, such as the Spice Finch (*Lonchura punctulata*) show an intermediate condition, between fluffing and ruffling, part of the plumage being fluffed and part of it ruffled.

These rather unspecialised responses are not restricted to the finches. The Carrion Crow (*Corvus corone*) threatens with all its body feathers ruffled. In this case, the autonomic response is performed simultaneously with somatic posturings of the wings and tail.

Hingston (1933) gives many more examples of generalised feather erections occurring during agonistic encounters, e.g., "Everyone knows how the fowl fluffs herself when rushing on a cat in defence of her chicks. Owls put up a tremendous show of threat, spreading their feathers so profusely, they look twice the normal size." Hingston, whose almost encyclopaedic knowledge of the whole range of animal displays is as admirable as his obsessional interpretation of them is deplorable, makes two generalisations as follows: Firstly, "All birds ruffle the body feathers in anger"; and secondly, "Were I to give all my examples of feather-fluffing under anger it would mean a list of every bird whose fighting behaviour I have seen recorded." It is clear from these two statements that he is not distinguishing between fluffing and ruffling, and in scanning his many excellent examples the reader must be warned of this fact.

For the sake of brevity, we will now pass on to the more specialised displays and these will be categorised according to the region of the body where the particular exaggeration has occurred.

Plate 6. Above: The Cutthroat Finch; The pair immediately before a bout of courtship activity. Below: The courtship of the Cutthroat Finch. The courting male (on the right) has assumed the striking ruffled posture which is characteristic of this species. Although the ruffling is fairly generalised there is nevertheless slight specialisation in the differentiated belly region.

Crests. In many species, too numerous to mention, differential ruffling may take place, in which only the crown feathers are erected. The latter are not modified in many cases, but in others they are brightly marked or coloured and the erection displays these markings or colours. In still others, the crown feathers are not only brightly marked, but also enlarged into an erectile crest. In the latter cases, the normal pilo-erection movements involved in ruffling may be unaltered, but the elongated crown feathers, when raised in the normal way, nevertheless give a violently exaggerated display. In some species, such as the Sulphur-crested Cockatoo (*Kakatoe galerita*) (see Fig. 8), the enormous crest fans out so far forward that it seems as if the pilomotor

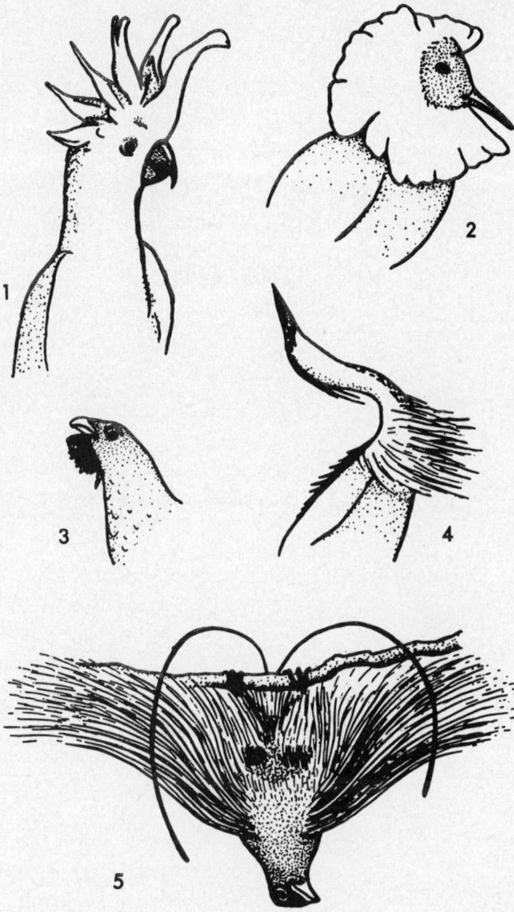

Fig. 8. Various specialised forms of feather signals. (1–5 respectively) crest, ruff, chin-growth, throat-plumes, flank-plumes.

movements themselves must also be exaggerated. In some species the crest has become rigidly fixed, leaving the bird in a permanently crown-ruffled state.

It would be pointless to include here a list of crested species, since there are so many and they are so well known. Suffice it to say that there is hardly any group of birds which does not possess certain members which are crested.

Ruffs. In a number of species, such as the Great Crested Grebe (*Podiceps cristatus*), or the Ruff (*Philomachus pugnax*) itself (see Fig. 8), there is a much enlarged erectile frill or circle of feathers around the neck, which is expanded in frontal displays.

Ear-tufts. A number of Owls and various other species possess elongated erectile outgrowths of feathers on either side of the head, which form discrete tufts. Even when not erect, these structures are usually clearly visible. Hingston reports that "Eagle Owls, when enraged, erect them so conspicuously that they look like black pointed horns."

Beards, or chin-growths. In some species there is a tuft of elongated feathers placed immediately beneath the beak and, in the display of the Capercaillie (*Tetrao urogallus*), for instance, the head is thrown back and these are fully erected (see Fig. 8). Hingston states that the Purple Fruit-Crow also has such a "beard."

Breast or throat plumes. Many species possess brightly marked breast patches which are differentially erected in display. One only need mention the well-known threat display of the Robin (*Erithaceus rubecula*) (Lack, 1943), in which it ruffles its red breast. In certain birds, however, notably the Herons, the feathers of the throat region are drawn out into very long erectile plumes (see Fig. 8).

Flank plumes. A fantastic development of patches of flank-feathers can be observed to have taken place in a number of species of Birds of Paradise. These "paradise plumes," which spring from the sides of the body, just beneath the wings, are dramatically erected and even vibrated as the display climax of these species (see Fig. 8).

Rump patches. A number of the more elaborate displays involve the erection of a bright rump patch, as, for example, in the courtship display of the Golden Pheasant (*Chrysolophus pictus*), where there is a fluffing out of a large and well differentiated rump region.

"Eye" tufts. Scattered throughout the bird world there are a number of unique forms of highly specialised feather signals, which will be represented here by the following example: The Superb Bird of Paradise (*Lophorina superba*) possesses an extremely elaborate display which, amongst other things, involves the erection of "two tiny patches of shining green feathers between the eyes." These "it raises in a manner so subtle that they catch the light and are transformed into a pair of scintillating green eyes, which, by their very brightness, mask the true eyes; they are so realistic that it is hard to believe they are no more than tiny spots of feathers" (Stonor, 1940).

Fragmentary as the above compilation may be, it nevertheless gives a clear picture of the immense variations and extreme developments, which specialised feather-signals may exhibit. For a further study, the reader is referred to the works of Hingston (1933), Stonor (1940) and Armstrong (1947).

It will be noticed that all the displays mentioned above take place in either agonistic or sexual contexts, during threatening, appeasing, pair-forming, greeting, or courting ceremonies. It will also be recalled that they have been grouped together here under the general heading of "feather-signals occurring under thwarting circumstances." It is essential therefore to point out here that recent ethological research (see Tinbergen, 1952; Hinde, 1953; Moynihan 1955; Morris 1954c and 1955) has shown that it is a general principle of agonistic and sexual ceremonies that they involve causally an intense arousal coupled with subsequent thwarting, the latter usually being the result of the simultaneous activation of incompatible tendencies. In the next section, a more comprehensive sketch of the total effect of thwarting situations will be given.

Before leaving the present section, there is still one aspect of feather-signals that must be dealt with. It was stated earlier that there is some evidence that fear may be linked predominantly with sympathetic and aggression with parasympathetic discharges. Now, the temperature regulatory mechanisms set in motion by the sympathetic system are those concerned with cooling and those set in motion by the parasympathetic with warming the animal. It would follow from this that in a bird in which fear predominated, the appropriate pilo-erection response would be ruffling and where aggression predominated fluffing should occur. What little evidence there is so far suggests that this is not so. Ruffling occurs in many cases where there is a marked tendency to attack, but only a weak tendency to flee. Also, fluffing is generally more characteristic of submissive birds than hostile ones. Among the Estrildine Finches, it is the males which are more aggressive towards their mates that ruffle while courting (e.g., Bronze Mannikin), whereas the "fluffed courters" are much less likely to attack their females (e.g., Bicheno Finch). But I do not feel that this necessarily points to a simple connection in birds between fear and the parasympathetic and hostility and the sympathetic. The whole problem is much more complicated than this. Undoubtedly the suddenness of the reaction is important and a difference would be expected between the autonomic response to a sudden fright, for example, on the one hand, and a state of subordinate submission, on the other, although both cases would involve intense fear and thwarted escape tendencies. Also, of course, sympathetic sleeking is associated with fear, and, in any case, all kinds of complicated combinations of sleeking, ruffling and fluffing can occur simultaneously in the same situation. In conclusion, it can only be stated that this whole question of the physioethology of thwarting urgently requires further investigation. The current, largely non-physiological, ethological approach to the problem of thwarting has been immensely fruitful, but the stage has now been reached where any further advance will almost certainly be dependent on more strictly physiological analyses.

CONCLUSION: THE GENERAL PROBLEM OF THWARTING AND THE EVOLUTION OF SIGNALS

Since the foregoing discussion of feather-signals has incorporated a somewhat new approach to the ethology of the responses that occur in thwarting conditions, it will perhaps be valuable here to review the more general aspects of the problem. For this purpose I propose to divide the reaction to thwarting into the Primary and Secondary Response. The Primary Response is subdivisable into the Somatic Response and the Autonomic Response in the manner already put forward. The Secondary Response consists of the alternatives to the Primary Somatic Response which may occur as a result of the failure of the latter to satisfy the requirements of the situation. They may arise in one of two ways. Either the thwarting situation is such that the animal is incapable of performing a Primary Somatic Response at all, or the latter is performed but fails in its task.

It is convenient at this stage to tabulate briefly the various types of Thwarting Stimulus-Situations and the Primary and Secondary Responses that are given to them. They are as follows:

Thwarting stimulus-situations

I. Absence of indispensible stimuli, following intense arousal.
II. Simple physical obstruction of aroused activity.
III. Simultaneous arousal of two or more incompatible tendencies.

Responses to above situations.

A. Primary Response.
- a) Somatic Response.
  1. "Perseverance" (applies to I and II above, when animal attempts to continue with aroused response despite circumstances).
  2. "Snap Decision" (applies to III above, when animal quickly responds to only one aspect of relevant stimulus-situation).
  3. Thwarted Intention Movements (applies to I, II, III).
  4. Ambivalent Posturing (applies to III).
  5. Alternating Ambivalent Movements (applies to III).
- b) Autonomic Response.
  1. Alimentary: Increase or decrease in salivation. Urination. Defecation.
  2. Circulatory: Pallor. Flushing. Vasodilation of sex organs. Fainting.
  3. Respiratory: Increase or decrease in respiratory rate or amplitude. Gasping. Sighing. Panting.
  4. Thermoregulatory: Sweating. Pilomotoric activity.
  5. Lacrimatory: Weeping.

B. Secondary Response.
- a) Displacement Activities.
- b) Redirection Activities.
- c) Regressive Activities.
- d) "Neurotic" Inactivity.

The three Thwarting Stimulus-Situations are self-explanatory and require no further explanation here. Also, little need be mentioned concerning the evolution as signals of Primary Somatic Responses 3, 4 and 5. The ritualisation of somatic posturing and locomotory intention movements has been dealt with fully elsewhere, especially in the works of Lorenz, Tinbergen and Daanje already cited.

A brief survey of the ritualisation of the various Primary Autonomic Responses, which appears to have occurred in many cases, is, however, highly relevant here. (Once again, only a concise selection can be attempted.)

The primary autonomic responses

Firstly, the alimentary effects: The most important of these are undoubtedly the many instances of the ritualised use of defecation and urination in mammals, as territorial marking systems. It is sufficient to cite as an example the elaborate territorial urinating of male dogs. Considering all that is known about the type of behaviour involved in territorial defence, it is highly probable that these methods of marking have evolved from the alimentary effects of the parasympathetic stimulation that occurs with thwarting.

Secondly, the circulatory effects: Parasympathetic vasodilation of various kinds of bare skin-patches in both birds and mammals appears to be of a quite widespread ritual significance. For example, Hingston states that "Birds with naked head or neck skin or fleshy head appendages, such as the turkey, jungle-fowl, bateleur eagle, either flush up in these areas, when angered, or cause the parts to become turgid" (p. 337). He also points out that man is not the only mammal to show a flushing of the face in emotional situations; it also occurs in the Bengal Monkey and the Mandrill, to give only two examples, and in both these cases their "hind ends flush in unison with their reddening faces."

Thirdly, the respiratory effects: Various forms of gaping and hissing may have evolved in this way. With gaping, however, there is a dangerous ambiguity, in that the opening wide of the mouth may also arise as a signal as a result of the ritualisation of intention movements such as biting. More important is the question of vocalisation. Spurway and Haldane (1953) have suggested that vocalisation may be "considered as a ritualisation of breathing." They refer to this ritualisation as being a modification of displacement breathing. Since it is the autonomic changes in respiration, which accompany the primary somatic response, which seem most likely to have given rise to this ritualisation, I cannot agree with them that it is necessary to invoke the concept of displacement activation here. This does not, however, really alter the significance of their suggestion. (The problem of displacement activities will be dealt with below.)

Another possible evolutionary trend originating in the respiratory disturbances that accompany thwarting is that leading to Inflation Displays. The latter are widespread in their occurrence, but we will consider only avian inflations here. Hingston states that: "Male Frigate-birds, when at courtship, blow their naked throats into scarlet distended bladders." Stonor records that the Sooty Grouse (*Dendragapus richardsoni*) has brightly coloured throat pouches which it inflates with air from the lungs when displaying. In the Sage Hen (*Centrocerus urophasianus*), the throat pouches are enormous and can be swelled out to almost half the size of the owner's body. Certain species of Bustard, especially the Australian Bustard (*Eupodotis australis*), possess huge throat pouches which, in the latter species, are so large when distended with air that they reach down to the ground. In a number of cases such pouches are not only brightly coloured, and act as visual signals, but also resound in various ways to produce auditory signals. Inflation displays are, of course, also widespread amongst the lower vertebrates, but these will not be discussed in the present paper.

Fourthly, the thermoregulatory effects: Sweating in mammals in thwarting situations may well have been the starting point for the evolution of much "scent signalling." It is not unlikely that scent glands have evolved from sweat glands and scent from sweat. Thus, the territorial behaviour of many mammals involving scent signals may have arisen in a fundamentally similar way to that involving defecation and urination. The difference between them would, of course, be that whereas the latter would be parasympathetic in origin, the former would be sympathetic. Pilomotor responses in birds need not be referred to again here, but similar reactions in mammals may be noted. Hingston has compiled a formidable list of pilo-erection responses in threatening mammals of many species. They include the following: A rather generalised hair erection is seen in dogs, wolves, jackals, mongoose, weasel, skunk, badger, bats and wild pigs, to mention only a few. More specialised hair erection is observed in the following: The lion spreads its mane expanding the region of dark hairs; the squirrel whisks its tail while spreading the tail-fur; the Giant Anteater erects a crest along the middle of its back and also its immense brush has its hairs spread out; Wart-hogs elevate their bristles into a crest. Amongst the most specialised forms of hair erection are those that occur in connection with the rump displays of certain ungulates, e.g., The Prongbuck "has a special circular muscle by which it can spread out the hairs on its rump into a pair of white flaming chrysanthemum-like discs" (Hingston, p. 59).

The widespread significance of ritualised autonomic signals in animal displays becomes immediately apparent when the above phenomena are grouped and considered in this way. In conclusion it needs hardly be mentioned that this whole field of study offers considerable possibilities for future ethological research.

The secondary responses

Turning now to the problem of the secondary responses, observations of thwarted animals reveal that they may frequently intersperse their primary somatic responses with alternative, apparently irrelevant, somatic activities. These are of several kinds.

Firstly, there may appear motor patterns which are typically seen as integral parts of instinctive patterns other than those being thwarted. These have been called Displacement Activities (see Tinbergen, 1939, 1940, 1952; Kortlandt, 1940; Tinbergen and van Iersel, 1947; Armstrong, 1947, 1950; Moynihan, 1953; Bastock, Morris and Moynihan, 1953; Morris 1954a). As regards their causation, it has been suggested that the "nervous energy" which is being denied an outlet by the thwarting circumstances is displaced into new channels and that, as a result of this, they "serve a function as outlets, through a safety valve, of dangerous surplus impulses" (Tinbergen, 1952).

Secondly, Redirection Activities (see Bastock, Morris and Moynihan, 1953) may occur. In these cases, the Primary Somatic Response is continued as far as the fixed motor pattern is concerned, but the orientation is changed. (Since this secondary orientation alters the whole significance of the response, I propose to include Redirection Activities here under the heading of Secondary Responses, despite the fact that the motor pattern itself is the same as the primary one.) Typically, an animal, which is prevented from attacking one individual or object, turns and vents its aggression on some different, inoffensive, individual or object, despite the continued presence of the former.

Thirdly, Regressive Activities may occur. This applies to special learning situations. When thwarted in psychological training apparatus, laboratory animals tend to show Secondary Responses which consist of a falling back to earlier behaviour patterns. That is, to patterns of responses given in the apparatus at an earlier stage of the training. This category of response has obvious implications for human behaviour.

Fourthly, with continued intense thwarting, a state of "neurotic" inactivity may develop. This may continue until the animal ceases to be responsive in any way to almost any stimuli and is then, of course, lethal. It has been suggested by Tinbergen (1952) that one of the possible functions of Displacement Activities is to prevent this from happening.

No more will be said about these last three types of secondary response, since they do not appear to be important from the point of view of ritualisation. Ritualised Displacement Activities, however, do play a very important role in social signalling, as Tinbergen has emphasized, and it is worthwhile therefore to conclude with a re-examination of the causal aspects of this form of Secondary Response.

In the past, a number of reasons have been put forward to explain why one particular Displacement Activity, rather than another, should occur in a given situation (see Tinbergen, 1952, p. 17–20). The three most important factors hitherto discussed are as follows: (1) The influence of the internal state of the thwarted animal. (2) The influence of the external stimuli present at the time of thwarting. (3) The influence of the initial posture of the thwarted animal. To give a simple, hypothetical, example: If an animal has its aggression thwarted, it may show displacement feeding rather than any other displacement activity because (a) It was rather hungry at the time its aggression was thwarted, or (b) because there was food nearby at that time, or (c) because the thwarted posture of biting the enemy was very similar to that of biting food.

Little or no experimental evidence has been presented for the first of these three. Räber (1948) has provided conclusive evidence for the second. He was able to show that whether Turkeys displacement fed or displacement drank was dependent of the relative availability of food and water. Tinbergen (1952) has discussed the third factor in some detail and states that "the posture in which an animal finds itself when a drive is thwarted might decide which activity will be used as an outlet." Although he points out that this has not yet been proved experimentally, he nevertheless cites a number of examples which can most plausibly be explained in this way.

I wish to add now a fourth causal factor which I consider to be of importance in determining which Displacement Activity will occur in a given situation, namely, the influence of the Primary Autonomic Responses. It seems highly likely that many Displacement Activities are selected because the appropriate stimuli for them are provided by the sympathetic or parasympathetic discharges which accompany thwarting. For example, violent pilomotor, sudomotor, or vasomotor activity may provide skin stimuli which provoke the animal to perform displacement scratching, wiping, shaking, or preening. Sleep is known to be intimately connected with parasympathetic activity and displacement sleeping (see for example, Makkink, 1936), may perhaps have been caused in this way.

It might be argued from the above that, if there is an autonomic causal link in a particular instance, then there can be no true Displacement Activity, because the autonomic response is primary and relevant to the thwarting situation. The stimuli that it provides, which determine the Displacement Activity, are relevant to the situation and therefore the so-called Displacement Activity is also relevant.

This might be contrasted with the other causal factors mentioned above, where the presence, at the moment of thwarting, of food, hunger, or the posture of feeding, is purely coincidental. It is not an essential effect of the thwarted aggression that the animal shall find itself hungry, near food, or in a feeding posture. (It can of course be a very common effect and thus lead to ritualisation, without actually being an essential one.) The autonomic responses on the other hand, are an integral, functional, part of the response to the thwarting situation. Secondary Somatic Responses occurring as a result of stimuli produced by this autonomic discharge are considerably less irrelevant to the situation than Displacement Activities arising in other ways. It might be suggested, on this basis, that autonomically stimulated Secondary Somatic Activities should not be called Displacement Activities, and that they should be separated from them conceptually. I think, however, that it would be unwise to do this at the present stage, when we know so little about the detailed causal mechanisms that operate in the production of any Secondary Response. For the present it is far better to leave the question open and to concentrate on the more detailed problems such as the more elaborate experimental analysis of the causal factors acting in special cases. In conclusion, it should however be pointed out that to call the Primary Autonomic Responses themselves Displacement Activities, as certain authors have done in various cases, is quite unjustified.

Summing up the foregoing, it can be said that there are a number of aspects of the response to a thwarting situation that have been specialised as social signals. Ritualisation has taken place, not only in connection with Thwarted Intention Movements, Ambivalent Reactions, and Displacement Activities, but also with Autonomic

Responses. It might be said, therefore, that there are three basic categories of signals arising from thwarting; namely, Primary Somatic Signals, Secondary Somatic Signals and Autonomic Signals. It is hoped that the present paper will serve to focus more attention on the latter category, the significance of which has been underestimated by ethologists in the past.

REFERENCES

Armstrong, E. A. (1947), *Bird Display and Behaviour*. London.

Armstrong, E. A. (1950), The nature and function of displacement activities, *Sympos. Soc. Exp. Biol.*, **4**, 361–387.

Baerends, G. P. (1950), Specializations in organs and movements with a releasing function, *Sympos. Soc. Exp. Biol.*, **4**, 337–360.

Bannerman, D. A. (1949), *The Birds of Tropical West Africa*, Vol. 7. London.

Bastock, M., Morris, D., and Moynihan M., (1953), Some comments on conflict and thwarting in animals, *Behaviour*, **6**, 66–84.

Cannon, W. B. (1929), *Bodily Changes in Pain, Hunger, Fear and Rage*. New York.

Cannon, W. B. (1932), *The Wisdom of the Body*. New York.

Daanje, A. (1950), On locomotory movements in birds and the intention movements derived from them, *Behaviour*, **3**, 48–99.

Davson, H. (1951), *A Textbook of General Physiology*. London.

Delacour, J. (1943), A revision of the subfamily Estrilidinae of the family *Ploceidae, Zoologica*, **28**, 69–86.

Gellhorn, E. (1943), *Autonomic Regulations, their Significance for Physiology, Psychology and Neuropsychiatry*. New York.

Gellhorn, E. (1953), *Physiological Foundations of Neurology and Psychiatry*. Minneapolis.

Goodwin, D. (1952), Recollections of some small birds, *Avicultural Mag.*, **58**, 24–29.

Hediger, H. (1950), *Wild Animals in Captivity*. London.

Hinde, R. (1953), The conflict between drives in the courtship and copulation of the Chaffinch, *Behaviour*, **5**, 1–31.

Hingston, R. W. G. (1933), *Animal Colour and Adornment*. London.

Hutt, F. B. and Ball, L. (1938), Number of feathers and body size in passerines, *Auk*, **55**, 651–657.

Huxley, J. S. (1922), Courtship activities in the Red-throated Diver (*Colymbus stellatus* Pont.); together with a discussion of the evolution of courtship in birds, *J. Linn. Soc. Lond.*, **35**, 253–292.

Kendeigh, S. C. and Baldwin, S. P. (1928), Development of temperature control in nestling house wrens, *Amer. Nat.*, **42**, 249–278.

Kortlandt, A. (1940), Wechselwirkung zwischen Instinkten, *Arch. Néerl. Zoöl.*, **4**, 442–520.

Lack, D. (1943), *The Life of the Robin*. London.

Langley, J. N. (1904), On the sympathetic system of birds, and on the muscles which move the feathers, *J. Physiol.*, **30**, 221–252.

Lorenz, K. (1935), Der Kumpan in der Umwelt des Vogels, *J. Ornithol.*, **83**, 137–213 and 289–413.

Lorenz, K. (1947), The companion in the bird's world, Auk, **54**, 245–273.

Lorenz, K. (1941), Vergleichende Bewegungsstudien an Anatinen, *J. Ornithol.*, **89**, 194–294.

Lorenz, K. (1954), *Comparative Studies on the Behaviour of the Anatinae.* The Avicultural Society, London.

Madsen, H. (1941), Hvad gor Fuglenes Fjer-Dragt Vandskyende? *Dansk Ornithol. Foren.*, **35**, 49–59.

Makkink, G. F. (1936), An attempt at an ethogram of the European Avocet (*Recurvirostra avosetta* L.), with ethological and psychological remarks, *Ardea*, **25**, 1–63.

Moore, A. D. (1945), Winter night habits of birds, *Wilson Bull.*, **57**, 253–260.

Morgan, C. T. and Stellar, E., (1950), *Physiological Psychology.* New York.

Morris, D. (1954a), The reproductive behaviour of the Zebra Finch (*Poephila guttata*), with special reference to pseudofemale behaviour and displacement activities, *Behaviour*, **6**, 271–322.

Morris, D. (1954 b), The courtship behaviour of the Cutthroat Finch (*Amadina fasciata*), *Avicutural Mag.*, **60**, 169–177.

Morris, D. (1954c), An analysis of the reproductive behaviour of the Tenspined Stickleback (*Pygosteus pungitius* L.), Doctor Thesis, Oxford. (Also, *Behaviour, Supplement*, **6**, in press.)

Morris, D. (1955), The function and causation of courtship ceremonies. Fondation Singer Polignac. Colloque Internat. sur L-Instinct, June 1954, in press.

Morris, D. (In preparation). The phenomenon of "Typical Intensity" and its bearing on the problem of ritualisation.

Moynihan, M. (1953), Some displacement activities of the Black-headed Gull, *Behaviour*, **5**, 58–80.

Moynihan, M. (1955), Some aspects of the reproductive behaviour of the Black-headed Gull (*Larus ridibundus* L.), and related species, *Behaviour, Supplement* **4**.

Moynihan, M. and Hall, F. (1954), Hostile, sexual, and other social behaviour patterns of the Spice Finch (*Lonchura punctulata*), in captivity, *Behaviour*, **7**, 33–76.

Räber, H. (1948), Analyse des Balzerhaltens eines domestizierten Truthahns (*Meleagris*), *Behaviour*, **1**, 237–266.

Spurway, H. and Haldane, J. B. S. (1953), The comparative ethology of vertebrate breathing. I. Breathing in Newts, with a general survey, *Behaviour*, **6**, 8–34.

Stonor, C. R. (1940), *Courtship and Display Among Birds.* London.

Sturkie, P. D. (1954), *Avian Physiology.* New York.

Tinbergen, N. (1939), On the analysis of social organisation among vertebrates, with special reference to birds, *Amer. Midl. Nat.*, **21**, 210–234.

Tinbergen, N. (1940), Die Übersprungbewegung, *Z. Tierpsychol.*, **4**, 1–40.

Tinbergen, N. (1948a), Social releasers and the experimental method required for their study, *Wilson Bull.*, **60**, 6–52.

Tinbergen, N. (1948a), Dierkundeles in het meeuwenduin, *De Levende Natuur*, **51**, 49–56.

Tinbergen, N. (1951), *The Study of Instinct.* Oxford.

Tinbergen, N. (1952), Derived activities; their causation, biological significance, origin, and emancipation during evolution, *Quart. Rev. Biol.*, **27**, 1–32.

Tinbergen, N. (1953), *Social Behaviour in Animals.* London.

Tinbergen, N. and van Iersel, J. J. A. (1947), Displacement reactions in the Three-spined Stickleback, *Behaviour*, **1**, 56–63.

Wetmore, A. (1921), A study of the body temperature of birds, *Smithsonian Misc. Coll.*, **72**, 1–52.

Wetmore, A. (1936), The number of contour feathers in Passeriformes and related birds, *Auk*, **53**, 159–169.

Remarks on the Original Sources of Displays

M. MOYNIHAN

Display behavior patterns, other than vocalizations, seem to have been derived, in varying combinations, from the following sources.

1. Obviously autochthonous and "intrinsic" activities. These are all very similar in basic nature, but they can be rather arbitrarily divided into three major groups.
   a) Intention movements of the drives producing the display.
   b) Higher intensity movements of the drives producing the display.
   c) Redirection activities belonging to the drives producing the display.
2. Apparently "extraneous" activities.

It has been suggested that certain supposedly "extrinsic" or "out of context" elements, very conspicuous in many displays, must have been derived from allochthonous or displacement activities. There is considerable evidence, however, that displacement activities are by no means as common as sometimes assumed, and that the supposedly "extraneous" elements in displays, as a group, are actually heterogeneous and quite varied in origin. Some are not really "extrinsic" at all. They have been derived from purely autochthonous and "intrinsic" reactions, particularly redirection movements; and their physical resemblances to some "extraneous" patterns are purely fortuitous. Many of the really "extrinsic" components, moreover, may have been derived from "associated" activities rather than displacement activities in the conventional sense. That is, they may have been derived from autochthonous patterns of drives other than those producing the display with which they have since been incorporated, autochthonous patterns that often occurred in close temporal conjunction with the ancestral form of the display.

There have been many attempts, within the last few years, to trace the evolution of individual behavior patterns and groups of patterns. This is particularly true of the "ritualized" activities or "displays", i.e., those peculiarly standardized and often exaggerated performances, including all vocalizations and many movements and postures, which have become specialized and modified as social signals or releasers.

Reprinted from the *Auk*, volume 72, pages 240–246 (1955).

Comparative studies of such performances have yielded results of interest to both ethologists and systematists; and the stage has now been reached when it is possible to begin to generalize these results, to draw some tentative conclusions about the sources from which some of the displays have been derived.

Tinbergen has already discussed these sources in an earlier review (1952) and some problematic aspects of their evolution have been also noted elsewhere, (e.g. in Bastock, Morris, and Moynihan, 1953); but the whole subject might, perhaps, be usefully reviewed and reassessed once again, as briefly as possible, in the critical light of some more recent information and conjecture.

The origin of vocalizations remains obscure (although the suggestions of Spurway and Haldane, 1953, are very interesting in this connection); but a good deal is now known about the nature of the elements that have been most frequently incorporated into ritualized movements and postures.

The commonest of these elements are "autochthonous" intention or low-intensity behavior patterns.

An "autochthonous" activity is one that is caused by its usual drive (see Kortlandt, 1940). Thus, for instance, an attack movement is said to be autochthonous when it is produced by attack motivation. The terms "motivation" and "drive" are used interchangeably here; as short-hand for "the complex of internal and external states and stimuli (usually or normally) leading to a given behavior." This usage follows Thorpe, 1951, in a somewhat altered form.

Many displays seem to have been derived from autochthonous intention movements alone. The "Aggressive Upright" threat display of many gulls is a good example of this type (see Moynihan, 1955). It is motivated by attack and escape drives; and it includes indications of advance and pecking (attack intention movements) plus indications of retreat or avoidance (escape intention movements), combined in a particularly standardized arrangement. Further examples, in other species, are described in detail by Tinbergen (1952).

Other displays have been derived from autochthonous movements of higher intensity. The "Swoop" and "Soar" displays of the Black-headed Gull, for instance, include attack and escape elements of much greater vigor and elaboration (Moynihan, 1955).

The most interesting displays, however, are those that would appear to have been derived, in part at least, from an "extraneous" or superficially "irrelevant" source, i.e., from the so-called "displacement activities" and some other behavior patterns that are sometimes confused with displacement.

The generally accepted definition of a displacement activity, as ethologists use the term, is "an activity belonging to the executive motor patterns of an instinct other than the instinct(s) activated" (Tinbergen, 1952). Such a reaction is supposed to occur, in most cases, when an instinct or drive is thwarted, when it is prevented from finding its usual expression. The "energy" of the blocked motivation is then supposed to "spark-over" somewhere in the central nervous system in a peculiar and as yet unexplained fashion, irrelevant act (see Bastock *et al.*, 1953). These apparently irrelevant acts, being caused by some drive other than their usual or normal one, are called "allochthonous."

The most plausible examples of displacement or displacement-like reactions, in this sense, are provided by certain "nervous" movements in man. It is well known that human beings may show unexpected "out of context" activities (e.g., yawning, scratching, playing with keys or other objects, etc.), under various conditions of stress and conflict.

It is also probable that similar reactions do sometimes occur in other animals. Thus, for instance, a male Three-spined Stickleback will show "fanning" (usually a parental activity) when its sex drive is thwarted during "courtship" (Tinbergen and van Iersel, 1947), and a Black-headed Gull will show preening and/or nest-building when its brooding drive is thwarted during the incubation period (Moynihan, 1953).

Many apparently irrelevant or "extraneous" movements of birds, therefore, have been interpreted as displacement activities of this sort. Such interpretations, however, are often obviously unwarranted. There is every reason to believe, in fact, that the great majority of the so-called "displacement activities" reported in the ornithological and ethological literature can be adequately explained without assuming the existence of any exceptional "spark-overs" in internal motivation. (It is true that any change in behavior of any sort, "normal" as well as "abnormal," may involve some type of internal switch; but this is usually much slighter, and/or less transitory, than the "spark-over" posited to explain most of the presumed displacements.) Of the many apparently irrelevant acts cited by Tinbergen (1952), for instance, it is probable that no more than a third, at best, are really displacement activities in the conventional ethological sense.

A few examples, from his list, should make this clear.

Many of the cited activities seem to be purely autochthonous, direct and usual reactions to internal and external stimuli, with only the most superficial and misleading appearance of unexpectedness or irrelevance. Thus, for instance, the "courtship" behavior of caged Willow Warblers, "when showing the inhibited migratory movements called 'migratory restlessness,'" can hardly be a typical displacement. These birds must almost certainly possess some activated "courtship" motivation. Similarly, the "song" of Skylarks after escaping from a Duck Hawk or Hobby is most unlikely to be allochthonous. The "songs" of many birds are hostile (i.e., produced by attack and escape motivation), and this situation is certainly one in which autochthonous hostility might even be expected as a general rule.

Other activities have been misinterpreted because of their peculiar orientation. These are the "redirection activities."

Redirection movements, like displacement, seem to occur when an instinct or drive is thwarted, and their physical form may be very similar to that of some displacement reactions; but the two types of activity are quite fundamentally and definitely different in nature, i.e. in internal causation.

Redirection movements can be defined as autochthonous activities of a drive directed toward an object or animal other than the one releasing and usually directing them (although the releasing object or animal remains available, or partly available, as a potential goal at the time) (see Bastock *et al.*, 1953). An example is provided by the behavior of a Prairie Falcon (cited by Bent, 1938), when disturbed at the nest by a human intruder. Both the attack and escape drives of this falcon were immediately activated; but they were largely incompatible, and the escape drive was strong enough to prevent the bird from venting its attack drive upon the real offending object, the actual disturber. The falcon then found an outlet for its thwarted attack motivation by pouncing upon some other birds, a Barn Owl and a Raven, which happened to pass by at a convenient moment. This sort of "unprovoked" attack upon an inoffensive scapegoat is the commonest type of redirection.

It is also the type of redirection that has been most frequently confused with displacement.

Thus, to give some more examples from Tinbergen's list, the pecking at the ground by many passerines during fights, and other hostile encounters, is probably redirected attack pecking rather than displacement feeding.

More complex are such performances as "grass-pulling" in Herring Gulls. When two of these birds become engaged in a territorial boundary dispute, one or both may begin to peck and pull, violently, at the nearby vegetation. Tinbergen notes that the violence of this performance is probably an expression of redirected aggressiveness, but he also believes that the activity includes an additional nest-building component (which must be displacement in these non-nesting circumstances). He bases this belief upon the fact that vegetation is used as nest-material, and, more important, that the vegetation pulled up by "grass-pulling" is usually thrown away with a sideways jerk of the head, a sideways movement also shown during the construction of an actual nest. This evidence of displacement is not, however, completely convincing. There is no obvious reason why redirected aggressiveness should not be vented upon vegetation as well as any other object; and the distinctive sideways jerking might be nothing more than an immediate and simple reaction to the presence of some nonedible material in the bill. In other words, there are no real indications that any internal motivation has "sparked-over" in the course of this performance.

Similar explanations might easily account for many of the other supposed displacement activities in which pecking, biting, pulling, or pushing movements are conspicuous.

Discounting such certain or probable autochthonous reactions, at least provisionally, the remaining list of probable or possible displacement activities is very greatly shortened. The majority of the remaining patterns, moreover, are comfort movements such as preening or scratching; just the type of reaction whose causation, whether displaced or not, is most difficult to determine with any degree of assurance. The real nature of many of them, if not all, is still highly dubious.

The fact that real displacement activities are apparently rarer than sometimes assumed is not without significance in connection with the origin of ritualized displays.

Many displays, particularly the most elaborate hostile and sexual performances, contain elements that appear to have been derived and modified from such "extrinsic" or "extraneous" contexts as sleeping, preening, or nest-building. There have been frequent suggestions that all of these "extraneous" elements must have originated, in the display situation, as allochthonous displacement activities. This theory may be quite correct, but we have very little evidence by which to judge it, and it is by no means the only possible explanation of the presence of these patterns in such peculiar circumstances.

Some of the supposedly "extraneous" elements in certain displays, of course, may be far more apparent than real. They may have been misinterpreted in the same way as some of the supposed displacement activities noted above. The pecking and throwing movements in the hostile and ritualized "choking" threat of the Black-headed Gull, for example, are rather misleading in much the same way as the similar movements in the "grass-pulling" of the Herring Gull (see Moynihan, 1955). They too should be classed as redirected attack, and they must always, therefore, have been strictly "intrinsic" as hostile reactions.

There are other components, however, in this and other displays, whose "extraneous" source is less easily questioned.

The ritualized "mock-preening" movements in the "courtship" of many male ducks (Lorenz, 1952), may be taken as representative of this group. They do seem to have been derived from real preening (although it is just barely possible that they might be modified forms of avoidance intention movements instead).

Assuming that they are indeed derivatives of preening, these movements may have been evolved from what was once displacement (conflict and thwarting are almost inevitable in "courtship" encounter); but such displacement preening, even if it did exist, was probably not the only source available. "Courtship" activities are also likely to provoke a considerable amount of purely autochthonous preening, during the performance itself or immediately afterwards, if only because vigorous activity of any sort is very apt to disarrange the plumage. Such ordinary preening may, therefore, have been the usual accompaniment of "courtship" on the males of the ancestral ducks.

It is extremely probable, moreover, that some conditioning would then result. In other words, the female ducks toward which such "courtship" was directed would then become conditioned to the associated movements, and autochthonous preening would thus acquire a "courtship" valence for them. This, in turn, would probably reinforce the connection between "courtship" and preening in the displaying males; as they would then become conditioned to the fact that preening movements had acquired a signal or symbolic function.

Granted the probability of this connection, it is easy to see that such associated preening might well become incorporated, by natural selection, into the actual "courtship" itself. (This would be a case of "neurophysiological emancipation," as Tinbergen has used the term. These preening movements, as they became incorporated into "courtship," with increasing ritualization, would be less and less motivated by preening drives, and more and more motivated by "courtship" drives.)

This hypothesis is apparently adequate to explain the origin of "courtship" preening in male ducks; and also, perhaps, the appearance of many other display patterns, some of them derived from very different sources, in many other groups of birds.

Some hypothesis of this sort might even seem to be more plausible than the alternative theory that would derive "extraneous" display elements from displacement activities. It might be preferable, primarily, because it is somewhat simpler; i.e., it does not need to assume the existence of some original displacement "spark-over," a type of "spark-over" which may be very rare.

A final point, in this connection, may help to put the matter in perspective.

The exact process by which an "extraneous" pattern is incorporated into a display is certainly difficult to imagine in detail; as we know relatively little about the immediate causal, internal, factors involved in ritualization. There is no reason to believe, however, that the incorporation of an autochthonous "extraneous" pattern would be much more complicated that the incorporation of an allochthonous one. The ritualization of an autochthonous "extraneous" element, in fact, must be essentially the same as that of any other autochthonous pattern in its proper context. The consequences of this basic similarity are obvious. The physical differences between "extrinsic" and "intrinsic" elements are rapidly reduced as ritualization progresses; and this, of course, is the very reason why the two components are so difficult to separate and analyze in many cases.

ACKNOWLEDGMENTS

I am indebted to many people for advice and comments; and to the U.S. Public Health Service for financial assistance during the writing of this paper.

REFERENCES

Bastock, M., Morris, D. and Moynihan, M. (1953), Some comments on conflict and thwarting in animals, *Behaviour*, **6**, 66–84.

Bent, A. C. (1938), Life histories of North American birds of prey, *Bull. U.S. Natl. Mus.*, **170**, viii + 482 pp., Smithson. Inst.

Kortlandt, A. (1940), Wechselwirkung zwischen Instinkten, *Arch. Néerl. Zool.*, **4**, 442–520.

Lorenz, K. (1952), Comparative studies of the behaviour of the Anatinae, *Avicultural Mag.*, London.

Moynihan, M. (1953), Some displacement activities of the Black-headed Gull, *Behaviour*, **5**, 58–80.

Moynihan, M. (1955), Some aspects of reproductive behavior in the Black-headed Gull (*Larus ridibundus ridibundus* L.), and related species, *Behaviour*, Supplement No 4 (in press).

Spurway, H., and Haldane, J. B. S. (1953), The comparative ethology of vertebrate breathing, *Behaviour*, **6** 1–34.

Thorpe, W. H. (1951), The definition of terms used in animal behaviour studies, *Bull. Anim. Behaviour*, **9**, 34–40.

Tinbergen, N. (1952), "Derived" activities; their causation, biological significance, origin, and emancipation during evolution, *Quart. Rev. Biol.*, **27**, 1–32.

Tinbergen, N., and van Iersel, J. (1947), "Displacement reactions" in the Three-spined Stickleback, *Behaviour*, **1**, 56–63.

*Comparative Studies of the Behaviour of Gulls (*Laridae*): A Progress Report*

N. TINBERGEN

This paper describes a number of displays of various gulls (*Laridae*) with special reference to the Herring Gull group, the hooded gulls, and the Kittiwake, and discusses their functions, causation, evolutionary origin and further evolution as signals.

After a sketch of the rationale of comparative behaviour studies (chapter II), the most common single displays and display sequences are described (chapter III). They are rather similar through the family, though many species-specific differences exist. The behaviour similarities strengthen the conclusion that the gulls are a monophyletic group.

Evidence is presented in chapter IV to show that the displays have signal function. A distinction is made between distance-increasing and distance-reducing displays. The reasons for the occurrence of more than one, yet no more than about six distance-increasing postures are discussed; it is argued that each display may well be adapted to deal with a distinct category of opponent: accidental trespassers require another repellent than intentional persistent intruders; and within each category actual trespassers are met in another way than potential trespassers. In addition, one particularly loud call is a typical long-distance threat. This call usually acts as advertisement in a double sense: It repels competitors and attracts unmated females.

Some of the distance-reducing (or appeasement) postures are used in agonistic situations, and still more regularly at the end of the pair formation or meeting ceremony.

The need for and the possibility of more systematic studies of the precise functions of displays is stressed.

Chapter V discusses aims and methods of analysis of the motivation of the displays. The value of "natural experiments" is stressed. The application of three independent methods leads to the conclusion that agonistic displays are ambivalent, i.e., the outcome of the simultaneous arousal of a tendency to attack and a tendency to fell: The relation between fluctuations in these tendencies and the displays shown is examined. There is a striking correlation between the motivation of these displays and the information they pass on to other individuals.

Reprinted from *Behaviour*, volume 15, pages 1–70 (1959). Original numbering of figures is retained.

Appeasement gestures always contain an element of fear; this tendency is in conflict with a tendency to stay, which can be, but usually is not part of the tendency to attack; it may be sexual attraction, attraction to a nest site, or attraction to a provider of food. The need for motivational analyses of the many different forms in which one posture can occur is stressed.

The similarity of motivation in the pair formation ceremonies in the different species is much greater than the formal similarity of the display sequences; the conflict theory of courtship is tested.

The origin of the displays (chapter VI) is varied. Some have clearly arisen as preparatory or intention movements of the patterns directly aroused by the situation ("autochthonous" movements); of these, some are redirected to inanimate objects. Others are derived from movements belonging to functional patterns not directly aroused by the situation ("displacement activities"); their various origins are discussed, and it is shown that they are second components of a dual movement, of which the first component is "autochthonous" in the above sense and facilitates the displacement activity.

In chapter VII some ultimate causes of evolutionary change are discussed, and a preliminary functional classification of alleged changes in displays is presented. It is argued that change has been enhanced by at least four different types of selection pressure: (1) towards improvement of signal function (conspicuousness); (2) towards increased intraspecific unambiguity; (3) towards increased interspecific unambiguity; and (4) as a corrollary of selection pressure in other functional systems. In some gulls, (4) may have made the major contribution to sexual isolation.

A preliminary classification is given of the postulated evolutionary changes in behaviour mechanisms.

The value of behaviour characters for taxonomic use is considered. After a discussion of adaptive and non-adaptive differences the relative validity of Lorenz's emphasis on the phylogenetic conservatism of displays is reconsidered, and the occurrence of convergent similarities is demonstrated.

INTRODUCTION

Stimulated by Heinroth's (1911) and Lorenz's (1941) studies of the behaviour of Anatidae, members of the Research Unit of Animal Behaviour of this Department started, in 1950, a programme of comparative studies of the behaviour of gulls. Foundations had already been laid by the work of Strong (1914), Portielje (1928), Goethe (1937) and Tinbergen (1936, 1953a) on the Herring Gull, and by Kirkman (1937) on the Black-headed Gull. The aims of our studies were: first, a description of the behaviour of as many species as possible; second, as complete a coverage as possible of the entire behaviour pattern of each species; and third, analyses of the functions, the causation and the origin of the displays, with the ultimate aim of understanding how they could have originated and diverged in the course of speciation. This wide scope of phenomena and of problems was judged necessary because it had become increasingly clear that an understanding of the origin and evolution of displays can be considerably enhanced by a wide knowledge of entire behaviour patterns and by some insight in the functions and the causation of displays.

The present paper attempts to give a summary of the results so far obtained, with a discussion of some problems of function, of motivation, of some evolutionary aspects, and of methods of study. It is based mainly on the following species:

1. The Herring Gull, *Larus argentatus* (the authors mentioned above; Moynihan, 1958b) and its close relatives the Lesser Black-backed Gull, *L. fuscus* (Paludan, 1951; J. M. and E. Cullen, in litt.), the Glaucous-winged Gull, *L. glaucescens* Veitch and Booth 1954; unpubl. observations of my own), the Western Gull, *L. occidentalis* (unpubl. observations of my own).
2. The Common Gull, *L. canus* (Weidmann, 1955) and its relative the Ring-billed Gull, *L. delawarensis* (Moynihan, 1956, 1958a and 1958b).
3. The Black-headed Gull, *L.* (subg. *Hydrocoloeus*) *ridibundus* (Moynihan, 1955a; Tinbergen and Moynihan, 1952) and other hooded gulls; Bonaparte's Gull, *L. philadelphia* (Moynihan, 1956; 1958a, b), Franklin's Gull, *L. pipixcan* (Moynihan, 1956; 1958a, b), *L. atricilla* (Moynihan, 1955; 1958b), Hartlaub's Gull, *L. novaehollandiae hartlaubi* (Tinbergen and Broekhuysen, 1954), Little Gull, *L. minutus* (Moynihan, 1955).
4. The Kittiwake, *Rissa tridactyla* (E. Cullen, 1957 and unpubl. observations; Paludan, 1955).
5. The Ivory Gull, *Pagophila eburnea* (Bateson and Plowright, unpubl.).

Almost all studies were conducted in the field; some observations made in zoological gardens have been added.

Films were made of the displays of the Herring Gull, the Lesser Black-backed Gull (N.T.), the Glaucous-winged Gull (N. Hancock, of Broadview, Ill.), the Western Gull (Dr. R. Boolootian of Los Angeles), the Black-headed Gull (N.T.), Hartlaub's Gull (Dr. G. J. Broekhuysen of Cape Town), the Kittiwake (N.T.) and the Ivory Gull (C. Plowright of Cambridge). Copies of all these films, as well as of a film of the Great Skua, *Catharacta skua* (G. Paris of Gouda, Holland) are in the Department of Zoology and Comparative Anatomy, Oxford. We have found these films of increasing value as the work progressed.

The illustrations in this paper were for the greater part drawn from these films and from still photographs.

Since the work thus deals with many species, and each species has a highly complicated behaviour repertoire; and because objective recording of movements is a much more laborious task than that of structures, even the initial descriptive phase of the work is an extremely time-consuming task, which can by no means be considered to have been completed even for those species we studied most intensely. Further, work on functions and causation is hampered by serious practical difficulties; and lastly, evidence on evolutionary problems is naturally indirect. For these reasons our work is still far from complete. Yet this progress report seems justified for two reasons. First, several conclusions have been reached which, while tentative, seem to me to be of some general interest; for instance, I believe that we have made definite progress with the analysis of the functions of displays, of some aspects of their causation, and of their evolutionary history; and I want to call attention to some ideas which might be checked in other animals. Secondly, I am convinced that his kind of work is still hampered by a certain lack of precision in our formulation of problems, of methods, and of conclusions. However much Lorenz (1935, 1941), Hinde (1955), Baerends (1950), Andrew (1956a, b), Marler (1955, 1956), Moynihan (1955b), Morris (1956, 1957), and others have contributed to theoretical clarification, and however much of great interest has been found already, we are still far from knowing exactly what we are doing; much of our work is guided by intuition rather than by conscious and systematic thinking. This is not surprising since this work is not only concerned with the causation of behaviour, where it is notoriously difficult to apply strictly biological methods of study, but since, in addition, the problems of causation, function and evolutionary history are so closely interrelated that the danger of circular arguments is great. Some space has therefore been given to attempts at reformulation of problems, methods and arguments with a view of further preparing the ground for future work.

I have tried to write this paper for the specialists as well as for the interested zoologists, ornithologists and psychologists. This gives the paper certain characters of a compromise.

I want to express my gratitude to the Cumberland County Council, the Farne Islands Committee, the National Trust, the Nature Conservancy, Sir William Pennington Ramsden, and the Earl of Verulam for permission to work in various gull colonies protected and managed by them; to the Ford Foundation, the National Science Foundation, the Nature Conservancy and the Nuffield Foundation for financial support: to Professor Sir Alister Hardy F.R.S. for encouragement and support of various kinds; to P. Bateson, Dr. E. Cullen, Dr. J. M. Cullen, G. Manley, Dr. M. Moynihan, Dr. A. C. Perdeck, C. Plowright, Dr. R. Weidmann and Dr. U. Weidmann for permission to use unpublished data; and to Dr. A. Cain, Dr. E. Cullen, Dr. J. M. Cullen and Dr. R. A. Hinde for reading and criticising the manuscript.

AN ATTEMPT AT A RATIONALE OF THE TAXONOMIC AND EVOLUTIONARY ASSESSMENT OF BEHAVIOUR CHARACTERS

Our comparative studies had two aims: (1) a mere description and listing of similarities and differences so as to arrive at a natural classification and to check the classifications given so far, and beyond this (2) an assessment of how and why these similarities and differences may have developed phylogenetically.

In the course of our studies it became gradually clear that much of the reasoning applied by us had often been vague, to a great extent intuitive, and less sophisticated than that which is now being developed in taxonomy (Cain and Harrison 1958). Partly because of this, and partly because behaviour studies meet with some difficulties not found in morphological studies, a critical examination and re-appraisal of our methods seemed necessary. For practical reasons I will discuss some general aspects of these methods in anticipation of the results: some readers might prefer to read the factual chapters first.

In comparison of behaviour, as in morphological comparison, similarities and differences in characters are described. For this purpose species-specific characters are singled out, i.e., properties which are relatively constant throughout the species* either in all members, or in males, females, or a certain ontogenetic stage, and which differ between species or larger taxa.

In describing differences in characters, it is of interest to know to what extent these differences are phenotypic, and to what extent, if at all, genotypic. This is particularly important in behaviour studies, since behaviour can in so many ways be moulded by the environment.

Evidence on this point is still relatively fragmentary. Without entering into a detailed discussion of behaviour ontogeny, it can be said that, on the whole, movements of the "fixed pattern" type (*Erbkoordinationen*, Lorenz, 1939) show a high degree of ontogenic stability in the sense of resistance against environmental modification, the song patterns of some song birds (Thorpe, 1958) being a relatively rare exception. The scattered reports of development of motor patterns in animals raised in abnormal environments particularly show that the type of movement on which the present study is mainly based, *viz.* displays, is highly environment-resistant (Goethe, 1955; Heinroth, 1928; Lorenz, 1939; Sauer, 1954). The situation is different with regard to other parts of the behaviour machinery. The selective responsiveness to specific releasing stimuli for instance is often subject to conditioning. Thus Bergman (1946) showed that Turnstones hatched under Redshanks ignored the alarm calls of adults of their own species and soon responded to their foster parents. On the other hand, Redshanks chicks hatched under Turnstones responded to the call notes of their own species; later they became conditioned to those of their foster parents. The specific responsiveness to some stimuli releasing food begging in young Herring Gulls (Goethe, 1957) and Black-headed Gulls (R. and U. Weidmann, in litt.) is the same in "naive" chicks as in chicks raised by the parents, though it can be changed at a later stage by conditioning (E. Cullen, 1957).

Other aspects of behaviour machinery have also been shown to be environmental-resistant: Thus the difference in mobility between chicks of the Kittiwake and of the Black-headed Gull is truly innate (E. Cullen, 1957).

While for our present purpose the characters compared can safely be assumed to be valid from this point of view, the need for more precise ontogenetic studies is obvious.

---

*For the present study, most gulls were studied in one or a few colonies, only, and our facts may not always be representative in detail of all populations of each species.

The variety of behaviour characters observed in gulls allowed a considerable widening of the range of taxonomically useful characters. This allowed a better assessment of their taxonomic affinity. Since all gulls so far studied are very similar in most of their behaviour their close taxonomic affinity was thus confirmed.

For phylogenetic interpretation, our method amounted, in a nutshell, to the application in detail of the same method by which, e.g., whales and bats are found to be mammals and thus to be more closely related to each other than whales are to fish and bats to birds. While for purely taxonomic purposes this is concluded from the fact that whales and bats share more characters with other mammals than with fish and birds respectively, phylogenetic interpretation depends on "weighting" of the various characters. This implies a refinement, through functional interpretation, in the crude use of sheer numbers of common or different characters. A study of function, or biological significance, can reveal the adaptive significance of (a) differences in single or a few characters between otherwise very similar species and thus strengthen the conclusion that such species are of common descent and have undergone recent adaptive radiation and (b) similarities of single or a few characters in otherwise different forms, and thus strengthen the conclusion that these similarities are due to convergence. The gulls being a monophyletic group, the present study is mainly concerned with adaptive divergence; but convergent traits shared with otherwise very different groups (analogous to the fish-like external shape of whales and fish) will also be briefly mentioned.

The weighting of characters on the basis of functional study has been done in the following way. Whenever, in a species which on total number of common characters is judged to be closely related to other species, the differences can be correlated to different functions and can thus be considered to be recently acquired, these differences can so to speak be "peeled off," and a core of characters revealed which are either different for (as yet) unknown reasons, or which are shared with the other species. If such cores are more similar between species than their total complexes of characters, their affinity is even more probably than before functional studies were undertaken; if the cores turn out to be more different than the total complexes, common descent of the species compared becomes doubtful. Apart from and beyond this, such functional studies allow conclusions about the particular types of selection pressure which must have been at work in adaptive radiation. Thus E. Cullen's study of the Kittiwake revealed that many behaviour (and morphological) peculiarities of this species have to do with its specialisation as a cliff breeder; the most striking differences between the various Galapagos Finches were found by Lack (1947) to be related to differences in feeding habits. When the Kittiwake is stripped of those peculiarities which can be recognized as cliff-breeding adaptations or their corollaries, its "core" of character so revealed is strikingly similar to those of most other gulls. At the same time, the nature of its specialisations is better understood.

Since description precedes functional interpretation, great care is needed in the naming of the characters. In the present study, which concentrates on displays, naming has been done in the following way. Within each species, a number of displays have been distinguished. This is possible in spite of the fact that each posture shows considerable variation, because there is, in most cases, a clear discontinuity between the displays. Further, each display, or rather the range of forms in which the same display can appear, is strikingly constant within the species. Third, in spite of the constant differences between species, there is, for each display, a degree of similarity between species often as striking as the morphological similarities.

Because the displays are at first described with regards to their formal appearance only, the names should be descriptive without any reference to either function or causation, for uncertainties in interpretation may lead to confusing changes of names as these interpretations change. The similarity between species allows the application of the same name to similar displays in different species. However, this is justified only if and when the same name can be assumed to indicate that the displays are not convergent or accidentally similar; application of the same name ought to indicate common descent, or perhaps parallel descent in many species. Strictly speaking therefore (since common descent is a matter of conjecture) the safest course would be to start by applying each name to one display of one species only. As the study proceeds, this procedure (which, while consistent, is of course pedantic and clumsy) can be simplified by allotting common names. This causes no great difficulty where interspecific differences are very small, such as in the Oblique, or in Facing Away, but with greater dissimilarity, such as between the Kittiwake's Bow-and-Moan and the Herring Gull's Mew Call, the uncertainty may be such as to warrant the use of different names, and thus avoiding prejudging the issue.

For these reasons it is important to be clear about the criteria used in the assignment of common names. Convergences or accidental similarity between displays is considered to be improbable when (1) their interspecific formal similarity is great (particularly when the displays are complex and "improbable"); (2) their distribution through the group is wide and continuous; (3) they are, in addition, found to have a similar motivation, and (4) a similar function throughout the group; and, finally (5) when they can be considered to have had the same origin. This latter aspect is particularly important in displays, for most, if not all are now recognised as being "derived" movements (Lorenz, 1935; Daanje, 1950; Tinbergen, 1952).

In practice, various difficulties arise. First, formal similarity, which per definition is not identity, can be great or small. Thus while there is not much doubt that the Upright of the Herring Gull and that of the Black-headed Gull deserve a common name, this is less certain of the Head-Flagging (Facing Away) of the Black-headed Gull and the Facing Away of the Kittiwake.

When the criterion of formal similarity cannot be applied without hesitation, but there are indications that function and causation may be very similar, such as is the case with the Vertical of the Little Gull and the Forward of the Black-headed Gull (which are both shown as the second posture in the meeting ceremony), uncertainty is best expressed by applying different names. This may or may not be temporary. Conversely, although there is a certain similarity between the Forward of the Black-headed Gull and "Jabbing" of the Kittiwake they appear in different circumstances and therefore must have different function and causation. For these reasons, naming is provisional and tentative, and may be subject to subsequent changes.

As to (5), movements such as the Upright of which the origin is clear, and Choking, about whose origin a relatively clear picture is gradually emerging, can be named with more confidence than the Mew Call Posture, about the origin of which there is as yet not much of a clue.

Apart from all these considerations a difficulty arises, as in comparative anatomy, whenever a posture seems to have been fused with other elements in some species while not in others. Thus the Oblique-cum-Long-Call of the Herring Gull is really a complex of three movements; while the Oblique-proper is very similar to that of many other species, it is preceded by a bending-down movement which is not found in the Kittiwake and the Black-headed Gull, and it contains the "Throwback" which is again not found in the latter two species, whereas just this element is still more conspicuous in the Common Gull. Similarly the Kittiwake's Bow-and-Moan contains one element which is very similar to the Mew Call Postures of other species, but the Bow may be an added element. The Long Call sequence of the Ivory Gull is peculiar in that the Oblique is preceded by a posture very much like the Forward of the Black-headed Gull.

This is one reason why the naming is no more than a crude approximation; the situation is rather similar to that in the comparison of the skulls of various vertebrates, and can be dealt with in the same manner.

Differences between postures receiving the same name may also be due to "ritualisation" of the movement itself. As will be seen, the Choking of the Kittiwakes is different from that of other gulls in that the mouth is usually wide open; this no doubt has to do with the fact that the inside of the mouth is a very bright orange, a device which presumably emphasises the conspicuousness of this posture. A similar functional explanation must probably be given to the spreading of the wings by the Great Skua when performing the Oblique-cum-Long-Call; this shows off the bright white patches on the wings. Without committing myself in the question which came first, the change in structure or the change of movement, I suggest that the correlation is functional.

Thus the evolutionary interpretation of these comparative behaviour data is beset with difficulties. Some of these cannot be removed—we cannot hope for direct evidence on the behaviour of the gulls of the past. This means that our reasoning has to be "probabilistic" (to use a term applied by Thieme, 1958, for the procedure of comparative linguistic studies); by increased precision of description, by extension of the number of species studied and of the characters used, by study of functional aspects, and by continuous scanning for inconsistencies or contradictions in the total set of conclusions the probability of the overall picture can be judged with increasing confidence. If the premises are false, inconsistencies are bound to appear; if uncertainties and inconsistencies tend to be reduced by more intensive and extensive study the premises will be strengthened. Even so, the gaps between present species may often be too wide to allow a full phylogenetic interpretation.

* * *

THE ORIGIN OF THE DISPLAYS

The description and the motivational analysis of the displays have already given us some hints about their evolutionary origin. In general it can be said that our gull studies have so far supported the conclusion accepted by various authors with regard to other species, *viz.* that displays are "derived" movements (Huxley, 1923; Lorenz, 1935; Tinbergen, 1952; Hinde and Tinbergen, 1958). While the derivation of some displays is rather obvious, we are still far from knowing the origin of all, and as long as this is so the question of origin deserves continued critical study in every single case.

The origin of some displays can be recognised, even without comparative study, by their similarity to the movements from which they are derived. Thus, a study of the Herring Gull alone leaves little doubt that the Aggressive Upright is in origin a mosaic of components of attack and escape behaviour, that Pecking-into-the-Ground is a redirected aggressive movement; and that Grass Pulling contains an element which can be recognised as "sideways building."

However, the origin of many displays cannot be recognised in this way, or often no more than a vague suspicion can be arrived at. This appears to be due to the fact, revealed by comparison of many closely related species that such derived movements have undergone ritualisation, which I will define as adaptive evolutionary change in the direction of increased efficiency as a signal. While the ritualisation of the displays of gulls will be discussed more fully in the next chapter, it is necessary to insert some general remarks here.

The idea of ritualisation is derived from two sets of facts. (1) Comparison of a given display in a number of closely related species allows one to arrange these species according to the degree of similarity of the derived movement to its supposed origin. Thus Cranes use a preening movement in their displays, in some species this is very similar to normal preening, in others it is more different; in the Manchurian Crane it is only recognisable as preening by comparison with other, in this respect more intermediate species (Lorenz, 1935). Similarly comparison has helped to recognise the inciting movements of female ducks as derived from a dual origin; the intention movement of pecking at another bird, followed by Chin-lifting, which is in many ducks an expression of fear, often with appeasement function (Lorenz, 1941). A third example is the wing-preening movements incorporated in the courtship of many male ducks; the Flag Touching of the Mandarin drake can be recognised as preening through less extreme forms shown by other species (Lorenz, 1941).

This alone however would not be convincing, if it could not be shown (2) that in general the less easily recognisable movements were functionally better suited than the others to the task of providing strong, conspicuous and unmistakable stimuli. This conclusion is based not only on a study of the displays, but also on their function and on an analysis of the type of stimuli animals usually respond to. Lorenz has pointed out long ago (1935) that "releasers" are so to speak the graphic representations, the crystallisations, of sign stimuli.

These two considerations suggest that divergence and evolution of displays are the result of a secondary adaptation to a new function of derived movements. As such there is a close parallel between them and, for instance, the secondary adaptation of the first pereiopod in many crayfish, originally a locomotor organ mainly, to the new function of catching and crushing prey, to which it has become adapted, for instance, in the Lobster.

So far, two main sources of derived movements have been recognised. (1) Movements immediately evoked by the situation (such as fighting movements evoked by the approach of an intruder into the territory) are performed either incompletely (intention movements) or with another orientation (redirected movements). Such movements are often called "autochthonous." (2) Movements belonging to patterns other than those immediately evoked by the situation and which are therefore usually unexpected and functionally out of context or irrelevant. For these reasons such movements, whatever their exact causation, are called collectively displacement activities. Evidence on the gulls in general confirms this, but they also add something to our knowledge.

I will first discuss some elements which are found in more than one posture, and then proceed to deal with the displays. The sequence will be the same as in preceding chapters.

Elements of displays

Elements common to more than one posture can be classified in fixed components and orientation components.

Fixed components

In at least the Oblique, the Upright and the Forward the position of the bill and head in the vertical plane varies with the relative preponderance of aggressive over avoidance tendencies; a very aggressive bird tends to point the bill down, a very timid bird less so or even points the bill up. The range over which this occurs varies from one species to the other; thus Hartlaub's Gull points the bill up more than the Black-headed Gull; an anxious Black-headed Gull may assume the Anxiety Oblique in which the bill is pointing up higher than usually in the Herring Gull or the Kittiwake.

Another fixed element which varies with the balance of tendencies is the position of the carpal joints. On the whole, in various postures, lifting of the carpal joints increases with aggression-escape conflict of increasing intensity. Again as we have seen, lifting of the carpal joints can be an element of the Oblique, the Forward, Choking, the Upright and the Mew Call Posture. Similarly, the Kittiwake opens its mouth in all agonistic postures.

Orientation components

Many aggressive postures of gulls show some orientation in the vertical plane. The Oblique of the Kittiwake and the Black-headed Gull is more vertical when it is evoked by a bird flying overhead than when it is on the same level. The Kittiwake's Oblique can even be directed far down when the opponent is lower on the cliff. The Kittiwake's Jabbing is always precisely aimed at the opponent. In the Oblique therefore the position in the vertical plane is controlled both by the attack-escape balance and by orientation.

Another source of variation is the orientation in the horizontal plane. When aggression is predominant the bill or the body axis tends to be pointed at the opponent, but when a tendency to move away from another bird (caused either by fear or by other tendencies) increases, the bill, the head and neck, or even the whole body turns away sideways. This is clear for instance in the Upright of the Herring Gull: with decreasing predominance of aggression, a bird tends to turn off sideways, and such a bird often walks up to another in a spiral, or even may walk around it in a circle. The Head-Flagging movement involves only the head in many postures of the Black-headed Gull; in the Common Gull it may involve the head alone, or the entire body, and Weidmann (1955) reports that all intermediates can be observed. The Forward of the Black-headed Gull often shows sideways orientation of the whole body; this is particularly clear in the pair formation sequence. Choking birds often turn their sides to each other, and in the Black-headed Gull it is not all uncommon to see two males turning round entirely until they face away from each other.

The effect of a growing avoidance tendency on the head is particularly interesting. While aggression makes a bird point the bill towards the opponent, preponderance of fear and perhaps other avoidance tendencies may result in four different movements which are all the opposite of the intention movement of pecking. (a) The head may be turned in the horizontal plane (Facing Away) (b) The neck may be withdrawn, while the bill still points towards the opponent (Hunched) (c) The bill may point upward (Bill-up Oblique, Bill-up Forward, Bill-up Upright, Upwards Choking). (d) The head can be bent down through the vertical until the bill points down and away from the opponent (Facing Away or Beak-Hiding of the Kittiwake).

These components may appear as parts of displays in their pure, nonritualised form, or they may themselves have undergone ritualisation. For instance, while Facing Away in the Kittiwake is beautifully oriented every time a Kittiwake shows it, in the pair formation displays of the Black-headed Gull the movement is fixed to the extent of having no orientation in the vertical plane, although in the horizontal plane the orientation is near perfect. As a result, no oriented downward movement of the head can be observed. It is possible that the head-down phase of the Herring Gull's Long Call is originally an aggressive movement; at the moment it does not seem to obey exactly the varying intensities of aggression, but is an integral part of the full Long-Call sequence.

The full displays

The Upright

The analysis of the form of the Upright shows its origin more clearly than that of any other posture. It must have originated as a mosaic of intention movements of attack, inhibited by escape tendencies. The two attack components recognised are (1) stretching the neck and pointing the bill down as a preparation to a downward peck, and (2) the lifting of the carpal joints as the preparation to a wing-beat. The posture seems to be little if at all ritualised.

Two comparative data support this derivation. First, the Kittiwake does not show the Aggressive Upright. Neither however does it have the downward peck as a part of its attack pattern. Second, the Great Skua, while having the Upright, never lifts the carpal joints in the Aggressive Upright. But neither does it fight with wing beating (Perdeck, 1958 and in litt.). This together with the fact that in other gulls the lifting of the carpal joints is the specific consequence of aggressive motivation, supports the derivation of this element of the Aggressive Upright.

The Oblique

This display, while showing great interspecific variation is, in all species studied, very dissimilar from any other display. Moynihan (1955a) suggested that it may have been derived from an inhibited forward-attacking movement. I agree that the presence of attack and escape tendencies, as revelated by a study of its motivation, accounts for (a) its orientation—the bill is aimed, at least in the horizontal plane, at the opponent; (b) the frequent lifting of the carpal joints; and (c) the fact that the Oblique is done while stationary, less often during an actual advance—if during an advance (as when it is adopted in the air) the attack is rarely driven home. All these elements however are present and in fact make up the Jabbing posture of the Kittiwake, the equivalent of which can be found in other gulls as well. But the Oblique has in addition features such as the stretching of the neck and the wide opening of the bill. These, I suggest, are the mechanical corollaries of the loud and long-drawn Long Call. It would thus seem that the Oblique is an aggressive advancing movement inhibited by fear, combined with and modified by the Long Call.

As described earlier, the Oblique is, in many species really a sequence of different postures. After an initial obliquely upward pointing phase (present in most though not all species) the head is bent down (most clearly in the Herring Gull group) then thrown up (either once as in the Herring Gull and—more emphatically—in the Common Gull, or repeatedly as in the Ring-billed Gull), after which a series of long calls is given during which the head gradually comes down to the obliquely forward position. Since, in general, pointing the head down reveals in most gulls a preponderant aggressive motivation, and head-up components a tendency to withdraw, it is possible that the phases of the polyphasic Obliques may have taken their origin in corresponding slight changes in motivation during the performance, and may then have been exaggerated during ritualisation (which in this posture seems to have promoted interspecific divergence, perhaps because it serves as song).

Of particular interest is the Throwback phase, since in some species it is remarkably similar to Head Tossing. Such species as Franklin's Gull suggest that the Throwback and Head Tossing have the same origin, and that their dissimilarity in the Herring Gull is the consequence of later divergence between the song, and the precopulatory display.

The Forward

Up till 1957 I assumed that the Forward of the Black-headed Gull and related species was homologous with the Jabbing of the Kittiwake, and was a slightly ritualised version of the intention movement of a forward-aimed peck. Moynihan (1958a) however pointed out that the Forward is more similar to the Low Oblique. Certainly the facts that the Forward is accompanied by a muffled form of the Long Call; that in Franklin's Gull and the Laughing Gull the Oblique in a complete Long-Call performance grades into a posture much like the Forward; and that the head plumage is not ruffled in the Forward as it is in the Jabbing and in the Gakkering performances of the Ring-billed and Franklin's Gull would indicate that the Forward is more likely to be derived from the Low Oblique.

The Little Gull's Vertical Display (Moynihan, 1955a) is probably the homologue of the Forward. Although its form is rather different from the Forward, it is not so different from the Low Oblique from which the Foward seems to be derived. It differs from the Bill-up Upright, with which it has a certain formal similarity by the extreme vertical position of the neck, and by the fact that, like the Forward of the Black-headed Gull, it is accompanied by a muffled version of the Long Call. Further it appears in the situations in which the Black headed Gull assumes the Forward. Finally, the Bill-up Upright is known in the Little Gull, where it is extremely similar to the Bill-up Upright of the other gulls.

Choking

The origin of Choking has been the subject of controversy. While Portielje (1928) and Kirkman (1937) pointed out that in the Herring Gull and the Black-backed Gull respectively, the movement was rather similar to that of regurgitation, I aruged (1953a) that in the Herring Gull it is in several respects more similar to the movement of depositing nest material. Since it appeared to be causally and functionally out of context (it is caused by a conflict between aggression and fear, and it is not a fighting movement but one belonging to another functional pattern), I have called it displacement depositing of nest material. Since then, however, Manley (pers. comm.) has observed the Black-headed Gull more closely, and has seen that at least in some situations (particularly when a male Chokes at the nest site in the presence of a female) regurgitation is a common part of Choking and he considers that Choking must be derived from the regurgitation movement.

Comparative data so far known seem to suggest the following interpretation. Choking in all species, begins with bending down and pointing the bill down. Further, it is always stimulated by the nest site. Thirdly, the initial bending-down movement is taxonomically much more widespread than the full Choking movement: neither Skuas (Perdeck, 1958; Pitelka, in litt.), nor terns (Tinbergen, 1931; Baggerman *et al.*, 1956; Bergman, 1953; J. M. Cullen, unpubl.) nor most waders (von Frisch, unpubl.) show anything resembling rhythmic Choking, but many if not all of them stand in a bent posture over the nest or scrape in situations similar to that in which a pair of gulls perform Choking (though not in boundary conflicts). Finally, while in the Kittiwake Choking closely resembles depositing of nest material, in other gulls the rhythmic repetition is much more pronounced in Choking than in depositing, and in the Black-headed Gull Choking is often followed by regurgitation. Further, the Black-headed Gull often alternates the rhythmic Choking call with the Mew or Crooning Call.

In all species where Choking occurs, except in the Kittiwake the rhythmic repetition of the down-up movements of the head is much more prolonged in Choking than in either depositing or regurgitation, though both too often involve some such movements. This prolonged repetition must be a result of ritualisation; rhythmicity is a common aspect of derived releasers (Morris, 1957).

All these facts combined render it likely that Choking is really a combination of two movements. The primary, taxonomically widespread movement is bending down over the nest site. This is very similar to what all these birds do prior to settling down on the eggs and to turning them. It therefore is most probably an incubation movement, which (like all incubation movements, Beer, unpubl.) appears well before the eggs are laid. This is supported by the fact that the posture is facilitated by the presence of the nest or scrape.

Once in this position, a bird which finds itself in a conflict situation (as it is when in the presence of either a female or an opponent) is likely to perform any movement which is facilitated by this initial movement. It is becoming clear from various studies that this type of facilitation (whether by external stimuli, or by proprioceptive stimuli, or by "central nervous commands" has not been analysed) may dictate which movement shall follow (J. M. Cullen, unpubl.; Lind, 1959). Both depositing and regurgitation begin with a head-down posture, and thus might be equally ready to appear. Is it possible to decide which of the two is the most likely origin of the rhythmic second phase of the Choking?

The fact that in the Kittiwake Choking resembles depositing so closely is due to the fact that depositing in this species involves very long series of down-up head movements. But this is doubtlessly due to the stickiness of the nest material, while the prolonged rhythmic Choking, found in all species, must be considered an aspect of ritualisation. Thus the similarity, in the Kittiwake, between depositing and Choking has probably developed under the influence of two different kinds of selection pressure and so must be discounted as an argument relevant to the origin of Choking.

J. M. Cullen (in press) has reported that the Black and Brown Noddy (*Anous stolidus* and *A. tenuirostris*) make a movement resembling Choking. The fact that these species are the only known species of terns which build nests and carry material to them and which regurgitate confirms the derivation from one of the two movements, but does not allow a decision between them.

The Great Skua does not have Choking (Perdeck, 1958). Neither does it carry material to its nest (it merely scrapes and builds sideways), but it does feed the female and the young by regurgitation. This points towards depositing as the origin of Choking, and pleads against regurgitation.

It seems most likely therefore that the rhythmical movements of Choking have been derived either from depositing or from regurgitation, or even from both, since either or both are facilitated by the initial bent posture. It is quite possible that it is in principle incorrect to say that it is derived from either one or the other, it may rather have been derived from the down-up element common to both. The prolonged rhythmicity must have developed as a further ritualisation of the initial rhythmicity.

As G. Manley (in litt.) points out, this view implies that Choking has originated in situations near the nest site. It is also widely used as a threat posture in boundary disputes. While it is true, and consistent with our view, that even in hostile encounters it is done almost exclusively on the territory and not far less on communal grounds, hostile Choking is often done many yards away from the nest. Here the initial posture may have to do with the general tendency to bend the neck and head down which normally appears in aggressively motivated Gulls, and this posture (although taken up for different reasons than at the nest site in the presence of a female) may facilitate the rhythmic component just as looking into the nest does. This suggests that hostile Choking has developed through "friendly" Choking.

In some species hostile Choking and friendly Choking show some interesting differences. Hostile Choking in the Black-headed Gull is often done with the wings spread; this occurs just after a sudden dash at an intruder. This is observed only very rarely in friendly Choking, and probably only in males which show a high level of aggressiveness in other ways as well. Neither does hostile Choking as a rule, develop into regurgitation. Noble and Wurm (1943) report that the Laughing Gull ruffles its plumage during hostile Choking which it does not do in friendly Choking. These are elements which probably have to do with the different situations in which the two postures occur, and thus with differences in motivation; in friendly Choking sexual tendencies may contribute which are absent in hostile Choking.

Upward Choking

While it seems plausible to assume that Upward Choking has been enhanced by selection favouring an appeasement movement clearly different from the preceding postures in the meeting ceremony, the origin of the upward posture of the head is obscure. As I pointed out, there is a tendency to point the bill either up, sideways, or deep down, or a tendency to withdraw the neck, in all postures where a tendency to move away is present, or at least a tendency not to point the bill at another bird (which is an aggressive component). This may have made the Kittiwake point its bill up at the end of a meeting ceremony, and the Choking might thus continue with the bill pointing more and more up. At the same time, the bill is closed; this correlates with the waning preponderance of the aggressive tendency, which latter tendency in the Kittiwake is always expressed by opening the mouth.

Another possibility is that the Choking is derived from a swallowing movement; this would be in line with the idea that Choking itself is originally regurgitation. However, normal swallowing does not usually involve an upward movement of the head. It seems to me most probably therefore that the upward tendency is essentially the same as that in several other postures, and indicates a growing preponderance of non-aggressive motivation.

The Mew call

The origin of this call and posture is unknown.

Pecking-into-the-ground

In 1953 I concluded, from the obvious similarity of the Grass Pulling of Herring Gulls to collecting of nest material, that it was "displacement collecting of nest material." I pointed out, however, that the movement was combined with actual attack movements: The pecking was as vigorous as in actual attack and much more vigorous than even seen in actual collecting; and if the bird got hold of a firmly rooted plant, it would pull at it as hard as it pulled at an opponent once it has got a grip on its bill, tail, or wing, I described that by writing that actual attack movements were superimposed on the displacement activity. A. Kortlandt pointed out to me (in litt.) that one could just as well and perhaps even better reverse this statement, and say that the displacement activity is superimposed upon attack movements. Our later observations on other gulls suggest that Kortlandt's view is correct. Many species of gulls give vigorous pecks at the ground in hostile and pair formation situations, and in the latter such pecks occur when the bird shows other signs of aggressiveness. Even in the Herring Gull group Grass Pulling usually begins with a vigorous peck. But the Black-headed Gull never or very rarely handles the nest material after such a peck; it fairly regularly performs bill wiping, extremely similar to the movement by which it cleans the bill after feeding in muddy soil or after regurgitation, when dirt or food sticks to the bill.

This points to a similar interpretation as that suggested for Choking: Grass Pulling is a dual movement, made up of an initial component (pecking down) which has a wide taxonomic distribution, and a subsequent displacement activity which is different in different species, but which in each case is facilitated by the initial posture. The initial movement is clearly recognisable as an attack movement. It is however "redirected" (Bastock, Morris and Moynihan, 1953; Moynihan 1955b): Instead of attacking the opponent or partner, the bird attacks the ground. This may seem rather odd, but it is less so when one remembers (a) that it is only little more extreme than the redirected attacks aimed at birds which themselves did not arouse the bird's aggressiveness and which were tolerated until another bird provoked (and yet inhibited) aggression, and (b) that we ourselves, when under the control of irrational anger and simultaneous inhibition (by fear, or by social pressure) do something very similar; we bang the table with out fist, or kick a chair. Allowing for the different motor patterns of attack in Man and gulls, I submit that the deeper causation (inhibited irrational anger) is the same in both cases. In Pecking-into-the-Ground therefore we have an example of an "autochthonous," but complete and redirected movement, followed by a movement originally belonging to another system (nest building or feeding) which is facilitated by the initial posture.

I cannot suggest an explanation why the Black-headed Gull should follow the initial movement by bill wiping, and the Herring Gull by sideways building, but it may be no accident that this difference runs more or less parallel to that in Choking. Nest building movements seem, for some as yet unknown reason, to be more readily available in the Herring Gull group than in the Black-headed Gull, which is more inclined to regurgitate and to wipe its bill (which is often associated with regurgitation).

In the Kittiwake Pecking-into-the-Ground is usually done on the nest, and it is often followed by picking up nest material. A bird starting to peck while on the top of the cliff may even fly to the nest and take up nest material there.

The Hunched

The striking features of the Hunched are its withdrawn neck and the orientation of the bill, which is usually aimed at the other bird. It seems probable that withdrawing the neck is the intention movement of total withdrawal, or alternatively of withdrawing the vulnerable face.

Head Tossing

The fact that this occurs in both the large young and the female when soliciting food suggests at least one step towards a solution: In the female, it may be nothing, originally, but an infantile movement which reappears at a later stage of life. Certain observations on human behaviour (see e.g. Prechtl, 1950) have shown that infantile movements under certain conditions occur in adult life, and I suggest therefore that this may have been the origin of the Head Tossing in the female. The Upward component may have to do with a fear tendency and with aiming at the bill of the male (parent), which is the source of the food.

The response of the male to the female's Head Tossing is also interesting: It feeds the female in exactly the same way as it feeds the young. This is another example of a movement moved forward in time, which in this case extends much further back in time than the bending movement which initiates Choking; another interesting point is that in the incubation period, which is between the period of intensive courtship feeding and parental feeding, it practically disappears, though to what extent this is due to lack of stimulation I could not say. The phenomenon of courtship-soliciting and courtship feeding is of course widespread in birds as well as in other animals (see Meisenheimer, 1921; Lack, 1940); in birds it is, however, less widespread than feeding of the young, which therefore must be considered primary. According to the definition of displacement activities given above, both Head Tossing by the female and courtship feeding must be diagnostically labelled as such: the detailed reasons why such irrelevant movements occur in the various examples given may of course be different in each case and further analysis is certainly needed; but the use of the distinction between intention movements and redirected movements on one hand and displacement activities on the other hand seems obvious and useful. A further point of interest is that so many displacement activities appear to depend on the primary occurrence of an intention movement which facilitates the performance of the "displacement activity."

Facing Away

We have seen that Facing Away signifies a fear component which is in conflict with a tendency to stay which is more general than a specifically aggressive tendency. Weidmann's observations on the Common Gull, which show many gradations between mere Facing Away and completely walking off or flying away strengthens the conclusion that Facing Away originated as an intention movement of turning in order to flee.

In conclusion, I would like to stress that it is clear that in the absence of historical documentation the interrelations with so many other kinds of evidence, certain hypotheses can be rated as more probable than others on the basis of the degree of consistency with known facts. Extension of these studies over more species and more intensive work on motivation and function will widen the range of facts which can be checked for inconsistencies. It is worth pointing out that work on the evolutionary origin of taxonomically related human languages is faced with very much the same situation and tries to arrive at conclusions in essentially the same way as the comparative ethologist (Thieme, 1958).

SOME EVOLUTIONARY ASPECTS

Interspecific similarities

The facts so far known show that the general behaviour, the organisation of the breeding colony, and the individual displays are very similar in all gulls. Taxonomically these characters considerably expand the total range of characters and thus this similarity strengthens the case for grouping all the gulls together and setting them off from other groups. Phylogenetically the similarities in general behaviour patterns, in the form of displays, in their causation and function and in their probably origin, together with their wide and often continuous occurrence in the whole group strongly support the conclusion that the gulls are a monophyletic group.

Within the groups, the similarity is on the whole greater between the species of sub-groups which have already been recognized as such on the basis of morphology. Thus behaviourally the large gulls form a homogeneous group, from which the hooded gulls and the Kittiwake are each set off sharply; within the large gulls, the Common Gull (and the Ring-Billed Gull) are to be separated from the Herring Gull group. Within the hooded gull group, the Black-headed Gull and its closest relatives Hartlaub's Gull, the other Silver Gulls and probably Franklin's and Bonaparte's gull are a homogeneous set rather different from the other hooded gulls. The present paper will not deal with the taxonomy; a revision by Moynihan is in preparation.

If the gulls are a monophyletic group, it must be concluded that the individual displays are each the result of divergent evolution started from a common root, i.e., that they are the "same" in the sense in which Cain and Harrison (1958) use this term*. They have each descended from either a display already possessed by the common ancestor, or have developed parallel after the ancestral species had broken up.

The more widely occurring, the more continuously distributed, and the more similar a display is throughout the group, the more probable I consider this common origin to be. Thus all the Obliques are undoubtedly of common descent, and so are all forms of Jabbing, of Choking and of Facing Away. It is also very probable for the Upright; but it is less certain for the Mew Call and Posture; and the common origin of the Forward of the Black-headed Gull and the Vertical of the Little Gull is still less certain.

Functional studies, which, together with the establishment of occasional, discontinuous occurrence, would give us a lead to possible convergences, have not so far revealed any clear case of this, although for instance the Large Hooded Gull (*L. ichthyaetus*) might well be either a Herring Gull which has developed a hood, or a hooded gull which has increased its size, and in either case could be expected to have developed some behavioural convergences.

Intra- and interspecific differences

As described earlier these are both considerable. Our study of the functional aspects has allowed an assessment of the adaptive significance of these differences. This assessment leads to the conclusion that four classes of selection pressure have been influential in the evolution of the displays: (1) intraspecific differences due to selection pressure on each display separately; (2) intraspecific differences due to selection towards interdisplay distinctness; (3) interspecific differences directly selected for; and (4) intraspecific differences developed as a consequence of pressure promoting differences in other functional systems.

1. As argued previously there is reason to suppose that certain postures derive some of their peculiarities from pressure favouring conspicuousness as such. Since

---

* Deliberately avoid the term "homologous" although of course it is often used in the same sense.

each posture has developed its own way of achieving this, the increase in conspicuousness has led to intraspecific divergence without selection for distinctness per se. Thus the Oblique-cum-Long-Call is directly adapted to the function of giving a loud, long-distance, advertising call, and this has indirectly (so to speak "accidentally") enhanced its difference from other calls. Choking has developed its prolonged rhythmicity as a means of making the movement more conspicuous and this itself has helped to make it strikingly different from other postures. The Forward may well have become so pronounced and so frequent in the hooded gulls as a correlate to the development of the dark facial mask. The Kittiwake opens its mouth widely in all threat postures as a correlate to the bright orange inside of the mouth: the Great Skua spreads its wings during the Oblique as a correlate to the possession of white wing patches. (By pointing this out, no pronouncement about the evolutionary priority of either changed movement or changed colour is intended—although occasional spreading of wings, occasional opening of the mouth, and occasional adoption of a "Low Oblique" is much more wide spread taxonomically than the bright colours mentioned, and in this sense the original movement is of course older than the bright colour). Alarm calls have developed a staccato quality and rhythmicity.

All these changes are to be considered mere improvements of movements functioning as signals; having started from different origins their increased conspicuousness has been achieved in different ways, and thus their distinctness from other signals has been increased as a by-product of selection for conspicuousness alone.

2. Other differences however must be understood as the outcome of selection pressure favouring differences between displays for the sake of their distinctness. This is perhaps most obvious where two sets of postures serve diametrically opposite functions. Thus all distance-increasing postures of a species may as a group show striking differences from all distance-reducing postures. For instance, all threat postures of the Kittiwake involve wide opening of the mouth, but in all friendly postures the mouth is shut even in those postures (such as Head Tossing) where other species might open the bill. We must assume that the need for displaying the orange mouth in hostile postures has a corollary the need for concealing it in non-hostile signalling. A still more specific example seems to be the intraspecific difference which can as a rule be observed between the appeasement postures at the end of the meeting ceremony and the preceding postures.

In general, the need for unambiguity may well have enhanced the mutual inter-display (intraspecific) distinctness in all displays; one of the means by which this seems to have been achieved is the development of a "typical intensity" (Morris, 1957) which has rendered intermediates between postures relatively rare.

3. The interspecific differences too may have developed either as a consequence of direct pressure towards distinctness, or in a more indirect way. The most likely direct pressure can be expected to have been influential with pair formation displays, for distinctness of these must promote sexual isolation. There is some evidence of this, although the situation is by no means as clear as, e.g., in the Anura studied by Blair (1955, 1956) or the grasshoppers analysed by Perdeck (1957),

where striking differences in song have been developed. The fact that gulls show far less, pronounced differences of this type is no doubt due to the circumstance that indirect pressure has already made some advertising displays of gulls sufficiently different (see below). However, it is possible that some differences may have to be explained as results of direct selection for interspecific distinctness. For instance, the Long Calls of most species are strikingly different. This is even true within the Herring Gull group, where for instance the Long Call of the Western Gull and the Glaucous-winged Gull are different from that of the Herring Gull, whereas all other displays of these three species are identical.

However, the entire problem of sexual isolation among gulls is complicated, since it seems possible that other factors are responsible as well, such as wing-tip patterns (which seem in general to be interspecifically different within each fauna), size, habitat, and mantle colour; this certainly requires further study. Species-characteristic selection of breeding habitats, which on the whole separates all species, is often overruled by gulls (and terns) being strongly attracted to any gathering of white birds, and occasional visits of individual gulls to colonies of other species are by no means rare: I have myself observed a Herring Gull in an Iceland Gull colony which went through some threat and pair formation displays; in a colony of Hartlaub's Gulls I observed a male Grey-headed Gull (*Cirrhocephala cirrhocephalus*) which displayed to a female Hartlaub's Gull; European Herring Gulls and Lesser Black-backed Gulls often breed in mixed colonies (even with occasional interbreeding); Kittiwakes have known to breed occasionally in Black-headed Gull colonies.

4. Some, and perhaps many, differences in displays seem to be the outcome of changes in other behaviour patterns. The best known example of this is the Kittiwake (E. Cullen, 1957). Mrs. Cullen has shown that many of the Kittiwake's aberrant traits are related to its cliff breeding habit. Her conclusions can be summarized by the following list, which enumerates and groups the characteristics which she relates to cliff breeding.

A. Relaxation of other ways of defense

1. Tameness
2. Alarm call rare
3. Predators not attacked
4. Chicks not camouflaged
5. Defecation on nest rim
6. Egg shells not carried away

Breeding habitat: Narrow ledges on very steep cliffs = protection from predators.

B. Protection against falling off cliff

1. Strong claws and foot musculature
2. Female lies down during copulation
3. Deep nest cup
4. Two eggs

5. Immobility of chicks
6. Chicks face "wall"
7. Facing Away in chicks; black neck band

C. *Fighting*

1. No Upright
2. Special fighting technique ("twisting")

D. *Formation of pairs*

1. Choking as song
2. (Upward Choking)

E. *Nest building*

1. Collecting of mud; trampling it down
2. Stealing of nest material
3. Guarding the empty nest

F. *Nest sanitation*

1. Incomplete regurgitation

G. *Parent-infant-relationships*

1. Parents lack food-call
2. Chicks lack food-call
3. Parents do not know chicks individually

The list contains five aspects in which the displays of the Kittiwake are peculiar. First it is the only species which uses the Choking as song, although it has the Oblique. Cullen argues that this may have come about through the fact that the Kittiwake's territory is nothing but the small ledge on which the nest is built. The Kittiwake therefore performs the entire pair formation sequence actually on the nest or the future nest site. We have seen that in all gulls Choking is strongly facilitated by the nest site, and this may originally have favoured Choking as the first threat display to be shown upon the approach of other birds. How this phenotypic effect has been incorporated as an "innate" (environment-resistant) character is of course an open question.

Now, in turn, the fact that the first part of the Kittiwake's meeting ceremony consists of Choking, Long Call, and Choking once more may, as we have seen have been responsible for the specific appeasement posture "Upward Choking"; the demand for intraspecific distinctness may have made this appeasement gesture so different from that of other gulls. (The fact that the large gulls follow Choking by Facing Away may have to do with the fact that the Choking is not shown as invariably as by Kittiwakes, nor do they necessarily meet very near a nest site.)

Another difference is the absence of the Aggressive Upright in the Kittiwake. This no doubt is due to the fact that the movement from which it has been derived, the peck-from-above, is also absent; instead, the Kittiwake pecks straight at the opponent without attempting to bring the bill above him first, and its bill fighting consists mainly of an attempt to throw the other bird off the cliff by twisting.

Thirdly, the alarm call has a greatly heightened threshold, as have all the direct anti-predator responses.

Fourthly, the Kittiwake lacks the "pumping" movement by which the young of other gulls attract their parents' attention when begging; the Kittiwake has no need of this, since the parent alights on the nest anyway.

A fifth example is the occurrence of Facing Away in Kittiwake chicks, which tends to prevent fights for food between the chicks. Other gulls have not got this appeasement gesture in the chicks; when fed a young just runs away from the others.

The evidence therefore suggest that, while many displays have undergone specialisation in the direction of their improvement as signals per se, resulting in increased conspicuousness, their divergence in evolution seems to be adaptive in three other respects as well: there must have been pressure towards intraspecific unambiguity, pressure directly favouring interspecific unambiguity and indirect pressure through other behaviour, either displays or nondisplays.

This analysis of the adaptive character of divergence now allows us to "strip" species of many of the effects of adaptive radiation and thus, by way of final check, compare the "cores" of what must be assumed to be older characters. As argued earlier these cores should be more similar than the total complex of characters if the species have descended from a common ancestor. This is actually so in the gulls. When the Kittiwake's Choking is stripped of the adaptive peculiarity (mouth open) it becomes more, and not less similar to that of other species. The Head Tossing, in which no adaptive differences have been discovered so far, is on the whole only very slightly different throughout the group. Once it is realised that the jerkiness of Head Flagging in the Black-headed Gull and its perfect orientation may have to do with the emphatic "not-showing" the brown mask (which is shown off well by the Forward), and that the presence of Facing Away in Kittiwake chicks is a corollary of its peculiar nesting habitat, this display is much more homogeneous throughout the group than it seemed before a functional appraisal was made. Thus "peeling off" the obviously adaptive differences strengthens still more the conclusion, already drawn from a mere description of taxonomic characters, that the gulls must be a monophyletic group and that the displays which were given a common name are actually the same.

Convergences with other species

The preceding has some repercussions upon the use of behaviour characters in taxonomy. Lorenz (1935, 1943) has stressed that displays are often more useful as taxonomic characters than other behaviour. The close similarity of many displays throughout the Laridae corroborates that they can be useful. However, it is easy to overrate this; as the above review of trends promoting evolutionary change emphasises. Considering the relatively small differences in, for instance, feeding behaviour and nest building, one could even argue that the displays have behaved rather less conservatively than these patterns. On the average, the truth is probably that displays are neither more nor less conservative than other behaviour patterns.

The recognition of the functional significance of many interspecific differences in displays may also lead to the recognition of the adaptive nature of similarities between distantly related species, and thus warn against the danger of overlooking possible convergences. For instance, Marler (1955) has shown that the striking similarities in the alarm calls of various small song birds are due to common adaptation to the function of being audible and yet not allowing localisation to the predators of the size which usually prey upon these species. The following illustrates in a slightly different way that similarity of display need not be due to common derivation, unlike (to quote Lorenz's example, 1953) the similarities between words of human languages, such as "father-Vater-vader-padre-père." Appeasement and related gestures, such as the more "defensive" threat postures may be very similar in species not at all closely related. In birds which attack with the bill (which itself is doubtlessly a very old character, retained by many different groups), threat postures are often derived from intention movements of this, and involve pointing the bill at the opponent. It is therefore not surprising that postures such as the Forward are found in many birds. Defensive threat postures and appeasement postures, as we have seen, often involve something like doing the opposite of the intention movements of pecking. Now this can in principle be done only in four ways. (1) The head can be turned in the horizontal plane, as in the Facing Away of the Black-headed Gull. (2) It can be turned up in the vertical plane as in the more anxious forms of the Oblique and the Upright, the Upward Choking of the Kittiwake, and the Vertical of the Little Gull. (3) It can be turned down until the bill points backward, as is often seen in the Facing Away of the Kittiwake. (4) Still another way is extreme withdrawal of the bill while it is still pointing at the opponent, and this is typical of the Hunched.

I suggest that it is because of the limited number of possibilities of doing "the opposite" to showing preparedness to attack, that we find some curious similarities in the definsive threat and appeasement postures of widely separated species. Bill-up postures in the extreme form of a "stretch" are found in such different species as terns, gannets and the Great Tit (Fig. 16). Bending the neck and facing down or even backwards is found not only in Kittiwake but also in Jackdaws and Coots. Facing Away in a manner more similar than to that of the Black-headed Gull is shown by the Common Crane (Heinroth, 1928; Lorenz, 1935) and by various ducks (Lorenz, 1941). There can be little doubt that these movements, either as defensive threats or as appeasement postures (they can probably not be sharply separated) occur in many more species (see also Moynihan, 1955b). Nor need convergences or independent origin of similar displays be rare; it seems quite possible that we are only just beginning to understand their interest and therefore to look for them. Comparison between more widely separated species, with functional studies, would be required to pursue this. It would seem then, that the chance of convergent similarities in human speech is so much smaller because the number of possible words is so infinitely larger than the number of possible simple appeasement gestures.

Fig. 16. Head-up postures in (a) Hartlaub's Gull (Anxiety Upright); (b) Cape Gannet (2 males in unbalanced clash); (c) Great Tit; (d) Swift Tern. (*a* and *d* after Tinbergen & Broekhuysen 1954; *c* after Hinde 1952).

Evolutionary changes in behaviour "machinery"

Having reviewed and classified the alleged evolutionary changes from the point of view of survival value I will not try and classify them according to the changes in underlying causal organisation which must have been involved. Again, this will be based on the observed intra-and interspecific differences of displays and on the intraspecific differences between displays as derived movements and their presumed origin. Naturally, since so little is known about the behaviour machinery of these displays, this analysis can at best do no more than distinguish between what may (but need not) be different types of change. I believe it is useful to stress the difference between this type of classification and that according to survival value and selection pressure.

1. The responsiveness of many movements and complex patterns to outside stimuli must have changed a great deal. This is true for instance of habitat selection: Each species selects its own breeding, feeding and roosting habitats. Selective responsiveness to conspecific individuals and categories, to nest material, to food, etc., is also different in each species. The critical distance to evoke territorial hostility is much smaller in the Kittiwake than it is in the Herring Gull, whereas the critical distance in winter is in each species merely proportional to the size of the species.

It is true that, since conditioning occurs so widely in gulls, the innate character of many of these differences is as yet uncertain, but in some instances there are at least indications that we have to do with hereditary changes, as mentioned earlier. Yet large-scale exchange of eggs between different species (to study the effect of the environment typical for other species) would be worth undertaking.

2. Purely quantitative shifts in the responsiveness of a movement or a pattern (threshold shifts of some kind) have occurred in both directions. Attacks on predators, escape from predators, and alarm call are either absent or only shown at the strongest provocation by the Kittiwakes, at least as long as they are on the breeding cliffs. The readiness to fly is generally much greater in small than in large species. Young Kittiwakes have strong inhibitions against leaving the nest, and, though less so, against wing flapping. In extreme cases movements have almost or even entirely disappeared, or new movements have appeared. Thus the Kittiwake has lost the aggressive pecking-from-above and with it the Aggressive Upright; but it has "gained" trampling as part of the nest building (E. Cullen, 1957). In the Great Skua lifting and spreading of the wings has been much facilitated in the Oblique. The same species lacks (i.e. must have lost) the wing-beating, and consequently the lifting of the carpal joints in aggressive displays.

3. The amplitude of certain movements or parts of movements has often been increased during ritualisation. Thus the Throwback during the Long Call is extreme in the Common Gull; the opening of the mouth in threat postures, especially Choking has been exaggerated in the Kittiwake. The Black-headed Gull has exaggerated the lifting of the carpal joints; the same species has increased the bending-down in Choking so that it stands in an almost vertical position. Changes of the types (2) and (3) together account for much of the "exaggeration and simplification" which so many derived movements have undergone during ritualisation.

4. Rhythmic repetition has been evolved or prolonged in several signals: the Long Call, Choking, the Alarm Call. This type of change may of course well be closely related to change in amplitude.

5. On a higher level of integration elementary displays have often been combined into one, either simultaneously or as an internally linked succession. Thus the Black-headed Gull has combined the Upright with Head Flagging (both movements which can occur separately even in this species) into one movement shown at the end of the meeting ceremony. The Ivory Gull has fused what would seem to be the Forward and the Oblique into one rigid series. Several species have added Head-Tossing to the Oblique, but while the Laughing Gull Head-Tosses after the full series of Long Call, the Ring-billed Gull Head-Tosses after each separate call. Pecking-into-the-Ground is successively integrated with Grass-Pulling and sideways building in the Herring Gull group, with billwiping in the Black-headed Gull. The Bow-and-Moan of the Kittiwake seems to be made up of the Mew Call posture and a bending-down element. The Herring Gull's Long Call has a bending-down component which may well have been added in this way, although it is also conceivable that a bending-down tendency, weak at first, has developed into the extreme bend we observe now.

On a still higher level of integration long sequences have been formed and more or less fixed; the various pair formation ceremonies are examples.

6. The motivation underlying a posture as a whole may often have changed. This is certainly true, as Morris (1957) has shown, in all those cases where a "typical intensity" has been established, i.e, where the range of motivational states in which a display is shown has been widened; this is particularly obvious in displays such as the Forward, the Oblique, and Facing Away, and it may often account for the discontinuity between displays of the same species, intermediate postures being rare and often being gone through quickly. It seems possible that many pair formation displays, while originally motivated by sexual attraction mixed with aggression and fear, may have become incorporated more firmly in the sexual pattern and lost part of the original agonistic motivation. This may be so even in those cases where signs of overt aggression and escape are still observable in the meeting ceremony. It also seems that a female approaching a male in the Hunched is motivated sexually mainly, and has very little if any aggressive tendency, nor is fear very obvious in this situation. As reported by Tinbergen and Broekhuysen (1954) the proportion of aggression and fear underlying the Forward is slightly different in the Black-headed Gull than in Hartlaub's Gull.

7. The speed with which movements are performed can change. Thus the Facing Away in the meeting ceremony of the Black-headed Gull is done much more suddenly than in the other gulls. Head-Tossing is repeated much more quickly by the Kittiwake than by the hooded gulls or the Herring Gull group. Extreme slowing down may result in "freezing": Good examples are the "Soar" of the Black-headed Gull and the Kittiwake, the wing spreading and lifting of the Great Skua during the Oblique, and the Vertical of the Little Gull.

On the whole these admittedly tentative and vaguely characterised types of change correspond well with those found in other groups (Blest, in press; Crane, 1957; Daanje, 1950; Hinde, 1955; Lorenz, 1941; Morris, 1957–1958). Yet there are some striking differences between gulls and some other groups. Feather raising, so conspicuous in, e.g., Gallinaceous birds, has not been used to any extent by gulls. Neither have conspicuous colour patterns developed except in a few cases, and gulls cannot compare with such groups as waders, ducks, or Gallinaceous birds. It is possible that this is because the white colouration of the ventral parts is of considerable survival value as aggressive camouflage (Craik, 1944; Tinbergen, 1953a); this, together with the absence of the need of breeding camouflage (which is found in the eggs and chicks) may also be the reason why colour dimorphism has not been developed. The lack of colourful signal structures may further be the reason why ritualisation has not been so extreme and has not led to such grossly distorted derived movements as in those more colourful groups. Further, although the smaller species indulge in a certain amount of aerial displaying, the larger species miss them almost entirely. Yet the hostile posture and the meeting ceremonies are both elaborate and often repeated. This may have to do with the need of a strong pair bond; it may also be due to the strong competition between the numerous members of a breeding colony. The size of the birds and their colonial nesting, both of which reduce predator pressure, may be the circumstance which have allowed the elaboration of the displays.

REFERENCES

Andrew, R. J. (1956), Some remarks on behaviour in conflict situations with references to Emberiza spec, *Brit. J. Anim. Behav.*, **4**, 41–45.

Andrew, R. J. (1956), Intention movement of flight in certain passerines, and their use in systematics, *Behaviour*, **10**, 179–204.

Andrew, R. J. (1957), The aggressive and courtship behaviour of certain Emberizinae, *Behaviour*, **10**, 255–308.

Assem, J. van den (1954), Waarnemingen over het gedrag van de Grotte Stern, *De. Levende Natuur.*, **57**, 1–9.

Baerends, G. P. (1950), Specialisations in organs and movements with a releasing function, *Symp. Soc. Exp. Biol.*, **4**, 337–360.

Baerends, G. P., Brouwer, R. and Waterbolk, H. Tj. (1955), Ethological studies on *Lebistes reticulatus* (Peters). I. An analysis of the male courtship pattern, *Behaviour*, **8**, 249–334.

Baggerman, B., Baerends, G. P., Heikens, H. S. and Mook, J. H. (1956), Observations on the behaviour of the Black Tern, *Chidonias n. niger* L., in the breeding area, *Ardea*, **44**, 1–71.

Bastock M., Morris, D. and Moynihan, M. (1953), Some comments on conflict and thwarting in animals, *Behaviour*, **6**, 66–84.

Beer, C. (Unpublished work on the incubation behaviour of the Black-headed Gull).

Bergman, G. (1946), Der Steinwalzer, *Arenaria i. interpres* L. in senior Beziehung zur Umwelt, *Acta Zool. Fennica.*, **47**, 1–151.

Bergman, G. (1953), Verhalten und Biologie der Raubseeschwalbe (*Hydroprogne tschegrava*), *Acta Zool. Fennica.*, **77**, 1–46.

Blair, W. F. (1955), Mating call and stage of speciation in the *Microhyla olivacea—M. carolinensis* complex, *Evolution*, **9**, 469–480.

Blair, W. F. (1956), Call difference as an isolating mechanism in south-western toads (Genus *Bufo*), *Texas Jour. Sci.*, **8**, 87–106.

Blest, A. D. (1959), The concept of "ritualisation," in *Modern Problems of the Behaviour of Man and Animals*, ed. W. H. Thorpe and O. L. Zangwell. Cambridge.

Cain, A. J. and Harrison, G. A. (1958), An analysis of the taxonomist's judgement of affinity, *Proc. Zool. Soc. London*, **131**, 85–98.

Craik, K. J. W. (1944), White plumage of sea-birds, *Nature*, **153**, 288.

Crane, J. (1957), Basic patterns of display in Fiddler Crabs (*Ocypodidae*, Genus *Uca*), *Zoologica N.Y.*, **42**, 69–82.

Cullen, E. (1957), Adaptations in the Kittiwake to cliff-nesting, *Ibis*, **99**, 275–302.

Cullen, J. M. (Unpublished work on the behaviour of the Shag).

Cullen, J. M. (In prep.). Observations on "Noddies" (*Anous stolidus* and *A. tenuirostris*). *Ibis*.

Daanje, A. (1950), On locomotory movements in birds and the intention movements derived from them, *Behaviour*, **3**, 48–99.

Drost, R. (1955), Neue Beitrage zur Soziologie der Silbermöwe, *Larus a. argentatus*, *Acta XI Congr. Intern. Ornithol.*, 564–569.

Frish, O. von. (Unpublished work on the behaviour of waders).

Goethe, F. (1937), Beobachtungen und Untersuchungen zur Biologie der Silbermöwe auf der Vogelinsel Memmertsand, *J. Ornithol.*, **85**, 1–119.

Goethe, F. (1955), Beobachtungen bei der Aufzucht junger Silbermöwen, *Z. Tierpsychol.* **13**, 402–433.

Goethe, F. (1957), Das Herabstarren, eine Ubersprungbewegung bei den Lariden, *Behaviour*, **11**, 310–317.

Heinroth, O. (1911), Beitrage zue Biologie, namentlich Ethologie und Psychologie der Anatiden, *Verh 5. Intern. Ornithol. Kongr.*, Berlin 1910, 589–702.

Heinroth, O. and Heinroth M. (1928), *Die Vogel Mitteleuropas*. Berlin.

Hinde, R. A. (1952), The behaviour of the Great Tit (*Parus major*) and some other related species, *Behaviour Suppl.*, **2**, 1–201.

Hinde, R. A. (1953), The conflict between drives in the courtship and copulation of the Chaffinch, *Behaviour*, **5**, 1–31.

Hinde, R. A. (1954), The courtship and copulation of the Greenfinch, *Behaviour*, **7**, 207–232.

Hinde, R. A. (1955), A comparative study of the courtship of certain finches, *Ibis*, **97**, 706–745; **98**, 1–23.

Hinde, R. A. (1956), Ethological models and the concept of "drive," *Brit. Jour. Philos. Sci.*, **6**, 321–331.

Hinde, R. A. and Tinbergen, N. (1958), The comparative study of species-specific behavior, In *Behavior and Evolution*. New Haven, 251–258.

Huxley, J. S. (1914), The courtship habits of the Great Crested Grebe (*Podiceps cristatus*); with an addition to the theory of sexual selection, *Proc. Zool. Soc. London, 1914*, 491–562.

Huxley, J. S. (1923), Courtship activities in the Red-throated Diver (*Colymbus stellatus* Pont.); together with a discussion on the evolution of courtship in birds, *J. Linn. Soc.*, **35**, 253–292.

Iersel, J. J. A. and Bol, A. (1958), Preening of two Tern species. A study on displacement activities, *Behaviour*, **13**, 1–88.

Kirkman, F. B. (1937), *Bird Behaviour*. London and Edinburgh.

Lack, D. (1940), Courtship feeding in birds, *Auk*, **57**, 169–179.

Lack, D. (1947), *Darwin's Finches*. Cambridge.

Lack, D. (1954), *The Natural Regulation of Animal Numbers*. Oxford.

Lind, H. (1959), The activation of an instinct caused by a "transitional action," *Behaviour*, **14**, 123–135.

Lorenz, K. (1935), Der Kumpan in der Umwelt der Vogels, *J. Ornithol.*, **83**, 137–213; 289–413.

Lorenz, K. (1939), Vergleichende Verhaltensforschung, *Verh. Deutschen. Zool. Gesellsch.*, 1939, 69–102.

Lorenz, K. (1941), Vergleichende Bewgungsstudien an Anatinen, *J. Ornithol.*, **89** (Festschr. Heinroth), 194–294.

Lorenz, K. (1953), Psychologie und Stammesgeschichte. In: G. Heberer, *Die Evolution der Organismen*. Stuttgart, 131–172.

Lorenz, K. (1955), Morphology and behavior patterns in closely allied species, *First Conf. Group Proc. Macy Found.*, 168–220.

Marler, P. (1955), The characteristics of some animal calls, *Nature*, **176**, 6.

Marler, P. (1956), The behaviour of the Chaffinch, *Behaviour Suppl.*, **5**, 1–184.

Meisenheimer, J. (1921), Geschlecht und Geschlecter im Tierreich, Jena.

Miller, N. E. (1944), Experimental studies of conflict. In: J. Mc V. Junt: *Personality and the Behavior Disorders*. New York.

Miller, N. E. and Murray, E. J. (1952), Displacement and conflict: learnable drive as a basis for the steeper gradient of avoidance than of approach, *J. Exp. Psychol.*, **43**, 227–231.

Morris, D. (1954), The reproductive behaviour of the River Bullhead (*Cottus gobio* L.) with special reference to the fanning activity, *Behaviour*, **7**, 1–32.

Morris, D. (1956), The function and causation of courtship ceremonies. In: Grasse: *Instinct*. Paris, 261–287.

Morris, D. (1956), The feather postures of birds and the problem of the origin of social signals, *Behaviour*, **9**, 75–113.

Morris, D. (1957), "Typical intensity" and its relationship to the problem of ritualisation, *Behaviour*, **11**, 1–12.

Morris, D. (1958), The comparative ethology of Grassfinches (Erythrurae) and Mannakins (Amadinae), *Proc. Zool. Soc. London*, **131**, 389–439.

Moynihan, M. (1955a), Some aspects of reproductive behaviour in the Black-headed Gull (*Larus r. ridibundus* L.) and related species, *Behaviour Suppl.*, **4**, 1–201.

Moynihan, M. (1955b), Some remarks on the original sources of displays, *Auk*, **72**, 240–246.

Moynihan, M. (1956), Notes on the behavior of some North American gulls I, *Behaviour*, **10**, 126–178; (1958a), II., *Behaviour*, **12**, 95–182; (1958b), III., *Behaviour*, **13**, 113–130.

Noble, G. K. and Wurm, M. (1943), The social behavior of the Laughing Gull, *Ann. N.Y. Acad. Sci.*, **45**, 179–220.

Paludan, K. (1951), Contributions to the breeding biology of *Larus argentatus* and *Larus fuscus*, *Vidensk. Medd. Dansk. Naturh. Foren.*, **114**, 1–128.

Paludan, K. (1955), Some behaviour patterns of *Rissa tridactyla*, *Vidensk. Medd. Dansk. Naturh. Foren.*, **117**, 1–121.

Perdeck, A. C. (1957), The isolating value of specific song patterns in two sibling species of grasshoppers, *Behaviour*, **21**, 1–75.

Perdeck, A. C. (1958), Jaarverslag van het Vogeltredstation over 1957, *Limosa*, **31**, 93–106.

Portielje, A. F. J. (1928), Zur Ethologie bzw. Psychologie der Silbermöwe, *Larus a. argentatus* Pont, *Ardea*, **17**, 122–149.

Prechtl, H. (1950), Zur Entstehung von Wackeltics, *Österr. Zs. Kinderheilk. und Kinderfürs*, **4**, 362–364.

Sauer, F. (1954), Die Entwicklung der Lautausserungen vom Ei ab schalldicht gehaltener Dorngrasmucken (*Sylvia c. communis* Latha.) im Vergleich mit spater isolierten und mit wildlebenden Artgenossen, *Z. Tierpsychol.*, **II**, 10–93.

Scott, J. P. and Fredericson, E. (1951), The causes of fighting in mice and rats, *Physiol. Zool.*, **24**, 273–309.

Strong, R. M. (1941), On the habits and behavior of the Herring Gull, *Larus argentatus* Pont., *Auk*, **31**, 22–50; 178–200.

Thieme, P. (1958), The Indo-European language, *Sci. Amer.*, **199**, 63–78.

Thorpe, W. H. (1958), The learning of song patterns by birds, with special reference to the Chaffinch *Frinquilla coelebs*, *Ibis*, **100**, 535–571.

Tinbergen, L. (1937), Feldeobachtungen an Zwergmöwen, *Larus minutus* Pall., *Limosa*, **10**, 12–21.

Tinbergen, N. (1932), Zur Paarungsbiologie der Flusseeschwalbe (*Sterna h. hirundo* L.), *Ardea*, **20**, 1–17.

Tinbergen, N. (1936), Zur Soziologie der Silbermöwe, *Larus a. argentatus* Pont., *Beitr. Fortpfl. Biol. Vogel*, **12**, 89–96.

Tinbergen, N. (1952), "Derived" activities; their causation, biological significance, origin and emancipation during evolution, *Quart. Rev. Biol.*, **27**, 1–32.

Tinbergen, N. (1952), A note on the origin and evolution of threat displays, *Ibis*, **94**, 160–162.

Tinbergen, N. (1953a), *The Herring Gull's World*. London.

Tinbergen, N. (1953b), *Social Behaviour in Animals*. London.

Tinbergen, N. (1954). The origin and evolution of courtship and threat display, In; *Evolution as a Process*. London, 233–251.

Tinbergen, N. (1955), Psychology and Ethology as supplementary parts of a science of behavior, *First Conf. Group Proc. Macy Found.*, 75–167.

Tinbergen, N. (1957). The functions of territory, *Bird Study*, **4**, 14–27.

Tinbergen, N. (1958). *Die Silbermöwe*. Berlin.

Tinbergen, N. and Broekhuysen, G. J. (1954), On the threat and courtship behavior of Hartlaub's Gull (*Hydrocoleus novae-holandiae hartlaubi* (Bruch), *Ostirch*, **25**, 50–61.

Tinbergen, N. and Hall, M. F. (1954), Hostile, sexual and other social behavior patterns of the Spice Finch (*Lonchura punctulata*) in captivity, *Behaviour*, **7**, 33–76.

Tinbergen, N. and Moynihan, M. (1952), Head Flagging in the Black-headed Gull; its function and origin, *Brit. Birds*, **45**, 19–22.

Tugendhat, B. (Unpublished work on conflict behaviour in sticklebacks).

Veitch, E. J. and Booth, E. S. (1954), Behaviour and life history of the Glaucous-winged Gull, *Walla Walla Coll. Publ. Biol. Sci.*, **12**, 1–39.

Weidmann, R. (Unpublished work on the behaviour of the Black-headed Gull).

Weidmann, U. (1955), Some reproductive activities of the Common Gull, *Larus canus* L., *Ardea*, **43**, 85–132.

B. STUDIES OF TRADITIONAL BEHAVIOR

As noted at the outset of this section, the study of traditional and culturally-transmitted behavior is a relatively new endeavor. Collected below are some studies that do more to open possibilities than to give a comprehensive viewpoint. The rather anecdotal observations of tits opening milk bottles has been appended by a recent note by Fisher and Hinde for this volume. Klopfer's experimental study of observational learning seeks to tap the abilities of the tits to learn traditional behavior from one another. Then a similar paper on a food tradition in mammals is presented. The collection concludes with an elegant study of traditional singing behavior in sparrows.

The Opening of Milk Bottles by Birds

JAMES FISHER AND R. A. HINDE

In 1921 birds described as tits were observed to pry open the waxboard tops of milk bottles on the doorsteps in Swaythling, near Stoneham, Southampton, and drink the milk. This is the first known record of an act which has now become a widespread habit in many parts of England and some parts of Wales, Scotland, and Ireland, and which has to date been practised by at least eleven species of birds.

The spread of the habit is interesting, because of the problems of behaviour involved. How far did the individual birds learn the habit from each other, or invent it for themselves? If most of them learnt it, by what process did they do so? How did, and how do, they detect the presence of food inside the bottle?

Proper answers to these questions can be obtained only from carefully controlled experiments on birds of known history—but the Research Committee of the British Trust for Ornithology supported the writers in their view that useful information could be derived from the collection of facts about the spread of the habit from members of ornithological societies and from the general public.

Questionnaires were distributed therefore to members of the British Trust for Ornithology. These were filled in by 126 members, giving 141 records of bottle-opening by species of tits. (Records of other birds are noted at the end of this paper.)

Reprinted in part from *British Birds*, volume 42, pages 347–357 (1949).

Eighteen replies were received from members of local Natural History Societies, and 43 as a result of appeals in the *British Medical Journal* and the *Lancet*. Thus about 200 records were obtained from ornithologists (mostly amateur) or from people with a scientific training. Over 200 further useful replies were obtained as the result of paragraphs in the daily or Sunday press or by personal interrogation: These confirmed the results of the preliminary enquiry.

In 223 cases the observer was able to state the year in which he first noted the habit, and to state also that he would have been in a position to note the habit in the previous year had it been occurring at his house. The remaining observers provided evidence of when and where the habit was known, but could not state the first year in which it occurred in their district.

* * *

The occurrence of the habit is naturally limited primarily by whether or not milk is distributed in bottles in the district. Thus the fact that the habit is unknown in most of the country districts of Ireland is undoubtedly due to the fact that bottles are not yet widely used there. There is, however, good evidence in many cases that the habit may not develop in a district for several years after the use of bottles has become almost universal there—this is known to have been the case in at least 23 out of 30 districts for which the year in which milk bottles were introduced is more or less accurately known.

Most British tits, and certainly the three species known to open milk bottles—the Great Tit (*Parus major*), the Blue Tit (*P. caeruleus*) and the Coal-Tit (*P. ater*)—are resident, and do not normally move, even in winter, more than a few miles from their breeding place. A movement of as much as fifteen miles is exceptional. It is probable that most of the birds which do move more than a few miles are in their first year. It would seem, therefore, that new centres and records more than fifteen miles distant from any place where the habit has been recorded previously probably represent new "discoveries" of the habit by individual birds.

The distribution of the records is consistent with the view that this source of food was actually discovered *de novo* by only a small proportion of the tit population, and was then passed on in some way to other individuals. Thus before 1930 there were ten records of the habit, in nine separate vice-counties, but at the present time there are very few isolated records. The majority of observations made after 1930 are either from near the pre-1930 localities, or else from independent areas in which the habit has subsequently become widespread.

By 1947 there were 89 records from places more than fifteen miles from any other place where the habit had been recorded previously. Seventy-four of these records were from England and Wales, twelve from Scotland and three from Ireland. In England and Wales, then, it seems likely that the habit has arisen *de novo* on an average at least once per vice-county, and may have arisen more often than this.

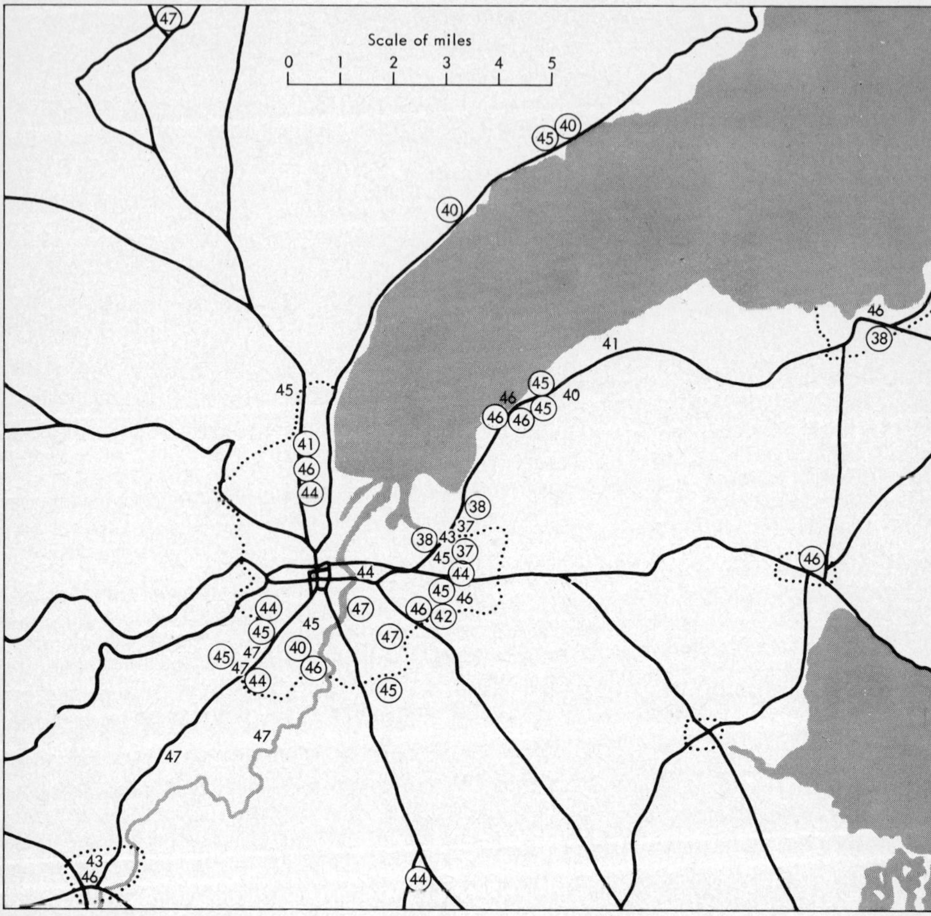

Fig. 1. Map of Belfast and District showing years in which opening of milk bottles by Tits was first recorded. Figures in circles indicate that the observer was present in the district in the previous year and so could have observed the habit if had occurred then. The dotted line indicates approximate boundary of the built-up area.

The spread of the habit from a number of centres is illustrated particularly well in East Anglia. The first record was at Lowestoft, Suffolk, in 1941; this was followed in 1942 by a further record from the same place. In 1943 there was another report from Lowestoft and three from Norwich. In 1945 there was a record from King's Lynn, and in 1946 a further Norwich record and four county records. In 1947 there were eight scattered county records and the habit was apparently widespread in most districts. In 1943 there was a record from near Ipswich, and by 1947 there were ten more records from within fifteen miles of the same place.

Reports from Ireland indicate a similar spread. Forty-seven records were obtained from Belfast and district: two for 1937, three for 1938 and forty-two for subsequent years. The two records for 1937 and two of the three 1938 records were all from points lying within an area of less than one third of a square mile in the northeastern suburbs of the city. There are indications of a fairly orderly spread from this area to other suburbs and neighbouring urban areas, as can be seen from the map (Fig. 1).

Similarly the whole of the spread in the London area, extending into parts of Essex and Herts, all Middlesex, and part of Bucks, Berks, Surrey, Sussex and Kent, can be derived, without anywhere jumping a gap of more than ten miles, from three points only—Dartford, Kent (1924), Chalfont St. Giles, Bucks (1925), and Richmond, Surrey (1929). We do not suggest that these were the only centres from which the habit spread—that seems unlikely—but the orderly way in which the spread of the habit can be traced from them, through the suburbs and garden-suburbs, branching down main roads, and filtering through the housing estates, wherever there are houses with small gardens, is extremely suggestive.

It might be argued that the pattern of distribution of the observations reflects, simply, the distribution of observers. In densely populated areas there are, of course, more observers, but there are also more milk bottles. Further, there are many densely populated areas from which no record of the habit has yet been received, or from which negative records have been received. Moreover the time interval between the introduction of milk bottles and the first occurrence of the habit, referred to above, more than justifies us in rejecting this argument.

The records show that the areas in which the habit was known remained comparatively restricted until the early war years, and then increased more rapidly with each succeeding year. It would, of course, be expected in an enquiry of this nature that more records would be obtained for recent years than for earlier ones. On the other hand, the area in which the habit occurred, as well as the actual number of records, increased more rapidly each year; this does seem to support the view that, when the habit has been acquired by one tit, it can then be spread through the population by some form of imitation or learning.

The records show that even when the habit is known to be widespread in a given area it may still be unknown in some households. This is the case in parts of Belfast, for instance. This may be due to such factors as differences in the method of delivering milk or variation in learning ability among the different members of the tit population; or it is possible that the spread may be limited in some way by the structure of the tit populations. However, in spite of this it is true to say that once the habit has been recorded by one observer in a given district it usually becomes nearly universal there within a few years. In Coventry, for example, a few isolated cases were noted for the first time in 1940, and within a few years the habit had become very common in the area. In part of Llanelly the habit was first noted in 1939 at one house only amongst a group of about three hundred near a wooded area. By 1946, and possibly earlier, all these houses had had milk taken. Even when the habit is universal in one area, it may be unknown in neighbouring districts and several records have been received showing that it may be very common in one village and unknown in another only a few miles away. For example, a milk roundsman reported that in 1946 bottle opening had been common for some years in Little Aston, near Birmingham, but was unknown in Streetly and Sutton Coldfield near by.

Thus, although the nature of the investigation makes it impossible to supply any proof, it can be said that the records entirely support the view that the practice has been begun by comparatively few individual birds and that the vast majority of tits have learned it in some way from others.

Although the habit occurs throughout the year, a large proportion of observers record that it is more prevalent during the winter months than in the summer. This may be due to the increased need which the birds have for fats during severe weather, but many observers record that tits are much commoner in winter in urban or semi-urban areas than they are in summer, and this is undoubtedly also a factor of importance.

The bottles are usually attacked within a few minutes of being left at the door. There are even several reports of parties of tits following the milkman's cart down the street and removing the tops from bottles in the cart whilst the milkman is delivering milk to the houses.

The method of opening employed varies greatly. When the milk bottle is closed by a cap of metal foil the bird usually first punctures the cap by hammering with its beak and then tears off the metal in thin strips. Sometimes the whole cap is removed, sometimes only a small hole is made in it. Cardboard caps may be treated in a variety of ways. The whole top may be removed, or only the press-in centre, or the cardboard may be torn off layer by layer until it is thin enough for a small hole to be made in it; the milk may be taken through this hole or the bird may insert its beak in the hole and flick off the remainder of the top. The records show that several different methods may be used in any one district, and that more than one method may be employed by one individual. For example, Margaret Campbell watched a Great Tit tap out half the small central disc and take a few sips of milk. The bird was then disturbed, but returned in a minute or two. This time it alighted on another bottle "on which the stopper was slightly crooked; it made no attempt to peck but inserted its beak under the raised part and flicked it off in one movement. . . ." It is therefore quite certain that the process which has been learnt is the whole business of obtaining milk from milk bottles, and not any particular technique for opening bottles.

If the cap is removed whole, it is often carried away and held in the claw while the semi-solid cream is pecked away from the under surface. Many observers have found quite large piles of cardboard stoppers under neighbouring trees or behind hedges. Two observers placed their milk bottle tops outside daily and found that the tits still pecked at them. When the birds drink from the bottle direct, up to $1\frac{1}{2}$ in or 2 in of milk may be taken. Several cases are known of Blue Tits being found drowned head first in the bottle, presumably because they tried to drink too deep and lost their balance.

In areas where the opening of milk bottles is common, it is often a considerable nuisance to the inhabitants, and milkmen are instructed to invert tins or jars over the tops of the bottles. This procedure is almost invariably effective, but in some cases, where tin lids were used, less success was achieved. Thus, one observer records that a paint tin lid 6–7 in in diameter was removed by the tit perching on its edge. Another observer saw a Blue Tit remove a flat stone from the top of a bottle three times in succession. At another house a Blue Tit managed to reach the milk in spite of a tea cloth which had been spread over the bottles. Of course, these are not necessarily instances of insight learning, but they do represent persistent attempts to reach the milk when the top of the bottle was no longer visible to the bird.

In many areas bottles containing milk of different grades are distinguished by having caps of different colours. Although no reference was made to the subject in the questionnaires, no less than fourteen observers, who had milk of more than one type delivered at their house, reported that the tits attacked only bottles with one particular type of stopper, and four others record an almost invariable preference for one type. There seems to be no rule as to which colour is preferred—in some districts gold, in others silver, red, blue, green or brown. In these cases it would thus appear that the differentiation on the part of the tits between bottles with different coloured stoppers is due to a habit formed perhaps on the first occasions when bottles were attacked.

Without performing proper experiments, it is impossible to decide which senses are of use to the birds in indicating the presence of food. Several of our correspondents have found that bottles filled with water, and even empty bottles, are still attacked, but this tells us nothing if the previous history of the birds is not known. Walter (1943), in an extensive series of experiments on a number of species, was unable to demonstrate a sense of smell in birds, and doubts the value of the work of earlier authors who claim to have done so. Dr. Lack has suggested to us that white objects may have a particular significance—white objects seem to be most effective as bait when trapping tits.

We received over 450 separate records of tits opening milk bottles in all, regarding openings at the same place by different kinds of tits (when known) as separate. The observers have not always identified the species of tit concerned, though they were often able to identify the openers as "tits" as opposed to other kinds of birds.

Exactly 400 identifications of the kind or kinds of tit concerned in opening bottles at a definite place were made, on which we have felt we could rely. Of these, 246 were of Blue Tits, 142 of Great Tits, 11 of Coal-Tits and one of a Marsh-Tit (*Parus palustris*). Of the Coal-Tit records, seven came from southeastern England, one from Somerset, one from South Wales and two from the Central Highlands. The single Marsh-Tit record was from St. Leonard's Sussex, in 1942, at a place where Great, Blue and Coal-Tits had been opening milk bottles since 1941. There is no record of the Willow-Tit (*Parus atricapillus*) opening milk bottles.

The habit is, however, by no means confined to the Paridae. Several other species have been seen drinking from bottles, but, whereas in some cases it is known that these species actually open the bottles for themselves, in others it is possible that they are merely drinking at bottles already opened by tits. The number of places from which records of these other birds opening bottles have been obtained are given below (species for which only one record has been received not included). Since some of the records are from non-ornithological observers, it is possible that some of the Hedge-Sparrow records refer to House-Sparrows.

	Seen to open bottles	Seen drinking from bottles which they did not necessarily open themselves
House-Sparrow	19	23
Blackbird	12	11
Starling	13	8
Robin	8	9
Chaffinch	3	9
Song-Thrush	4	2
Hedge-Sparrow	1	2

To what extent it would be correct to refer to the behaviour of those individuals which "invented" the habit for themselves as "insight learning" is a problem which it is justifiable to discuss only briefly on the present evidence. Thorpe (1943 and 1944) has discussed the ability of certain species to pull up food suspended by a thread; he takes the view that this behaviour probably represents some sort of insight learning and is not dependent on a fully formed inborn automatism. The opening of milk bottles can be divided into two parts:— (1) The recognition by the bird of the milk bottle as a potential supply of food; and (2) The technique of opening the bottle. As far as the second of these is concerned, we have seen that each individual may vary his technique from bottle to bottle, and, although the techniques may in some cases bear a superficial resemblance to those used by tits opening nuts, so that instinctive acts (in the sense used by Lorenz, 1937) may form a part of them, it is clear that the primary learning process is concerned with the recognition of the bottle as a food supply. Now it is possible that those birds which first drank from milk bottles without having previously seen others do so, drank from bottles which had already been opened, or from bottles in which the top was awry. One or two occasions of this type may have been enough to produce an association between the milk bottle and food. In this case the learning process would appear to resemble an advanced type of trial and error learning rather than insight learning. All that can be said, therefore, is that our present data are insufficient to prove the use of insight learning. On the other hand, if it is assumed that the first occurrences of the habit were not dependent on some accident such as a misplaced bottle top, then it would seem that in each district the milk bottles presented equal sensory clues to all birds which came near them, that only certain rather exceptional birds were able to profit by them, and that the subsequent learning of the habit by other individuals in the same district took place by some sort of imitation similar to that referred to by Buxton (1948), though we have no evidence as to the precise nature of the process. It is interesting to note that Thorpe and many other writers have commented on just such a variability of learning ability among individuals.

REFERENCES

Buxton, E. J. M. (1948), Tits and Peanuts, *Brit. Birds,* **41**, 229–232.

Lorenz, K. Z. (1937), The Companion in the Bird's World, *Auk,* **54**, 245–272.

Thorpe, W. H. (1943), A Type of Insight Learning in Birds, *Brit. Birds,* **37**, 29–31.

Thorpe, W. H. (1944), Further Notes on a Type of Insight Learning in Birds, *Brit. Birds,* **38**, 46–49.

Walter, W. G. (1943), Some Experiments on the Sense of Smell in Birds: studied by the method of Conditioned Reflexes, *Arch. Néerl. Sci.* (*C. Physiol.*), **27**, 1–72.

Further Observations on the Opening of Milk Bottles by Birds

R. A. HINDE AND JAMES FISHER

The opening of milk bottles by birds is now a well known habit throughout most of the British Isles. In a previous paper over 400 records of the habit were summarized, and the spread of the habit discussed (Fisher and Hinde, 1949). The information used in that paper was obtained largely through the help of the British Trust for Ornithology. Since its publication we have received many more records from the British Isles, and also a few from other countries. In the present paper the foreign records are summarized, and the processes involved in the spread of the habit are discussed.

FOREIGN RECORDS

Sweden

Mrs. I. Kramer recorded Blue Tits (*Parus caeruleus*) opening milk bottles at Göteborg. Inquiries made by her among the local milk distributors revealed that the habit was widespread there, and that "Magpies and Blackbirds" also raided the bottles. M. M. Rundstrom has recorded tits opening coffee cream bottles at Vendelso. S. Durango states that the habit is not known at Taby. Mrs. Tote Thoren (per Arne Olander) recorded Great Tits (*Parus major*) opening bottles at Stockholm. In reply to an appeal in *Fauna och Flora* (1950, p. 45), for which we are indebted to Prof. Hanstrom and Mr. Durango, five records of Great Tits opening milk bottles were received: Uppsala, Dec., 1949 (K. Thomasson), Saltsjöbaden, 1948 and 1949 (S. Hesselman), Göteborg since 1940 (Torsten Borgvall), Kil, July 1st to August 15th, 1949 (Arne Kelmentsson) and Lund (Gunilla Lindergard)—vide, *Fauna och Flora,* 1950, pp. 171–172, 256. Both cardboard and foil stoppers were attacked. Some of these correspondents also recorded Great Tits tearing various types of paper, as also did Henry Bark at Surahammar.

Reprinted in part from *British Birds,* volume 44, pages 392–396 (1951).

Denmark

F. Salomonsen states that the habit is well known, and "in suburban villa-quarters, almost a pest." The main birds and the Great Tit and Great Spotted Woodpecker (*Dendrocopus major*). The habit appears to have become more common in recent years. (See also *Dansk. Orn. Foren. Tidsskr.*, **37** (1943) p. 184). A. F. Davidson (*Countryman*, Winter 1948; 291) says that the opening of bottles by Great Spotted Woodpeckers in the Copenhagen area is apparently quite a general practice. Mrs. Ilse Runner, of Switzerland, has sent us a record of Great Tits opening milk bottles at Ormager 1919 and 1924. A request for information published in the *Dansk. Orn. Foren. Tidsskr.* brought two replies. One recorded Great Tits opening bottles at Helsingor (H. Cristrup) and the other stated that the habit was common along the south border of the Geel Forest, where Great Spotted Woodpeckers, and also Jackdaws (*Corvus monedula*) and Great Tits have been seen opening bottles (A. Hohn).

Holland

Enquiries made by Dr. L. Tinbergen by means of questionnaires distributed through popular nature magazines brought in a number of replies. Bottle-opening was recorded from at least 21 localities south of the Zuider Zee; the precise distribution of the records was probably influenced by the method of sampling. In all cases where the bird was certainly identified it was a Great Tit. We have also received one other record of a Great Tit opening a milk bottle in Amsterdam from G. Los. Dr. Tinbergen points out that milk bottles practically disappeared from Holland during the war, and only became reasonably common again in 1947–48. It is unlikely that many tits which had learnt the habit in pre-war years still survived at this date, so the habit must have originated in all the recorded localities since 1948. Since Great Tits are relatively sedentary birds, it seems certain that the habit was started in many different places and by many individuals. Dr. Tinbergen also noted that in Holland, as in England, there are many places where the habit is unknown although both milk bottles and tits are common.

Belgium

A. van Beneden published an article describing the habit as seen in England in Parcs Nationaux, but we have had no records from Belgium. J. E. Shaw states that suitable bottles are scarce in Brussels.

France

N. Mayaud made some enquiries in France, but was unable to discover any records—apparently milk bottles are scarce.

Switzerland

Records are being collected by D. Burckhardt and will be published separately.

U.S.A.

A photograph of a Steller's Jay (*Cyanocitta stelleri*) opening a milk bottle appeared in the *Seattle Times* for 6/7/37. (Information from E. R. Parrinder). V. B. Schaffer, who has lived in Seattle for 25 years, could not recall any other cases.

NOTES ON THE PROCESSES INVOLVED IN LEARNING TO OPEN A MILK BOTTLE

No experimental analysis of the behaviour involved in the opening of milk bottles has yet been made, but the discussion in the previous paper can be carried a little further with the aid of field observations. Previously the processes involved were considered in two parts—(1) the "recognition" of the milk bottle as a potential supply of food and (2) the technique of opening the bottle. It was pointed out that the actual opening of the bottle probably depended on innate motor patterns—that is on sequences of movements common to all members of the species, which do not have to be learned. Study of tits in the field has confirmed this view, for the movements used by tits in opening milk bottles are the same as those used when feeding on natural foods. For example, the hammering action with which foil caps are punctured is very similar to a motor pattern used in opening nuts, and the tearing action often used on cardboard is similar to a movement used in tearing bark from a twig.

We are thus primarily concerned with the "recognition" of the milk bottle. Dr. L. Tinbergen has suggested to us that the initial "discovery" of the bottle as a source of food may be a logical consequence of the feeding habits of tits. They appear to have an inborn tendency to inspect a great variety of conspicuous objects which contrast with their surroundings, and to test their palatability. If the outside of the object is unpalatable, they may attempt to open it by hammering—hammering seems to be evoked by the sound of a hollow object. (This is probably true of woodpeckers as well as tits.) In fact it seems that when tits are looking for food (though not for any particular type of food) objects with certain very general characteristics (such as "contrast with the background") may be sufficient to release a more limited type of searching behaviour (such as flying to the food). By a succession of similar steps (hammering, etc.), evoked by successively more specific stimuli, a reward may eventually be obtained. It so happens that milk bottles have sufficient characteristics in common with the natural foods for tits occasionally to "discover" them in the course of their normal searching.

When a bird has fed once or a few times at milk bottles, it will learn to return to them on later occasions (presumably by a form of trial-and-error learning—see Thorpe 1951a and 1951b). On these subsequent occasions the bird may come to respond to quite different stimuli from those which controlled its behavior the first time. For instance, it was shown in the previous paper that, if milk bottles with differently coloured stoppers are available, the tits will often concentrate their attacks on those having stoppers of one particular colour—presumably because the colour of the top was one of the characters which the bird learnt on the first visits.

In our previous paper we brought forward considerable circumstantial evidence to show that tits may learn from each other to drink out of milk bottles. Thorpe (1951a) and Armstrong (1951) class this with other cases reviewed by Roberts (1941) as "Local Enhancement." We may examine briefly the nature of the processes involved. Feeding, like other cases of instinctive behaviour, consists of a series of fairly complex forms of searching or "appetitive" behaviour leading eventually to the "consummatory act" of actual eating. Thus "flying towards the milk bottle" is one of the stages of the preliminary appetitive behaviour, and it can be analysed into a motor pattern (flying) and a steering or "taxis" component. Now when one bird, engaged in looking for food, sees another feeding on a milk bottle and flies towards the bottle two things have happened—a new stage in the appetitive behavior (flying) has been released, and the flying has been guided in a particular direction. These, then, are the effects of the local enhancement. There is no need to assume that any of the actions of the bird after it reaches the bottle on the first occasion are influenced by its view of the other bird feeding—they may occur simply as a result of the tit's inborn mechanisms as described in the preceding paragraphs. The change in the appetitive behaviour occurs before the bird has reached the reward, and so is apparently independent of it. Since it is "behaviour modified by experience" it comes within the definition of learning (Thorpe, 1951b). On the other hand it may also be regarded as innate behaviour, for it is possible (though unproven) that there is an inborn tendency to investigate objects similar to those on which other birds have been seen to feed. It is not profitable to pursue the discussion further without more precise experimental evidence.

The preceding discussion has been rather hypothetical, and the views advanced need to be verified. It does, however, serve to show that the habit formed by so many birds of drinking from milk bottles may not be so far removed from their normal feeding behaviour as would appear at first sight, and to suggest some of the mechanisms involved. In brief, it is suggested that:

1. The initial investigation of the milk bottle by the pioneers can probably be understood on the basis of the normal innate feeding behaviour.
2. The subsequent return of these individuals to milk bottles is probably due to trial-and-error learning.
3. The apparent copying of the behaviour by other individuals is probably due to another process, Local Enhancement, which modified the appetitive behaviour of feeding and is initially independent of reward.

We are grateful to Dr. W. H. Thorpe and Dr. L. Tinbergen for valuable advice given during the preparation of this paper, and to the correspondents mentioned in the first part of the paper for the records from abroad.

REFERENCES

Armstrong, E. A. (1951), The nature and function of animal mimesis, *Bull. Anim. Behav.*, **9**, 46–58.

Fisher, J. and Hinde, R. A. (1949), The opening of milk-bottles by birds, *Brit. Birds*, **42**, 347–357.

Roberts, D. (1941), Imitation and suggestion in animals, *Bull. Anim. Behav.*, **3**, 11–19.

Thorpe, W. H. (1951a), The learning abilities of birds, *Ibis*, **93**, 1–52, 252–296.

Thorpe, W. H. (1951b), The definition of some terms in animal behaviour studies, *Bull. Anim. Behav.*, **9**, 34–40.

SOME COMMENTS ON THE RE-PUBLICATION OF TWO PAPERS ON THE OPENING OF MILK BOTTLES BY BIRDS

The re-publication of papers over fifteen years old requires some re-assessment of their content. We would like to make the following points:

1. Our earlier conclusion that "the records entirely support the view that the practice has been begun by comparatively few individual birds and that the vast majority of tits have learned it in some way from others" was based on evidence where circumstantial nature we were at pains to emphasize. We would now put the emphasis differently. While the second half of the conclusion is probably valid, the practice must have been started by many individuals. For example in the second paper we quoted the late Dr. L. Tinbergen's evidence that the habit started in many different places when milk bottles were re-introduced in Holland after the war; and it seems clear that in many areas in this country it is confined to local pockets. We should now put as much emphasis on local spread, but less on geographic spread.

2. We described the movements used in opening the bottles as "innate motor patterns." While they are indeed common to all members of the species, the role of learning in their development is an open issue. It is now known that learning is involved in the development of the comparable movement used in transferring objects (such as milk bottle tops) from beak to foot (Hinde and Tinbergen, 1958; Vince, 1964). In any case the use of the adjective "innate" for a type of behaviour is open to a number of objections, and it is better reserved for those differences between types of behaviour which are not environmentally determined (e.g. Schneirla, 1952; Lehrman, 1953; Hinde, 1966). For similar reasons we would no longer use the terms inborn or innate to describe other aspects of behaviour, as we did later in the paper.

3. In the second paper we ascribed the copying of the habit by other individuals to "local enhancement" (Thorpe, 1951, 1963). We now feel that we should emphasize that, while this term is useful for a descriptive category, it has little explanatory value. Furthermore, the manner in which we analyzed the behaviour involved now seems to us a little naive. We would no longer refer to it as "instinctive behaviour," and we believe that the distinction between appetitive and consummatory behaviour depends on a number of characters which are not necessarily correlated with each other (Hinde, 1953). Furthermore, in separating the act of flying to the bottle into the motor pattern of flying and the steering component, we should not wish to imply that these depend on different stimuli.

4. Finally, we must express our surprise that the editors should wish to resurrect these papers under the title of *Readings in Animal Behavior*. They seem to us to display some of the worst faults of ethology at that time, such as the tendency to regard descriptive labels as explanatory, and they fall very far short of the better papers which were then being published. Certainly they are on the quite different level from most of the work in comparable journals in recent years. Furthermore, they must not be used as evidence that ethologists are concerned wholly or even mostly with "field studies": if that was ever the case, it is certainly so no longer.

REFERENCES

Hinde, R. A. (1953), Appetitive behaviour, Consummatory act, and the hierarchical organisation of behaviour—with special reference to the Great Tit (*Parus major*), *Behaviour*, **3**, 189–224.

Hinde, R. A. (1966), *Animal Behaviour: a Synthesis of Ethology and Comparative Psychology*. London, New York: McGraw-Hill.

Hinde, R. A. and Tinbergen, N. (1958), The comparative study of species-specific behavior, From *Behaviour and Evolution*, Eds. Anne Roe and G. G. Simpson. New Haven: Yale Univ. Press.

Lehrman, D. S. (1953), A critique of Konrad Lorenz's theory of instinctive behavior, *Quart. Rev. Biol.*, **28**, 337–363.

Schneirla, T. C. (1952), A consideration of some conceptual trends in comparative psychology, *Psych. Bull.*, **49**, 559–597.

Thorpe, W. H. (1951), The learning abilities of birds, *Ibis*, **93**, 1–52, 252–296.

Thorpe, W. H. (1963), *Learning and Instinct in Animals*. Methuen.

Vince, M. A. (1964), Use of the feet in feeding by the Great Tit (*Parus major*), *Ibis*, **106**, 508–529.

A NOTE BY THE EDITORS

We felt it a valuable addition to this historical anthology to include the comments by Hinde and Fisher on their early papers. We do wish to point out, however, that we do not agree that their papers "fall very short of the better papers which were then being published." On the contrary, the discovery of behavior that is at least partially transmitted by "culture" rather than biological inheritance, was a stimulating point to ethology (as the following papers help demonstrate). And whose works of two decades ago are not naïve in some respect? Indeed, the development and maturation of ethology as a young science are two of the things we wish most to demonstrate by this collection of readings. It is a tribute to Hinde and Fisher, and in fact to all the authors who allowed older papers to be included here, that they so well understand the importance of a student's viewing intellectual progress in the discipline that they feel no false embarrassment in having old words see new print.

Observational Learning in Birds: The Establishment of Behavioral Modes

PETER H. KLOPFER

In the greenfinch, *Chloris chloris*, single birds can learn a food discrimination task more rapidly than pairs of birds, though fewer of the latter are fearful of approaching the discriminanda. This behavior can fail to be maladaptive only in species with relatively conservative food habits, or of a solitary nature. In the more inquisitive and exploitative great tit, on the other hand, the birds suffered no disadvantage when trained in pairs.

Thus, additional evidence is provided to support the view that species characteristic behavior may be determined by imitative processes and the establishment of traditions, as well as by innate differences in structure and behavior.

GENERAL

There are essentially two methods by means of which the behavior that is characteristic of a species may be maintained from one generation to the next. On the one hand, the appropriate behavior can be built into the organism. Anatomic peculiarities may restrict it to special feeding techniques, or, neural circuits may limit its responses to particular stimuli. More generally, such structural determinants of behavior may produce slightly different degrees of efficiency in different environments so that learning and competitive processes assist in the final determination of the normal behavior pattern. On the other hand, the appropriate behavior may develop solely as the result of imitative propensities and lead to the establishment of traditional behavior. We can summarize these processes, all of which lead to the maintenance of species-specific behavior, as follows:

1. Characteristic behavior is largely endogenous or self-differentiating (Ewer, 1957).
2. Minor morphologic differences produce different efficiencies in different environments, leading to selection for the most appropriate behavior and/or environment.
3. Imitation of congeners leads to the establishment of traditions (observational learning).

Social learning and traditions

Instances of social facilitation, which may appear similar to observational learning, are common to many classes including birds. Examples, as well as a careful definition of the phenomenon can be found in Thorpe (1956). True observational learning, however, requires that the effect of the observation persists for an interval after the removal from the scene of the animal being observed. Theoretically, the effect should persist

Reprinted from *Behaviour*, volume 17, pages 71–80 (1961). Original numbering of figures is retained.

for at least as long an interval as that which can be interposed between any other type of training and retesting cycle. Social facilitory effects are much more transitory. In the, now proverbial, instance of the satiated hen which is stimulated to begin feeding upon being placed with a hungry hen, the former will promptly stop feeding after her hungry partner has been removed. This is an instance of what is meant by a transitory effect.

Few documented instances exist where a true tradition can be said to exist, apart from *Homo* species, although studies of primates (cf. Koehler, 1921) have certainly suggested this possibility. King (1955) has convincingly reported the existence of traditions in the black-tailed prairie-dog, *Cynomys ludovicianus*. Both the characteristic social signals and the boundaries of the coterie territory appear to be stable features that are transmitted from one generation to the next in a nongenetic fashion. Even more recently, Miyadi (1958) has shown how isolated colonies of Japanese monkeys (genus unidentified) may develop strikingly different food preferences. Once established, these preferences can remain stable through ensuing generations. In one colony, a particular animal began washing sweet potatoes, one of the provided food items. The behavior was then imitated by this animal's mother, playfellows, then its siblings, until it was shown by most members of the group. Significantly, the older animals are distinctly less prone to accept behavior patterns of younger beasts than conversely.

Whether similar phenomena can be demonstrated in nonmammalian vertebrates has remained unclear. Klopfer (1957) has shown that, at least in a few instances, a duck (*Anas platyrhynchos*) can learn to avoid a particular pattern as the result of observations of avoidance responses by a congener. It was suggested that this type of learning could explain the basis of certain instances of Batesian mimicry as well as the persistence of food, forage site, and habitat preferences of certain species.

Species-specific behavior

It is clearly important for sympatric species to maintain the integrity of their respective niches and the imitation by the young of parental behavior would provide an easy and satisfactory explanation, especially for situations such as those described by MacArthur (1958) where a group of anatomically similar warblers have nearly divided the environment between themselves. However, a number of complicating variables must be dealt with if traditions are to be thought to play a role in the determination of species-specific behavior. In the first place, birds which habitually move in mixed species flocks cannot be subjected to observational influences if interspecific behavioral differences exist. Alternatively, such influences might be limited in effectiveness to a brief time early in the life of the individual while its observations were restricted to its parents. Secondly, solitary species should be much less likely to have their characteristic behavior established by observation. We might even expect that if such species were experimentally placed in a situation where they could be influenced by the responses of their congeners, such influences would prove to be unimportant. Finally, we might expect that birds with highly conservative and restricted habits would show a different type of observational effect than more tolerant species. This assumption rests on the fact that an observer can be influenced by one of two aspects of the observed birds' performance. If the latter tries an unpalatable food and then rejects it, the observer may be led to associate either the initial feeding or the final rejection with the food. We may call the former acceptance- and the latter avoidance-learning. An exploitative, inquisitive species (such as the great tit, *Parus major*) would find itself in a difficult situation if it were capable only of acceptance-learning, for then it would continually repeat its congener's errors. For a conservative species, which would not be prone to try new foods, this would make little difference. The greenfinch, *Chloris chloris*, for instance, seems to fall into this latter category. Apparently, in the greenfinch, the sight of a conspecific feeding on a particular (unpalatable) object can serve to over-ride both the effects of observing a subsequent rejection of the object as well as the personal experiences with the object of the observer itself (Klopfer, 1959). Whether these suggestions have any general validity remains open to experimentation and discussion. However, we have obtained some preliminary indications that the role of observational learning may indeed differ between species, though it is not yet possible to determine whether all the differences are as predicted. The data have been summarized below in the hope they will stimulate other workers to take up this problem. A detailed analysis of the cues used by some of these birds in their learning has been reported elsewhere (*ibid*).

METHODS

In order to eliminate the disturbances associated with captive conditions and artificial patterns, birds were housed in large wire aviaries, about 6 feet on each side, provided with suitable nesting sites, perches, feed, and water, *ad libitum*. Into each cage was placed a pair of adult birds, one of each sex, or a single bird. In the latter instances the cage was partitioned so as to reduce its volume by one-half ($6 \times 6 \times 3$ feet). The cages were placed along the edges of a wood in the normal habitat for the species. Into each cage there was placed at 24 hour intervals one sprig of box (*Buxus* species) and one of ivy (*Hedera* species) for each bird. To the leaves of one sprig were glued sunflower seeds and to the other sunflower seeds whose kernels had been replaced with moist aspirin, a nonvolatile, odor-free, and relatively nontoxic substance known to be unpalatable to all of the species tested. For half of the birds the ivy represented the unpalatable food source, and for the others it was the box. Neither plant is normally preferred by these species. The seeds were replaced at 24 hour intervals.

Scoring was accomplished simply by counting the number of seeds taken from each sprig after a 24 hour interval. The graphed scores represent the difference between the number of palatable and the unpalatable seeds taken; thus the higher the score, the better the discrimination. Since single birds had available half the number of seeds (6 of each type) as did the pairs (12 of each type), their scores were doubled in order to make direct comparisons of learning speed possible. On the basis of earlier tests, it was possible to define a learning criterion as a total score of 25 or more accumulated in three consecutive days. Once this score had been attained, performances no longer dropped to a nondiscriminatory level. Discrimination at this level is significant at the .99 level as determined by the sign test (Dixon and Massey, 1951). Where birds took two or fewer seeds they were considered to fall in the same category as those which made no responses at all. In point of fact, it was very rare that fewer than all or at least 5/6 of the seeds were taken, if any were taken at all.

The two species examined were the greenfinch, *Chloris chloris*, and the great tit, *Parus major*. Since the latter is a far bolder and more explorative creature than the former, both the time and the number of trials required to attain criterion are compared. This follows from the assumption that the more cautious greenfinch could require longer before attempting to feed from the sprigs while actually requiring fewer trials to learn the discrimination.

Since motivational levels, hence learning speed, are clearly influenced by temperature, day length, and other environmental variables, and since the ten sets of experiments conducted for each species could not be run simultaneously, analysis of the data requires a test allowing paired comparisons of single and paired birds. Since one pair and one single bird were always tested simultaneously, the Sign Test provides a simple and suitable means for assessing significance.

RESULTS AND DISCUSSION

It has previously been shown (Klopfer, 1959) that the speed and possibly the manner in which adult greenfinches learn to feed from one but not from another pattern varies with the experience of the partner. A single male can learn the required discrimination rapidly; a single male being observed by a naive male requires somewhat longer. A male which has previously observed a partner making correct responses may do as well as a solitary male, but no better, while a male who has observed the entire learning sequence of another (including its initial errors) will do very poorly, and through its own errors lead the formerly trained partner to "forget" the discrimination. An analysis of the relevant cues showed that the most potent stimulus apparently was the sight of the other bird feeding. This stimulus served to overcome any learned tendency to avoid the object being fed upon.

The present, more natural, experiments (Fig. 21–24) fully substantiate the conclusions drawn. The paired greenfinches, in 9 cases out of 10, required longer to learn the discrimination, than the single birds, often failing to learn it at all in the time allowed. In all instances but one, both members of a pair were seen to feed from the sprigs. The exception proved the rule, for in this case the responding member of the pair learned as quickly as the single control.

The superiority of the single birds does not mean that pairs need always be at a disadvantage. A great many more single birds were tested than appear in the data summaries herewith presented. Those excluded were the birds who failed to approach

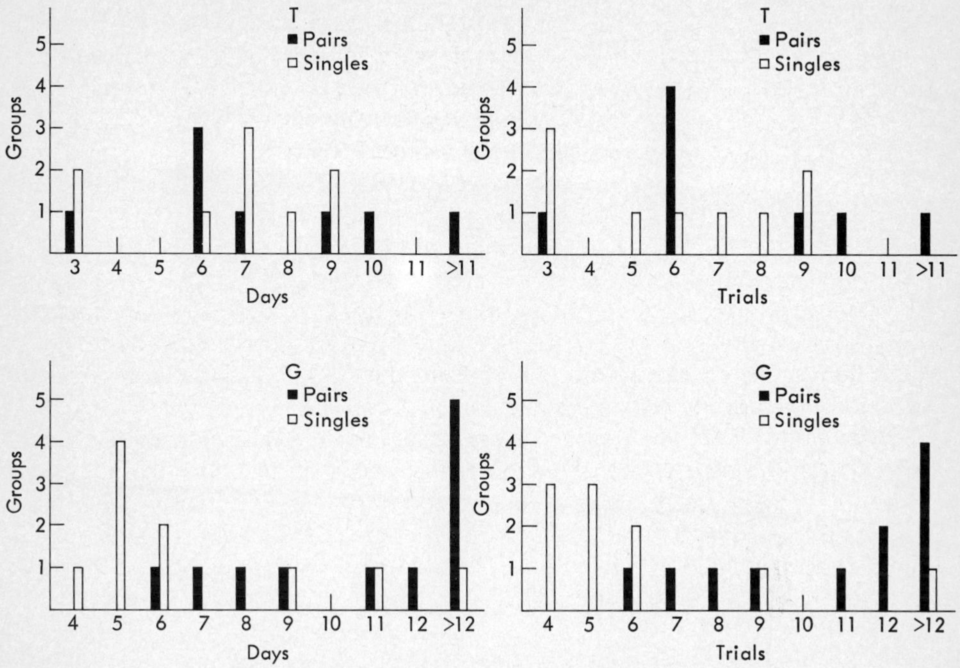

Figs. 21–24. Cumulative totals for great tits and greenfinches: T = great tit; G = greenfinch; Ordinate: number of pairs or single birds; Abcissa: number of days or trials required to attain criterion.

the sprigs altogether. This was true of 6 out of a total of 9 females and 2 of a total of 10 males. Thus, for this species, the same processes which retard learning rates in social situations also enlarge that part of the bird's Umwelt which is explored. As will be seen, the bolder great tits do not require social stimulation to try new foods or feeding sites, and far fewer individuals fail to respond. Of 11 single great tits, none failed to feed from the sprigs, while only a single pair out of 10 did not respond. Comparing the nonresponders among single birds, we have 8/19 for the greenfinches as against 0/11 for the tits. That is to say, in order to find 11 responders, almost twice as many greenfinches as great tits had to be tested. However, it is not possible to say whether this is a statistically significant difference. As one might expect from these results, in the great tits, paired birds do not show themselves inferior to single birds. Indeed, no significant differences in learning speed between the two groups, pairs and singles, exist among the tits. This accords well with the gregarious and inquisitive nature of this species.

REFERENCES

Dixon, W. J. and Massey, F. J. (1951), *Introduction to Statistical-Analysis.* McGraw-Hill.

Ewer, R. F. (1957), Ethological concepts, *Science,* **126**, 599–603.

King, J. (1955), Social behavior, social organization, and population dynamics in a Black-tailed Prairie-dog town in the Black Hills of South Dakota, *Contributions from the Laboratory of Vertebrate Biology*, No. 67, University of Michigan.

Klopfer, P. (1959), Social interactions in discrimination learning with special reference to feeding behavior, *Behaviour*, **14**, 282–299.

Klopfer, P. (1957), Experiments on emphatic learning, *American Naturalist*, **91**, 61–63.

Köhler, W. (1921), *The Mentality of Apes*. London.

MacArthur, R. (1958), Population ecology of some warblers of northeastern coniferous forest, *Ecology*, **39**, 4, 599–619.

Miyadi, D. (1959), On some new habits and their propagation in Japanese Monkey bands, *Proceedings International Zoological Congress*, London, 857–860.

Thorpe, W. H. (1956), *Learning and Instinct in Animals*. Harvard.

Social Life of Japanese Monkeys

DENZABURO MIYADI

Japanese monkeys live in troops of between 30 to 200 individuals, which form a social group integrated in part by vocal communication. Behavioral traits may be culturally transmitted within troops. For instance, some troops eat eggs, while others do not; most troops do not allow males to mount females outside of the breeding season, although one troop does; paternal care of babies is seen in only one troop; and so on. The propagation of washing of sweet potatoes before eating was studied in the Kojima Island troop, where it was initiated in 1953 by a one-and-one-half-year-old female. It spread to her mother and playfellow, then to her sisters and brothers, and finally to all members of the troop save the old males. The importance of behavioral distinctiveness of individual troops cannot be overlooked as an important aspect of primate behavior. [Editor's Abstract.]

Japanese monkeys have a more northerly range than any other nonhuman primate. They are about the size of the Gibraltar monkeys, found at the entrance to the Mediterranean Sea, and have beautifully bright faces and very short tails.

They forage in the forest in troops of from 30 to 200, collecting the leaves, fruits, and insects which are their principal foods. The number of species of plants in the diet of a troop may be as high as 200, and the troop's foraging area is estimated to cover about 3 to 5 square kilometers.

The wild Japanese monkeys are extremely shy, and it is very difficult to catch even a glimpse of them. Hardly any photographs had been taken of them in nature

Reprinted in part from "Social Life of Japanese Monkeys," D. Miyadi, *Science*, vol. 143, pages 783–786, 21 February 1964.

before the beginning of our study in 1958. But at our secret approach to a troop we were surprised at the variety of their voices coming from the depth of forest. Thus we came to regard vocalization as one of the most important clues to their natural history and undertook to list in our field notebooks, every vocal sound heard. Other clues were the droppings, footprints, and fragments of food plants they left scattered about.

To study the vocalization more fully we needed to ascertain the situation of the troop before and after each cry and to identify the behavior of the monkey emitting a certain vocal sound as well as the influence of the sound on another monkey or the reaction of the troop as a whole. Thus it was necessary to lure the monkey troop to an open place to observe it more clearly at shorter distances.

For this purpose we tried setting out sweet potatoes and wheat grains on rocks in a deep, pathless forest of Kojima Island in Kyushu, where the monkeys were accustomed to pass. They ate the sweet potatoes, and by and by the troop's wanderings became influenced by the food we set out. Next, the number of these provisioning points was gradually reduced, finally to a single point: the sandy beach of the small island. On the rock of this beach, large quantities of sweet potatoes and wheat grains were set out. At first the monkeys were wary of a fisherman's cottage on the beach and were unwilling to come down to the food. The suspicion and fear they felt toward human beings, however, were no match for their greed. Once they realized that we would do them no harm, they gradually grew bolder, and soon, unobserved, we had a complete view of them.

After this success our first task was to distinguish the monkeys of the troop by giving each one a name. In this way we learned about the membership of the troop and also about the relationships among individuals and the social status of each individual, so that eventually we were able to grasp the social structure and the lineage of the troop as a whole.

* * *

Some cultural habits and their propagation

Animal behaviors might be grouped in two classes, those innate to a species and those learned during the life of an individual. The latter are often acquired by trial and error and usually vanish with the death of the individual without being transferred to the offspring.

However, in a higher animal species whose social form is a permanent troop, behavior acquired by one individual may be imitated by other members to become a new habit of the troop, preserved there even when the individual dies. Such preserved behavior is accepted by newborn animals without hesitation and appears early in their development as if it were an innate habit. Thus, these behaviors are beyond the level of the individual and may be said to have been socialized or to have reached the cultural level. It is not always easy to decide whether a certain behavior which all individuals of a troop possess is innate in an individual or encultured in a group. For distinguishing the two, comparative study of the behaviors of different troops or of the process of new-habit formation may be useful.

Members of the Primate Research Group at Kyoto University as well as members of the Japan Monkey Center have found some differences in the habits of Japanese monkeys belonging to different troops. While the monkeys of the Mt. Arasiyama troop have an egg-eating culture, those of the Syodosima troop in the Inland Sea do not. The monkeys of the Mt. Atago troop near Tokyo are indifferent to unhulled grains of the rice plant and to soya beans even when they pass along footpaths between the fields where these are cultivated, in contrast to monkeys of many other troops who do considerable damage to these plants. In most troops the mounting of a female by a male is "taboo" except in the breeding season, but this is not so in some exceptional troops. Paternal care of babies by leader and subleader males in the delivery season is seldom seen in Japanese monkeys except among those of the Mt. Takasaki troop. These differences in habits according to troop can hardly be assumed to be innate.

When we lured monkey troops to artificial feeding places, at first the monkeys ignored new foods, such as candies and boiled rice, but they became gradually accustomed to them and established new food habits. Newborn monkey babies do not distinguish between artificial and natural foods, and the infants were quick to start eating the new food, although the mother interfered with this adventure. After a while, however, the mother imitated her children and the behavior spread to others, until the whole troop had acquired a new habit. The speed of acquisition differed considerably with age, and elderly individuals proved the most conservative. There were also some personal differences. While the first male of the Mt. Takasaki troop was reluctant to try a new kind of food, that of the Mt. Minoo troop near Osaka was quick in learning, and he was imitated by other members of his troop very quickly. Established habits are handed down quickly from mothers to babies or from older monkeys to younger ones, but along the reverse route new habits are acquired very slowly.

In the troop of Kojima Island we discovered another kind of new habit: the washing of sweet potatoes in sea water before eating them. This hygienic habit was initiated in 1953 by a $1\frac{1}{2}$-year-old female, and it was imitated by other monkeys. It was first learned by the mother and playfellow, then by sisters and brothers. At present, almost all monkeys except the old males have this behavior, which seems to have been established as a new cultural habit of this troop. These new habits may spread to other troops through solitary males who happen to move from one troop to another.

The benefits of troop formation to both the individual and the species are manifold. Among them may be counted the acquisition of new habits as part of the culture of the troop. Such acquisition may be significant in the social evolution of higher animals.

The social life of the Japanese monkey has been studied for about 12 years through observation of more than 20 natural troops fed at various points throughout the country. However, no sign of a troop division was noticed until 1958 and 1959, when division took place in two troops in completely different ways. So far as we know, there is no other record of the multiplication of a natural troop.

In Japanese monkey society, two distinct parts may be distinguished, the central and the peripheral. The central part consists of one or several leader males, all the females (both mature and immature), and babies and infants less than 2 years old. At the periphery are the subleader males and young males, who have a different status from the leaders and are not allowed to enter or eat in the central part. While the males must leave the central part as they grow up, the females remain around the leader males throughout life, except in the breeding season when they consort with peripheral males.

One of the troop divisions occurred when the population of the Mt. Takasaki troop swelled to about 500 (when we started to feed this troop in 1952 it had about 200 members and six leaders). Perhaps because of the widening area covered by the troop, some adult females accompanied by their babies and infants came out of the central part to join the young males at the periphery. The first sign of division came when part of the troop left the feeding ground for the troop sleeping place more than an hour later than other members. Later the branching troop sometimes passed the night near the feeding ground while the rest of the troop went to its usual sleeping place at some distance. Division was complete when the branching troop chose its own sleeping place, distant from that of the main troop and from the feeding ground. Both the leader and the subleaders of the original group remained with the main troop, contrary to our expectation that some of the subleaders might take the initiative in division. Six young males played important roles in the branching, and one of these became the leader.

The other observed case of troop division did not proceed in this peaceful way. It seemed to be the result of a decline in rank of the troop's third leader, whom we had named Saburo. Saburo was initially surpassed in rank by the fourth leader, Siro, who later became the first leader. A few months after Saburo had been mounted by a young male named Take, there was a report of unusually severe fighting. The next day Saburo and one-fourth of the troop did not appear at the feeding ground. A search on the mountain revealed the new troop foraging in the forest, and it never again appeared at the feeding ground. The detached troop has two leaders: Saburo, the former declining leader, and Kuro, promoted from subleader status. After the division, the new troop observed the forms of social etiquette and manners according to rank, such as the down-motion of the tail in the presence of a superior male, more strictly than the original group. It is not clear how the 50 monkeys chose the separate troop. Was there a secret plot among them?

While change in social ranking among leaders is common, the declining leader does not always desert the troop to become a solitary, or try to establish a new troop. In one of the troops we observed there was a retired leader, very old, who still posted himself in the central part, although he was not much concerned with the management of the troop.

The Japan Monkey Center has also succeeded in organizing a new troop with monkeys collected from different natural troops. When a newcomer was added to the artificial group, he or she was recognized at once and warmly accepted by individuals who had come from the same natural troop several months before. The monkey troop is undoubtedly a society of mutual acquaintance; each monkey knows every other monkey—its rank, status, mother-child relationship, and so on. Some of the infants, for example, are ranked high because they are the children of influential mothers. In these senses, each monkey troop has its own troop peculiarity and cultural trend, and each member of the troop differs from the other members in personality and life history.

Thus, nonuniformity is a fundamental characteristic of individuals as well as of troops of the Japanese monkey, and case-by-case observation, with identification and naming of individuals and comparison of troops, is essential in primatological research.

Culturally Transmitted Patterns of Vocal Behavior in Sparrows

PETER MARLER AND MIWAKO TAMURA

Male white-crowned sparrows have song "dialects," acquired in about the first 100 days of life by learning from older males. In the laboratory an alien white-crowned sparrow dialect can be taught. Once the song is established further acoustical experience does not change the pattern. White-crowned sparrows do not copy recorded songs of other sparrow species presented under similar conditions.

The white-crowned sparrow, *Zonotrichia leucophrys*, is a small song bird with an extensive breeding distribution in all but the southern and eastern parts of North America (Banks, 1964). Ornithologists have long remarked upon the geographical variability of its song. Physical analysis of field recordings of the several vocalizations of the Pacific Coast subspecies *Z. l. nuttalli* reveals that while most of the seven or so sounds which make up the adult repertoire vary little from one population to another, the song patterns of the male show striking variation (Marler and Tamura, 1962).

Each adult male has a single basic song pattern which, with minor variations of omission or repetition, is repeated throughout the season. Within a population small differences separate the songs of individual males but they all share certain salient characteristics of the song. In each discrete population there is one predominant pattern which differs in certain consistent respects from the patterns found in neighboring populations (Fig. 1). The term "dialect" seems appropriate for the properties of the song patterns that characterize each separate population of breeding birds. The detailed structure of syllables in the second part of the song is the most reliable indicator. Such dialects are known in other song birds (Armstrong, 1963).

The white-crowned sparrow is remarkable for the homogeneity of song patterns in one area. As a result the differences in song patterns between populations are ideal

---

Reprinted from "Culturally Transmitted Patterns of Vocal Behavior in Sparrows," P. Marler and M. Tamura, *Science*, Vol. 146, pages 1483–1486, 11 September 1964.

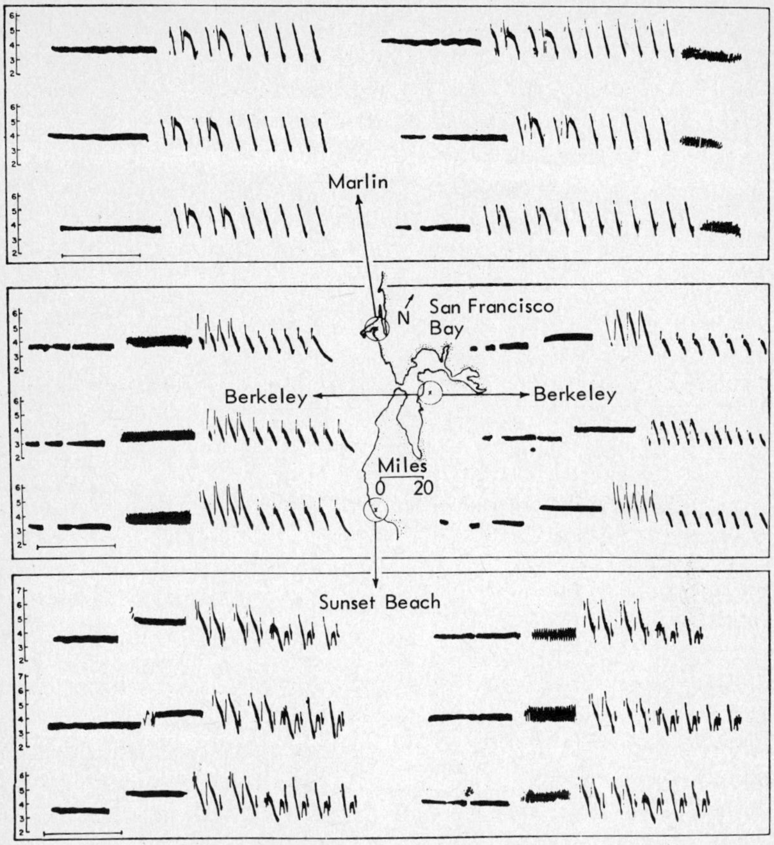

Fig. 1. Sound spectrograms of songs of 18 male white-crowned sparrows from three localities in the San Francisco Bay area. The detailed syllabic structure of the second part of the song varies little within an area but is consistently different between populations. The introductory or terminal whistles and vibrati show more individual variability. The time marker indicates 0.5 second and the vertical scale is marked in kilocycles per second.

subjects for study of the developmental basis of behavior. If young male birds are taken from a given area, an accurate prediction can be made about several properties of the songs that would have developed if they had been left in their natural environment. Thus there is a firm frame of reference with which to compare vocal patterns developing under experimental conditions. Since 1959 we have raised some 88 white-crowned sparrows in various types of acoustical environments and observed the effects upon their vocal behavior. Here we report on the adult song patterns of 35 such experimental male birds. The several types of acoustical chamber in which they were raised will be described elsewhere.

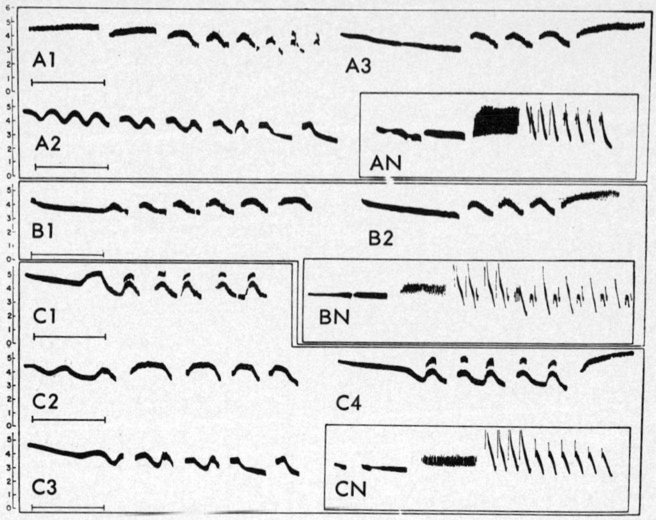

Fig. 2. Songs of nine males from three areas raised together in group isolation. (A1 to A3) Songs of individuals born at Inspiration Point, 3 km northeast of Berkeley. (B1 and B2) Songs of individuals born at Sunset Beach. (C1 to C4) Songs of individuals born in Berkeley. The inserts (AN, BN, and CN) show the home dialect of each group.

In nature a young male white-crown hears abundant singing from its father and neighbors from 20 to about 100 days after fledging. Then the adults stop singing during the summer molt and during the fall. Singing is resumed again in late winter and early spring, when the young males of the previous year begin to participate. Young males captured between the ages of 30 and 100 days, and raised in pairs in divided acoustical chambers, developed song patterns in the following spring which matched the dialect of their home area closely. If males were taken as nestlings or fledglings when 3 to 14 days of age and kept as a group in a large soundproof room, the process of song development was very different. Fig. 2 shows sound spectrograms of the songs of nine males taken from three different areas and raised as a group. The patterns lack the characteristics of the home dialect. Moreover, some birds from different areas have strikingly similar patterns (A3, B2, and C4 in Fig. 2).

Males taken at the same age and individually isolated also developed songs which lacked the dialect characteristics (Fig. 3). Although the dialect properties are absent in such birds isolated in groups or individually, the songs do have some of the species-specific characteristics. The sustained tone in the introduction is generally, though not always, followed by a repetitive series of shorter sounds, with or without a sustained tone at the end. An ornithologist would identify such songs as utterances of a Zonotrichia species.

Males of different ages were exposed to recorded sounds played into the acoustical chambers through loudspeakers. One male given an alien dialect (8 minutes of singing

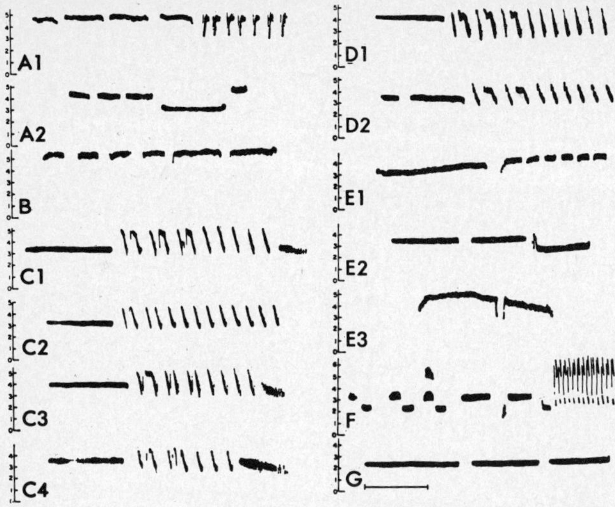

Fig. 3. Songs of 12 males raised under various experimental conditions. (A1 and A2) Birds raised in individual isolation. (B) Male from Sunset Beach trained with Marin song (See Fig. 1) from the 3rd to the 8th day of age. (Cl) Untrained. (C2 to C4) Trained with Sunset Beach songs; C2 at about 100 days of age, *C3* at 200 days, *C4* at 300 days. (*D1*) Bird from Sunset Beach trained with Marin white-crowned sparrow song and a Harris's sparrow song (see G) from the age of 35 to 56 days. (*D2*) Marin Bird trained with Marin white-crowned sparrow song and a song-sparrow song (see F) from the age of 6 to 28 days. (E1 to E3) Two birds from Sunset Beach and one from Berkeley trained with song-sparrow song from the age of 7 to 28 days. (F) A song-sparrow training song for *D2* and *E*1 to *E3*. (G) A Harris's sparrow training song for *D1*.

per day) from the third to eighth day after hatching, and individually isolated, showed no effect of the training. Thus, the early experience as a nestling probably has little specific effect. One of the group-raised isolates was removed at about 1 year of age and given 10 weeks of daily training with an alien dialect in an open cage in the laboratory. His song pattern was unaffected. In general, acoustical experience seems to have no effect on the song pattern after males reach adulthood. Birds taken as fledglings aged from 30 to 100 days were given an alien dialect for a 3-week period, some at about 100 days of age, some at 200, and some at 300 days of age. Only the training at the age of 100 days had a slight effect upon the adult song. The other groups developed accurate versions of the home dialect. Attention is thus focused on the effects of training between the ages of about 10 and 100 days. Two males were placed in individual isolation at 5 and 10 days of age, respectively, and were exposed alternately to the songs of a normal white-crowned sparrow and a bird of a different species. One male was exposed at 6 to 28 days, the other at 35 to 56 days. Both developed fair copies of the training song which was the home dialect for one and an alien dialect for the other. Although the rendering of the training song is not perfect, it establishes that the dialect patterns of the male song develop through learning from older birds in the first month or two of life. Experiments are in progress to determine whether longer training periods are necessary for perfect copying of the training pattern.

The training song of the white-crowned sparrow was alternated in one case with the song of a song sparrow, *Melospiza melodia,* a common bird in the areas where the white-crowns were taken, and in the other case with a song of a Harris's sparrow, *Zonotrichia querula.* Neither song seemed to have any effect on the adult patterns of the experimental birds. To pursue this issue further, three males were individually isolated at 5 days of age and trained with song-sparrow song alone from about the ninth to thirtieth days. The adult songs of these birds bore no resemblance to the training patterns and resembled those of naive birds (Fig. 3). There is thus a predisposition to learn white-crowned sparrow songs in preference to those of other species.

The songs of white-crowned sparrows raised in isolation have some normal characteristics. Recent work by Konishi (in preparation) has shown that a young male must be able to hear his own voice if these properties are to appear. Deafening in youth by removal of the cochlea causes development of quite different songs, with a variable broken pattern and a sibilant tone, lacking the pure whistles of the intact, isolated birds. Furthermore, there is a resemblance between the songs of male white-crowned sparrows deafened in youth and those of another species, *Junco oreganus,* subjected to similar treatment. The songs of intact juncos and white-crowns are quite different. Konishi also finds that males which have been exposed to the dialect of their birthplace during the sensitive period need to hear themselves before the memory trace can be translated into motor activity. Males deafened after exposure to their home dialects during the sensitive period, but before they start to sing themselves, develop songs like those of a deafened naive bird. However, once the adult pattern of singing has become established then deafening has little or no effect upon it. Konishi infers that in the course of crystallization of the motor pattern some control mechanism other than auditory feedback takes over and becomes adequate to maintain its organization. There are thus several pathways impinging upon the development of song patterns in the white-crowned sparrow, including acoustical influences from the external environment, acoustical feedback from the bird's own vocalizations, and perhaps non-auditory feedback as well.

Cultural transmission is known to play a role in the development of several types of animal behavior (Etkin, 1964). However, most examples consist of the reorientation through experience of motor patterns, the basic organization of which remains little changed. In the development of vocal behavior in the white-crowned sparrow and certain other species of song birds, we find a rare case of drastic reorganization of whole patterns of motor activity through cultural influence (Lanyon, 1960). The process of acquisition in the white-crowned sparrow is interesting in that, unlike that of some birds (Nicolai, 1959), it requires no social bond between the young bird and the emitter of the copied sound, such as is postulated as a prerequisite for speech learning in human children (Mowrer, 1958). The reinforcement process underlying the acquisition of sound patterns transmitted through a loudspeaker is obscure.

REFERENCES

Armstrong, E. A. (1963), *A Study of Bird Song*. London: Oxford Univ. Press.

Banks, R. C. (1964), *Zool.*, **70**, 1. Univ. Calif. Berkeley Publ.

Etkin, W. (1964), *Social Behavior and Organization Among Vertebrates*. Chicago: University of Chicago Press.

Konishi, M. (in preparation).

Lanyon, W. (1960), Animal Sounds and Communication, *AIBS Publ. No.* 7, 321, W. Lanyon and W. Tavolga, Eds. (American Institute of Biological Sciences, Washington, D.C.); Thorpe, W. H. (1961), *Bird Song. The Biology of Vocal Communication and Expression in Birds*. London: Cambridge Univ. Press; Thielcke, G. (1961), *J. Ornithol.*, **102**, 285; Marler, P. (1964), in *Acoustic Behaviour of Animals*. R. G. Busnel, Ed. Amsterdam: Elsevier, 228.

Marler, P. and Tamura, M. (1962), *Condor*, **64**, 368.

Mowrer, O. H. (1958), *J. Speech Hearing Disorders*, **23**, 143.

Nicolai, J. (1959), *Z. Tierpsychol.*, **100**, 93.

ACKNOWLEDGMENTS

M. Konishi, M. Kreith, and J. Mulligan cooperated generously in locating and raising the birds and conducting the experiments. W. Fish and J. Hartshorne gave invaluable aid in design and construction of soundproof boxes.

We thank D. M. Konishi and Dr. Alden H. Miller for reading and criticizing this manuscript. The work was supported by a grant from the National Science Foundation.

EPILOGUE

We have seen in this volume how the statement of ethological concepts by Lorenz was challenged, creating a climate of active investigation concerning the function and historical origins of behavior. In the second volume, further empirical investigations are presented that are directed towards the questions of how behavior is controlled by external stimuli and internal variables, and how behavior develops in the individual. With such investigations exemplified, the second volume then leads to conclusions about ethological theory, thus completing the cycle of thesis, antithesis and synthesis.

BIOGRAPHICAL SKETCHES

JOHN T. EMLEN, JR.

John T. Emlen, Jr. grew up in a Quaker background in Philadelphia. An overwhelming enthusiasm for bird study and natural history, fostered by sympathetic parents and teachers, threatened to interfere with scholastic activities until he entered Haverford College in 1927. Haunting the bird room of the Philadelphia Academy of Natural Sciences, he was invited to participate in summer expeditions to the West Indies and Central America and encouraged to visit museums and seek out professional ornithologists. At Cornell University where he received the Ph.D. degree in 1934, he fell under the influence of Arthur Allen, Albert Wright, William Hamilton and others who broadened his sights and introduced him to the scientific method. Here he met and married Virginia, daughter of Physicist Ernest Merritt, and together they traveled to Wisconsin to work briefly with the inspiring Aldo Leopold, before moving on to a teaching and research position at the University of California at Davis. Then Emlen turned his efforts to practical problems with the swarming rat populations of the city of Baltimore. Here Dr. Curt Richter, psychobiologist at Johns Hopkins University introduced him to new methods of research and steered him into studies of population growth and regulation.

A call to the University of Wisconsin in 1946 offered new freedom in research and an opportunity to return to teaching and graduate student guidance.

To Emlen, field studies of free-ranging animals in nature should be the starting and finishing point for all behavior research. Behavior, he believes, is functionally meaningful only in the contexts in which it evolved, and should be removed to the controlled laboratory experiment only after its role in the broad natural context has been intensively examined. Concerned that naturalistic studies are being neglected in modern ecological and ethological research, Emlen has carried his studies and led his students from the tropical forests of Africa to the desert of Death Valley and the frigid icefields of Antarctica. Included have been projects on distance navigation in penguins, olfactory orientation in petrels and social behavior in swallows, blackbirds, rhinoceroses, macaques, and gorillas.

I was born of parents and reared in the usual way. I successfully prevented my schooling from interfering with my education until I went to Harvard, where Ernst Mayr and Donald Griffin interested me in making the pursuit of biology life's work. Although my research is presently on frogs, as a child I detested the slippery creatures; my interest in living things really grew from the enjoyment of watching birds. The greatest influence on my scientific career came from my graduate major professor and current coauthor: I was his first doctoral student, at Duke. My early work was observational studies of birds in the field, but under the gentle brainwashing of Peter Klopfer I learned as well as I'll ever learn how to experiment, and my interest in the ontogenetic development of behavior bloomed. I studied imprinting and the development of begging in gull chicks, and was further prepared as a postdoctoral fellow with Daniel Lehrman. When I went to the University of Maryland to teach, my research interests drifted increasingly toward problems of visual perception in vertebrates, and I spent an exciting three years interacting with a bright faculty in behavior studies there. Since coming to the University of Wisconsin my interests have remained in sensory aspects of behavior, especially phototaxis and color vision in frogs.

JACK P. HAILMAN

Robert A. Hinde was born October 26, 1923 in Norwich, England. He was educated at Oundle School, St. John's (Cambridge), London, and Oxford. He is a Fellow, St. John's College, Cambridge, Royal Society Research Professor, and the Honorary Director of Medical Research Unit on the Development and Integration of Behaviour.

His publications include *Animal Behaviour*, Second Edition, (McGraw-Hill, New York) 1970 and numerous scientific papers.

ROBERT A. HINDE

PETER H. KLOPFER

A student of G. Evelyn Hutchinson, Dr. Klopfer has interests which include ecological problems relating to community structure and faunal diversity, and psychological problems of maternal-filial attachments and early learning. His major publications are *Behavioral Aspects of Ecology* (Prentice-Hall), *An Introduction to Animal Behavior: Ethology's First Century* (with J. P. Hailman, Prentice-Hall), and *Habitats and Territories: A Study of the Use of Space by Animals* (Basic Books). He has also completed a child's primer on behavior (which is still seeking a publisher). Major nonacademic activities include work with the American Friends Service Committee, the Carolina Friends School (an important experiment in elementary education), and sailing and cross-country racing. At Duke, he is professor of zoology, Director of the Field Station for Animal Behavior Studies, and Associate Director of the Primate Facility. His work is supported by grants and a Research Scientist Award from the National Institute of Mental Health and of Child Health.

KONRAD Z. LORENZ

"At the age of twenty-four I read the works of Max Hartmann and have never since forgotten the method of how to deduce a general law from the observation of separate special cases and how to verify them by keeping an open eye for comparable further special cases to verify the deduction."

PETER R. MARLER

Peter Robert Marler was born February 24, 1928 in London, England. He was educated at University College, University of London, B.Sc. 1948, Ph.D. 1952 (Botany) and University of Cambridge, Ph.D. 1954 (Zoology).

He holds numerous honors including Guggenheim Fellow 1964–65; President, Animal Behavior Division of the American Society of Zoologists 1968–69; President, Animal Behavior Society 1969–70; Fellow, American Association for Advancement of Science 1965; Fellow, New York Zoological Society 1969; Fellow, American Academy of Arts and Sciences 1970. He is currently a Professor at The Rockefeller University.

His research interests include behavior of animals, with special reference to the development of vocalizations in birds and primates, study of the processes of communication in animals, and field studies of social behavior.

DESMOND MORRIS

Desmond Morris began his career as a painter, holding his first one-man show in London in 1950. He entered zoology initially because of his fascination for biological shapes and patterns to be used in his paintings but after a single exposure to a lecture by Niko Tinbergen became irreversibly imprinted on comparative ethology and has since devoted much of his time to studies of vertebrate behaviour, covering a wide range of species. Passing from research on fish, to birds, and then to mammals he finally concentrated his attention on the human species, at one stage returning to the world of painting as Director of the Institute of Contemporary Arts. He is now occupied in making further studies of human ethology and divides his time between the Zoology Department at Oxford University and his house in Malta. He is still actively painting and pursuing his interest in primitive art and archaeology.

Courtesy of American Museum of Natural History

THEODORE C. SCHNEIRLA (1902–1968)

T. C. (Ted) Schneirla was Michigan born and bred, having earned his bachelor and doctoral degrees at the University of Michigan in Ann Arbor. His notebooks and sketches, made during studies under John Shepard during the early 1920's, became the basis for his now-classical *Principles of Animal Psychology*, co-authored with Norman Maier. Ted actually completed his dissertation while teaching at New York University from 1927–1930, after which he went to Karl Lashley's laboratory in Chicago for a year. It was there that Ted met Norman Maier and planned the text published in 1935, a book later reprinted (with added material by both him and Maier) as a 1964 paperback.

After Chicago, Ted Schneirla returned to New York, where he remained throughout his career, first at New York University (establishing lifelong ties there), and then ultimately at the American Museum of Natural History. Ted's first field trip to Barro Colorado Island in the Panama Canal was made in 1932, thus initiating his lifelong studies of army ants, which studies led to many lectures, papers and films. Indeed, we last saw Ted in Panama in 1966, when all three of us were engaged in field work at Barro Colorado.

In 1943 Schneirla was brought to the Museum as Associate Curator in the Department of Animal Behavior by Frank Beach, whom he eventually succeeded as Chairman. Retaining an adjunct professorship at N.Y.U., Schneirla cultured a number of graduate students, and eventually built a thriving research group at the American Museum. His interests ranged widely through animal behavior, from ontogeny to social organization.

Ted is probably best known for his critical analyses of certain concepts of early European ethology, and for his proposals of a complicated theoretical formulation about ontogenetic development and "stimulus intensity." At the time of his death, Ted had very nearly finished a long manuscript summarizing his lifelong studies of army ants, a book that may yet see print in a posthumous volume. Probably, though, Ted's major impact on the science will be judged as his very great personal influence upon the thinking and activities of a number of important workers in the discipline.

NIKO TINBERGEN

b. The Hague, Netherlands—April 15, 1907 D. Phil., M.A., F.R.S. 1962
Reader in Animal Behavior, Oxford University since 1960
Educ.—Leiden, Vienna, Yale
Lecturer, Leiden U. 1936
Professor of Experimental Zoology, Leiden 1947
Lecturer in Animal Behavior, Oxford 1949 Publications:
Eskimoland 1935
The Study of Instinct 1951
The Herring Gull's World 1953
Social Behaviour in Animals 1953
Curious Naturalists 1959
Contributions to German, British, and American journals

ABCDEFGH|7 9 8 7 6 5 4 3 2